Biosemiotics

Volume 13

Aims and Scope of the Series

Combining research approaches from biology, semiotics, philosophy and linguistics, the field of biosemiotics studies semiotic processes as they occur in and among living systems. This has important implications and applications for issues ranging from natural selection to animal behaviour and human psychology, leaving biosemiotics at the cutting edge of the research on the fundamentals of life.

The Springer book series *Biosemiotics* draws together contributions from leading scholars in international biosemiotics, producing an unparalleled series that will appeal to all those interested in the origins and evolution of life, including molecular and evolutionary biologists, ecologists, anthropologists, psychologists, philosophers and historians of science, linguists, semioticians and researchers in artificial life, information theory and communication technology

More information about this series at http://www.springer.com/series/7710

Ekaterina Velmezova • Kalevi Kull
Stephen J. Cowley
Editors

Biosemiotic Perspectives on Language and Linguistics

Springer

Editors
Ekaterina Velmezova
Centre for Linguistics and Language Sciences / Department of Slavic and South Asian Studies
University of Lausanne
Lausanne, Switzerland

Kalevi Kull
Department of Semiotics
University of Tartu
Tartu, Estonia

Stephen J. Cowley
Centre for Human Interactivity and the COMAC Cluster
University of Southern Denmark
Slagelse, Denmark

ISSN 1875-4651 ISSN 1875-466X (electronic)
Biosemiotics
ISBN 978-3-319-20662-2 ISBN 978-3-319-20663-9 (eBook)
DOI 10.1007/978-3-319-20663-9

Library of Congress Control Number: 2015955149

Springer Cham Heidelberg New York Dordrecht London

Printed on acid-free paper

Springer International Publishing AG Switzerland is part of Springer Science+Business Media (www.springer.com)

Contents

Language, Linguistics: Life, Biosemiotics…

Kalevi Kull and Ekaterina Velmezova

Abstract Since mostly human modes of action take on a symbolic aspect, and since there are many semiotic (meaning making) systems without any symbolic signs, the application of purely linguistic models in biology is mostly incongruent. However, there exist many common features between human language and other (non-human) sign systems, and even the developed linguistic universe remains internally connected to pre-linguistic expressive forms. Therefore, at least this role of biosemiotic phenomena and processes in the functioning of human language is worth paying attention to, as manifested by the contributions to this volume.

Keywords History of ideas • Semiotics • Linguistics • Human sciences • Non-human sign systems • Biology

The idea of this book arose during the 12th Gatherings in Biosemiotics (Tartu, July 2012), which included a special session entitled "Language and Life: The double interface".[1] At that time, reading Donald Favareau's review "Twelve years with the Gatherings in Biosemiotics" published in a book describing this series of annual meetings,[2] we paid attention to the fact that he mentioned very few linguists who had given talks in the Gatherings between 2001 and 2011. They are (in the order of joining the biosemiotic gatherings): Tuomo Jämsä, Stephen Cowley, (psycho)linguist Joanna Rączaszek-Leonardi, Natalia Abieva, Prisca Augustyn, and Angelo N. Recchia-Luciani.

Even if the choice of designations (are they "linguists"? or "philologists"? or maybe "philosophers [of language]"?) can sometimes alter the interpretation of facts (*nomina sunt odiosa*), this rather insignificant rate of *linguists* interested in

[1] Cf. Cowley 2012.

[2] Favareau 2012.

K. Kull (✉)
Department of Semiotics, University of Tartu, Tartu, Estonia
e-mail: kalevi.kull@ut.ee

E. Velmezova
Centre for Linguistics and Language Sciences / Department of Slavic and South Asian Studies, University of Lausanne, Lausanne, Switzerland
e-mail: ekaterina.velmezova@unil.ch

E. Velmezova et al. (eds.), *Biosemiotic Perspectives on Language and Linguistics*, Biosemiotics 13, DOI 10.1007/978-3-319-20663-9_1

biosemiotics provoked not only the question about the possible reasons for this state of affairs, but also a wish to contribute to the improvement of this situation. In particular, we were inspired not only by the example of Thomas A. Sebeok, a linguist as to his basic education and one of the founders of modern biosemiotic research (his intellectual heritage is discussed in several articles of this book), but also by the idea that, in the future, for biosemioticians it would be of use as well to become more knowledgeable at least in these aspects of linguistics in which the two fields may overlap. Let us also refer to Juri Lotman's words going back to 1990: "I think that zoosemiotics should become part of linguistics, or linguistics part of zoosemiotics; let us not argue about the priority, but it seems to me that a zoologist ought to be a linguist, and maybe a linguist ought to be a zoologist".[3]

This is how the project of the volume "Biosemiotic perspectives in linguistics" was launched. The *general idea* of the book was to try to present new methods, directions and perspectives of studying human language in general and various languages in particular within the framework of biosemiotic models, or of studying language and languages simply with an interest in biosemiotics. In particular, the (potential) contributors to the volume were invited to answer the following questions:

- What can biosemiotics bring to linguistics (and vice versa)?
- What are the biosemiotic implications for language sciences?
- What are the biosemiotic groundings of language and how to study them?
- How has the interdisciplinary union of linguistics and biosemiotics contributed to the reconsideration of some linguistic concepts – such as *language* itself, *language* as *langue(s)* and *langage, syntax, (linguistic) sign, dialect, text, discourse, code* etc.?

Additionally, articles on historical backgrounds and intellectual premises of biosemiotic approaches to the study of language and languages were also welcome.

Several months after the Tartu conference, a letter was sent to around thirty scholars all over the world, with an invitation to write a contribution to this volume.

Per aspera...

Some of the difficulties that we have encountered already from the very outset of the project allow us to ask once again the question about the reasons of this not so intensive cooperation of linguists with biosemioticians.

Of course, *prima facie*, already a difference in the objects of study in the case of linguistics and biosemiotics seems significant. According to Ferdinand de Saussure's classical work, linguistics (even in his seemingly narrow definition) is a part of

[3] Cf. Kull 1999b, p. 125.

semiotics ("semiology"). Charles S. Peirce would not disagree. And if biosemiotics studies signs and sign systems in regard to all living organisms (including human pre-linguistic semiosis), it may come very close to providing a general theory of semiotics.[4] In this case, the very objects of biosemiotics and linguistics, in all their diversity, would correlate as general and specific. It explains that linguists can feel more easy staying in the "comfort zone" of their own object of study,[5] but hardly justifies the rather limited interest of biosemioticians in linguistics. Though, of course, the proper focus of biosemiotics lays at non-linguistic semiosis, so they therefore have an excuse. After all, biosemiotics can even be defined as pre-linguistic or non-linguistic semiotics.

But in our specific case, linguists had other reasons for being reserved. In particular, if at the beginning several colleagues were very enthusiastic about the whole project (proposing, for instance, such titles of contributions as "Biosemiotics of the nineteenth century? A view from the 'pre-Saussurean' linguistic tradition", or "German philosophy of nature as a source of inspiration for structuralism and biosemiotics" etc.), their enthusiasm did not find sufficient support from the current level of biosemiotic methodology, which, obviously, has not yet made itself clear enough in order to remove all fears of biologization (like those caused earlier by social darwinism, sociobiology or evolutionary psychology). One of the arguments was that drawing analogies between cultural phenomena and those going back to natural sciences would hardly constitute a reliable method leading to important *discoveries*.

Even if this criticism is hardly sound, it concerns a noticeable image of the biosemiotic community, and that is why it seems important to explicate its origins.

... (per historiam)...

To a certain extent, the roots of this criticism go back to the history of ideas, and even though so many works have already been written on the problems of relations between linguistics and biology that their simple enumeration would need a book, to recall some tendencies does not seem unnecessary even in this short foreword. In particular, certain modern linguists consider the ideas of late eighteenth and early nineteenth century Romanticism to have been forever left behind – specifically the ideas emphasizing the union of nature and culture (and therefore, also that of biology and the humanities). However, still not so long ago such ideas had their right to existence and seemed fruitful for linguistics; moreover, some "biologico-linguistic" ideas have not lost their value at present. Even if with the corresponding conceptions we are still not in the field of biosemiotics as such, it can be useful to bring to

[4] Cf. also Kull et al. 2009, p. 171 about biosemiotics aiming at general semiotics.

[5] Cf. the following observation of one of the participants of our project: "Linguists attending the biosemiotic *Gatherings* conferences always run the risk of being perceived as naïve or uninformed about the many layers of language and communication that the inadequate abstractions in the field of linguistics cannot address" (Augustyn 2012, p. 185).

mind some of these reflections, providing here only two striking examples and referring to the articles of this volume for other illustrations of the corresponding tendency: the history of an eternal dialogue between linguistics and biology is repeatedly reviewed in the contributions we publish.

From Biology to Linguistics

One of the best known attempts to establish an analogy between linguistics and biology on the *object* is immediately associated today with the so-called linguistic naturalism and the name of August Schleicher, Darwin's admirer for whom, at first glance, languages living organisms. This conviction resulted in the conclusion that linguistics is a natural science and its *methods* are, **in general outline**, almost the same as in other sciences of life. Other scholars working within the framework of the naturalist "paradigm" (as Max Müller) also shared this point of view. Languages could be considered as living organisms, among other things, in virtue of the belief that language evolution follows the laws which men cannot influence. The notion of *law* was transposed into linguistics from biology (even if Schleicher was convinced that Darwin's theory could be only in a very general way applied to the study of languages, the latter being too different from both plants and animals). From the above it appeared that the "life" of languages could be analogously divided into the same phases as that of living organisms (hence Schleicher's evolutionary typology), that both struggle for existence and natural selection are possible among linguistic phenomena (words, morphological constructions), etc. All this witnesses that these scholars moved, indeed, from analogy on object to analogy on *method*, trying to study the evolution of languages by analogy with the evolution of living organisms as understood by Darwinians.[6] Despite their visibly naïve character, some of these views have survived in linguistics until today (though in a less "literal" state). The image of the language family tree, worked out within the framework of the naturalist current, is still widely spread in linguistics, even if now it is completed by other models.

One of these "supplementing" models is connected with a similar analogy on method between the humanities and sciences of life which was drawn later. A manifest anti-Darwinian example of such analogy goes back to the 1920s in Russia. In the book *Nomogenesis; or, Evolution Determined by Law* (1922),[7] biologist and geographer Lev Berg set out a conception of evolution which was an anti-Darwinian one. Among other things, his idea of evolution was that of convergence of non-related species on the same territory, as opposed to Darwin's conception of evolution by divergence. This point of view has much influenced scholars who transposed Berg's model into linguistics; in the 1920s–1930s, Nikolai Marr, Roman Jakobson, Nikolai Trubetzkoy and others spoke (with different degrees of reliability in their discourses, and either completely rejecting the divergent model of language

[6] Cf. Velmezova 2014.

[7] English edition Berg 1969.

evolution [Marr] or only completing it correspondingly [Jakobson, Trubetzkoy]) about languages evolving by convergence, hence the idea of not only lexical but also morphological loans, of language unions, etc.[8]

Therefore as concerns the relations between the humanities and life sciences, or more precisely between linguistics and biology, even the method of "drawing analogies" used to be fruitful.[9]

However, the whole situation will receive a different light if one takes a position that both biosemiotics and linguistics are parts of semiotics which share the principal processes of meaning making as to their objects. Therefore there could also be much in common in their methodologies. Biology as seen from this perspective would not belong to natural sciences (at least in the earlier sense), and accordingly, the regularities that linguistic and biosemiotic descriptions may share will not be of the same kind as "natural laws".

Anyhow, already in the examples discussed above, the influence of biology on linguistics was not irreciprocal: both Darwin and Berg discussed linguistic examples in their works. But it is nothing in comparison with the interest of biologists in semiotics (and linguistics as a part of it) during several past decades.

From Linguistics to Biology

It was in the 1970s that application of linguistic principles in biology became particularly frequent among theoreticians.[10] In the decades that followed, simple language metaphors (like "DNA language" or "cellular language" of life, etc.) have become widespread in biology. Such linguistic metaphorics have been characterised as "spontaneous semiotics" in biology.[11] However, a more profound understanding of the relations between biology and linguistics yet has to be developed.[12]

Early attempts to redefine the relationship between linguistics and biology on a semiotic basis were made already in the 1970s. Conrad H. Waddington stated: "It is language [...] that I suggest may become a paradigm for the theory of General Biology".[13] Howard Pattee[14] spoke about linguistic and dynamic modes of description of living systems. Applying explicitly semiotic terminology, Belgian biochemist Marcel Florkin wrote: "We believe that in future develop-

---

[8] Cf. also Sériot 2014; Velmezova 2007.

[9] Cf. also Auroux (ed.), 2007.

[10] Jakobson 1971; Marcus 1974; Pattee 1972, etc.

[11] Hoffmeyer 2008, pp. 360–364; cf. also Markoš and Faltýnek 2011.

[12] Cf. Kravchenko 2013.

[13] Waddington 1972, p. 289.

[14] Pattee 1977.

ment, linguistic semiology will become based on molecular biosemiotics of the activities of the brain. We shall therefore use in the perspective of this subject several general concepts elaborated by de Saussure such as significant and signified, synchrony and diachrony, syntagm and system with the special meaning they have in molecular biosemiotics. It must be noted that in the mind of F. de Saussure these concepts arose from the consideration of existential (not psychological) aspects of natural science. [...] It is therefore fitting to situate these concepts in the most general context of semiotics, the general science of signification, of which linguistics and biosemiotics are special aspects".[15] Further, on February 1–2 1978, a conference "Biology and linguistics" was organized in Tartu, with the participation of several leading scholars in the fields of both semiotics and theoretical biology who worked at that time in the Soviet Union.[16] One of the conference sessions was titled "Biosemiotic research abroad". In addition, in the Tartu-Moscow semiotic school some projects relating linguistics and biology were carried out (works on aphasia, studies of relationships between neurobiological and semiotic brain asymmetry, etc.).[17]

In parallel, discussions on the applicability of linguistic models in animal communication were carried out within the framework of zoosemiotics.[18] A remarkable crystallization of biosemiotic ideas took place in the 1990s, particularly due to Thomas A. Sebeok's and Jesper Hoffmeyer's work.[19]

Overall, since only human modes of action take on a symbolic aspect,[20] and since there are many semiotic (meaning making) systems without any symbolic signs, the application of purely linguistic models in biology is mostly incongruent. However, there exist many common features between human language and other (non-human) sign systems, and "even the fully developed linguistic universe of expressive sounds remains internally connected to those pre-linguistic expressive forms".[21] Therefore, at least this role of biosemiotic phenomena and processes in the functioning of human language is worth paying attention to, as manifested by a number of contributions to this volume.[22]

---

[15] Florkin 1974, p. 14.

[16] The conference was organised by three research groups, starting to work in the direction of establishing connections between biology and semiotics. They were from St. Petersburg (leaded by Sergei Chebanov), Moscow (Alexander Levich, Alexei Sharov), and Tartu (Kalevi Kull with colleagues) (cf. Kull 1999b, p. 122).

[17] Cf. for instance Ivanov 1978; articles on these problems in Minc (ed.), 1983, etc.

[18] Cf. a review about the history of zoosemiotics in Maran et al. (eds.), 2011.

[19] Sebeok and Umiker-Sebeok (eds.), 1992; Hoffmeyer 1993 [1996]; for reviews of this tendency cf. Favareau 2010a; Kull 1999a.

[20] Cf. in particular Deacon 1997. It corresponds to Th.A. Sebeok's usage of the term *language* as referring to the sign system which is almost uniquely human.

[21] Hoffmeyer 2008, p. 274.

[22] Some earlier works on the relations between linguistics and biosemiotics were reviewed in the anthologies on biosemiotics (Favareau [ed.], 2010) and zoosemiotics (Maran et al. [eds.], 2011). In addition, we may mention the work, e.g., by Alexander Kravchenko (2006 and 2013), Terrence Deacon (2003), Stephen Cowley (2006), etc.

...ad astra

In this volume are presented contributions of both young researchers and eminent professors from several countries and continents, many of whom know each other personally and/or by research work. One third of the articles in this book are those written in collaboration, and in many contributions, there are references to the works (including the latest ones) of other participants of our project. Therefore the volume could be considered as a fruit of collaboration between researchers belonging to a very dynamic and rapidly developing international community of scholars; the variety of subjects discussed here[23] reflects different aspects of their activity.

The book contains four parts; articles within each of them are united by common subjects and/or problematics, even if this division is certainly relative.

The first part ("Theory and Theoretical Models") opens with an article by Donald Favareau and Kalevi Kull about "biosemiotics and its possible relevance to linguistics". This text can be considered as a general *theoretical* introduction to our volume, with its emphasis on the idea of meaning-making as one of the most important phenomena studied in both biosemiotics and linguistics. The question of the extraction of meaning as possible via semiosis and narration is discussed in the text written by Anton Markoš and Dan Faltýnek who at the same time remind us of a blurred nature of some central concepts in linguistics and biology – such as *language*, in the first place. The notion of *language* remains central also in the article by Susan Petrilli and Augusto Ponzio, where "language as primary modelling" and "natural languages" are confronted in a biosemiotic perspective. Morten Tønnessen's contribution deals with language and umwelt. In this article, relations between these two "entities" turn out opposite in comparison with a "traditional" view. In Jamin Pelkey's paper, the evolutionary aspect of language is emphasized with a particular insistence on a "deep congruence between linguistic and biotic growth".

From multifarious theories to *empirical and observational* work: that is how the following part of the volume could be described, in which we decided to put only one contribution: Stephen Cowley's article seems particularly important for our book because of specific case studies which are discussed in this text in the light of its author's theoretical theses.

Indispensable for our volume was the question of relations between biolinguistics and biosemiotics, to which the third part of the book is dedicated. In fact, what has often been labeled as biolinguistics[24] manifests mostly a quite separate approach from biosemiotics: biolinguistics studies the biological preconditions for (largely a computational model of) language, while biosemiotics focuses on the pre-linguistic

[23] Let us specify from the very beginning that some theses discussed in the contributions of this volume, or conclusions to which their authors come, were not always shared by the three editors of the book (whose views also sometimes diverged). We also gave (relatively) free hand to our authors as to their own right for spelling the words and terms with non-established orthography (*Umwelt* or *umwelt*? etc.), for putting (or not) into References works which they only mention (without quoting) in their contributions, etc. Likewise, each author could choose either British or American spelling for her/his contribution; among other things, this allowed us to avoid potential inconsistencies in quotations, etc.

[24] Berwick and Chomsky 2011; Di Sciullo and Boeckx (eds.), 2011.

sign processes. A major difference between these approaches lies in the view to biology: authors of works that use the label *biolinguistics* usually do not accept meaning making processes at the biological level.[25] In our book, biolinguistico-biosemiotic problems are taken up in three articles. Embracing both fields, biolinguistics and biosemiotics, Winfried Nöth discusses their common points and their differences in a meticulous overview. With a particular insistence on the history of the corresponding disciplines, major problems of relations between biology, linguistics, biolinguistics and biosemiotics are examined by Prisca Augustyn. Finally, turning from historical problems to those of current research, Piera Filippi discusses the evolutionary continuity between animals' communication systems and human language in light of the general question about the "evolutionary roots of human language".

In the fourth part of this book are gathered contributions on the history of biosemiotic and linguistic ideas in their interrelation. Being last, this part is far from being least not only as to the number of contributions it contains, but also because of the fact that historical questions are discussed, in one way or another, in the majority of texts presented in this book. It appears therefore that history and historiography of sciences provoke today much more enthusiasm from those interested in linguistics and biosemiotics than, for instance, any empirical research. *Quae sunt Caesaris, Caesari*: the first contribution of this part, that of Sara Cannizzaro and Paul Cobley, puts forward Thomas Sebeok's transition from linguistics to biosemiotics. The works of such "classical authors" as Ferdinand de Saussure (for linguistics) and Charles Darwin (for natural sciences) are discussed in the articles of Jui-Pi Chien, and Thomas Robert and Deana Neubauer, correspondingly. At last, we publish Ekaterina Velmezova's text about the Bakhtinian notion of *dialogue*, which is sometimes referred to in the context of biosemiotics studies. Without claiming any biosemiotic orientation of her work, E. Velmezova offers an overview of Mikhail Bakhtin's references to this concept, complex and evolving with time.

We hope that this book will offer an opportunity to look at numerous phenomena in a new way, therefore allowing their original interpretation. If one of the purposes of biosemiotics as an interdisciplinary research consists in bridging the gap between natural sciences and the humanities, as well as in redefining their relationship, this volume could be considered as a step in this noble direction. We hope to continue our work in the future, organizing a series of events and publications on the corresponding passionate problems and in this way favoring a cross-disciplinary exchange, a dialogue between specialists in several conventionally separated fields of knowledge.

[25] Asked about the definitions ("biolinguistics is the study of biological preconditions of language; biosemiotics is the study of pre-linguistic sign systems"), one of the current leaders in the field of biolinguistics, Cedric Boeckx, responded with the following: "Regarding your definitions, they seem fine to me, as far as definitions are concerned, and I believe the two fields have lots to teach to one another" (letter from Cedric Boeckx to Kalevi Kull [July 19 2013]). Cf. also Cowley 2006 about the differences between biosemiotic and biolinguistic approaches.

Acknowledgements We would like to thank all participants of this project. Kalevi Kull's research has been supported by the European Union through the European Regional Development Fund, grant IUT2-44.

References

Augustyn, P. (2012). Linguistics. In D. Favareau, P. Cobley, & K. Kull (Eds.), *A more developed sign: Interpreting the work of Jesper Hoffmeyer* (pp. 183–185). Tartu: Tartu University Press.

Auroux, S. (Ed.). (2007). *Le naturalisme linguistique et ses désordres*. Paris: SHESL [*Histoire Épistémologie Langage*, 29.2].

Berg, L. S. (1969). *Nomogenesis or evolution determined by law*. Cambridge: MIT Press.

Berwick, R. C., & Chomsky, N. (2011). The biolinguistic program: The current state of its development. In Di Sciullo, Boeckx (Eds.), 2011, pp. 19–41.

Cowley, S. (2006). Language and biosemiosis: Towards unity? *Semiotica, 162*, 417–443.

Cowley, S. (2012). Language and life: The double interface. In S. Rattasepp & T. Bennett (Eds.), *Gatherings in biosemiotics* (pp. 142–144). Tartu: University of Tartu Press.

Deacon, T. (1997). *The symbolic species: The co-evolution of language and the human brain*. London: Penguin.

Deacon, T. (2003). Universal grammar and semiotic constraints. In M. H. Christiansen & S. Kirby (Eds.), *Language evolution* (pp. 111–139). Oxford: Oxford University Press.

Di Sciullo, A. M., & Boeckx, C. (Eds.). (2011). *The biolinguistic enterprise: New perspectives on the evolution and nature of the human language faculty*. Oxford: Oxford University Press.

Favareau, D. (2010a). Introduction: An evolutionary history of biosemiotics. In Favareau (Ed.), 2010, pp. 1–77.

Favareau, D. (Ed.). (2010b). *Essential readings in biosemiotics: Anthology and commentary*. Dordrecht: Springer.

Favareau, D. (2012). Twelve years with the gatherings in biosemiotics. In S. Rattasepp & T. Bennett (Eds.), *Gatherings in biosemiotics* (pp. 64–72). Tartu: University of Tartu Press.

Florkin, M. (1974). Concepts of molecular biosemiotics and of molecular evolution. In M. Florkin & E. H. Stotz (Eds.), *Biochemistry* (Comprehensive biochemistry, Vol. 29A, pp. 1–124). Amsterdam: Elsevier.

Hoffmeyer, J. (1993 [1996]). *Signs of meaning in the universe*. Bloomington: Indiana University Press.

Hoffmeyer, J. (2008). *Biosemiotics: An examination into the signs of life and the life of signs*. Scranton: Scranton University Press.

Ivanov, V. V. (1978). *Čet i nečet: Asimmetrija mozga i znakovyx system*. Moskva: Nauka [Even and odd: Asymmetry of brain and of sign systems].

Jakobson, R. (1971). Linguistics in relation to other sciences. In R. Jakobson, *Selected writings, vol. II: Word and language* (pp. 655–696). The Hague: Mouton.

Kravchenko, A. (2006). Cognitive linguistics, biology of cognition and biosemiotics: Bridging the gaps. *Language Sciences, 28*(1), 51–75.

Kravchenko, A. (2013). *Ot jazykovogo mifa k biologičeskoj real'nosti: Pereosmysljaja poznavatel'nye ustanovki jazykoznanija*. Moskva: Rukopisnye pamjatniki drevnej Rusi [From a language myth towards the biological reality: Rethinking cognitive purposes of linguistics].

Kull, K. (1999a). Biosemiotics in the twentieth century: A view from biology. *Semiotica, 127*(1–4), 385–414.

Kull, K. (1999b). Towards biosemiotics with Juri Lotman. *Semiotica, 127*(1–4), 115–131.

Kull, K., Deacon, T., Emmeche, C., Hoffmeyer, J., & Stjernfelt, F. (2009). Theses on biosemiotics: Prolegomena to a theoretical biology. *Biological Theory, 4*(2), 167–173.

Maran, T., Martinelli, D., & Turovski, A. (Eds.). (2011). *Readings in zoosemiotics*. Berlin: De Gruyter Mouton.

Marcus, S. (1974). Linguistic structures and generative devices in molecular genetics. *Cahiers de linguistique théorique et appliquée, 11*, 77–104.
Markoš, A., & Faltýnek, D. (2011). Language metaphors of life. *Biosemiotics, 4*(2), 171–200.
Minc, Z. G. (Ed.), (1983). *Tekst i kul'tura*. Tartu: Tartu University Press. [*Trudy po znakovym sistemam/Sign Systems Studies*, 16]. [Text and culture].
Pattee, H. H. (1972). Laws and constraints, symbols and languages. In C. H. Waddington (Ed.), *Towards a theoretical biology 4: Essays* (pp. 248–258). Edinburgh: Edinburgh University Press.
Pattee, H. H. (1977). Dynamic and linguistic modes of complex systems. *International Journal of General Systems, 3*, 259–266.
Sebeok, T. A., & Umiker-Sebeok, J. (Eds.). (1992). *Biosemiotics: The semiotic web 1991*. Berlin: Mouton de Gruyter.
Sériot, P. (2014). *Structure and the whole: East, west and Non-Darwinian biology in the origins of structural linguistics*. Berlin: De Gruyter Mouton.
Velmezova, E. (2007). *Les lois du sens: la sémantique marriste*. Bern/Berlin/Bruxelles/Frankfurt am Main/New York/Oxford/Wien: Peter Lang.
Velmezova, E. (2014). Avgust Šlejxer, Maks Mjuller i… Prosper Merime o zakonax jazyka i jazykov. *Mirgorod, 1*(3), 45–60. [Linguistic laws as seen by August Schleicher, Max Müller and… Prosper Mérimée].
Waddington, C. H. (1972). Epilogue. In C. H. Waddington (Ed.), *Towards a theoretical biology 4: Essays* (pp. 283–289). Edinburgh: Edinburgh University Press.

Part I
Theory and Theoretical Models

On Biosemiotics and Its Possible Relevance to Linguistics

Donald Favareau and Kalevi Kull

Abstract Biosemiotics is the study of meaning-making in biological systems. It argues that all organisms are biologically semiosic systems. This provides for linguistics a firm basis to ground the problem of the origin of meaning and to build upon the findings of this field.

Keywords Biosemiotics • Linguistics • Biolinguistics • Biology • Meaning • Thomas A. Sebeok

Understanding Life from a Fundamentally Semiotic Perspective

According to a physicalist perspective that is widely accepted in the contemporary sciences, an organism is a chemical-metabolizing body capable of self-copying (via its genetic endowment), and growth. Due to random inexactness in copying and differences in reproduction outcomes between individuals, the structure of such organisms at the population level slowly changes over time. This phenomenon of differential reproduction is called evolution via natural selection, and along with the growth of the diversity of organisms, the complexity of such systems may become extraordinarily high. Most relevant from the biosemiotic standpoint is the fact that, according to the physicalist understanding, an ineliminable part of the metabolism constituting such systems are those substances that may, in small quantities, cause large changes. Such configurations often take the form of *signals* or *signal particles*, and in those cases where they move (or are moved) between bodies, it is often said that a biologically primary kind of "information transfer" occurs.

However, a biosemiotic approach to the same phenomena draws a radically different picture, together with many new implications for research. Living systems,

D. Favareau (✉)
National University of Singapore, Singapore, Singapore
e-mail: uspfdf@nus.edu.sg

K. Kull
Department of Semiotics, University of Tartu, Tartu, Estonia
e-mail: kalevi.kull@ut.ee

E. Velmezova et al. (eds.), *Biosemiotic Perspectives on Language and Linguistics*, Biosemiotics 13, DOI 10.1007/978-3-319-20663-9_2

including the bodily organization of an individual organism, are always understood to be a swarm of communicative processes. None of its particular *structures* can be found to be living per se. Rather, *life* (much like *conversation*, or *language*[1]) names the emergent *process* that occurs exclusively between the participants of the swarm. Life (again, like language) owes its very being to *mediatedness*, and is thus, by definition, fundamentally distributed. Lastly, and again like language, life, at every level which it is considered, is a sign process.

From a biological point of view, these two approaches can be seen as complementary. From the point of view of their research agenda and interdisciplinary applicability – which is the focus of our investigation here – these two approaches differ quite radically. For biosemioticians, semiosis assumes mediation; this can be found from the cellular level on upwards. Therefore, life is a communication process throughout all its levels. The resulting structures of organisms – like the resulting language of a culture – are themselves products of semiosis, which simultaneously work as scaffolding both for further communication and for further possible evolution. The latter can be described as the situation of problem-resolving processes manifesting due to certain incompatibilities in the system. Accordingly, all (decentralized and non-anthropomorphic) "decision-making" to be attributed to such behaviour is a collective and communicative phenomenon.

These short few words, both above and below, are meant to give readers just a taste of the deep parallels between "life" and "language" as emergent enactments of semiosis, and of the concerns that bind biosemiotics to linguistics. It is our hope that this brief contribution, and especially the larger volume of which it is a part, will help seed fertile new communication between our two fields.

---

[1] The authors wish to note at the outset that use of the word *language* in this text should not be read to indicate a view of "language" as a reified "thing in itself", over and apart from the actual interactions between humans that give rise to such reified notions. The authors of this article find it important to emphasize the centrality in biosemiotic thinking that languaged behavior, as a subset of semiosic behavior, inherits all the latter's definitional properties, and that thus it could not be otherwise that *sign* behavior of all sorts is grounded in situated, actually instantiated *action* at all points – just as both Charles S. Peirce and Jakob von Uexküll each independently observed, approximately 100 years ago. If there is a view of "language" that holds it to be either a "thing in itself" or, indeed, anything other than semiosic behavior in its fully enacted cycles of perception and action (*Funktionskreis*) and unceasingly dynamic and emergent sign-object-interpretant (the last relatum of which must always be an action or an enacted change) interaction by and between living agents, it is not one held by us, or any other biosemioticians who are sufficiently conversant in the works of Sebeok, Uexküll and Peirce. Moreover, and as with the higher-order term *semiosis*, we feel that readers of the present volume will be sophisticated enough in such matters as to clearly understand that our use of the word *language* here is referring to an enacted and emergent semiosic *process* that arises solely from the interactions of living beings, and not with some misguided, fundamentally *anti*-biosemiotic notion of a "thing in itself". We include this note here for so as not to be mis-read, and to remind our readers to understand the following uses of the term *language* as shorthand for "linguistically-aided semiosis" – with all the ineliminable interactivity and triadicity that sign action in the world involves.

Linguistics and Biology: The Relationship Redefined

The formal disciplines of linguistics and biology have borrowed models from each other for a long time, and therefore their interrelations are very rich. Some of the more famous examples from the side of linguistics include (1) August Schleicher's extrapolation of Charles Darwin's ideas on the divergence of species to proposals about the evolution of language, and to the building of phylogenetic trees of language families,[2] (2) Noam Chomsky's invocation of the idea of brain-based and genetically implicated program as a generator of a universal grammar,[3] (3) biolinguistics, the study of the biological preconditions of language[4]; and (4) the application of neo-Darwinian models of competition, population genetics and fitness dynamics, particularly from sociobiology, in explaining various phenomena of language dynamics.[5] Most notably, and as the review by Quentin D. Atkinson and Russell D. Gray[6] regarding the connections between historical linguistics and evolutionary biology shows, the main shared aspects of the fields up to now are the emphasis on historicity and the phenomenon of inheritance.[7]

It is remarkable to note that most of the examples cited above draw on biological models that are *not* principally concerned with the phenomenon of *meaning-making* or *communication* per se. Indeed, Darwin's model of natural selection itself does not include a consideration of the evolutionary role of communication; rather, its focus is simply on the reproduction of individuals and populations. The same is true for most of later models that have been transferred from biology to linguistics – they are based on an individual-centric vision that sees communication between individuals playing only a derivative role, if any, to the extra-individual "forces of evolutionary pressure" structuring such communicative ability.

Indeed, it is not surprising that, despite some enthusiastic initial support, most of the above biological models in linguistics have been met with substantive criticism from other linguists (who stress that linguistically-aided semiosis is shaped by culture, not by biology) and from humanists more generally.[8]

For up until very recently (and, arguably, for the most part, even now) biology in its mainstream has simply not taken communicative processes deeply into account, and communication per se has appeared as an epiphenomenon somewhat later in evolution and not at all as a fundamental, generative and causal attribute of life itself.

---

[2] Koerner (ed.), 1983.

[3] Chomsky 1965 and 2007.

[4] E.g., Di Sciullo and Boeckx (ed.), 2011.

[5] E.g., Ritt 2004.

[6] Atkinson and Gray 2005.

[7] In a few more recent works, Darwinian biology has been understood in an extended way, such as, for instance, in developmental linguistics (Andresen et al. 2014). Such work, however, has yet to transcend the inherited shortcoming of failing to incorporate the *semiosis* of the biological level within its primary analytical scope.

[8] E.g., Tallis 2011.

No science, however, consists in only its dominant or mainstream paradigms. Likewise, biology has always harbored within it approaches (often not so widely known) that see communicative meaning-making as a fundamental process for all living Beings, and that should therefore naturally be of help (1) in informing a fundamental understanding of sign systems as they appear in the natural world, and (2) for supplementing linguistic theory with a more general one that would extend over all kind of organic meaning-making processes. One such approach is nowadays being developed under the label of *biosemiotics*.[9]

Defining Biosemiotics, and Its Relation to Linguistics

Although there are many ways to define biosemiotics, we may say for the purposes of this article that biosemiotics is primarily interested in the study of all pre-linguistic sign systems.[10] Such systems are composed of the manifold meaning-making processes that pre-exist and coexist with language and include both internal and external cognitive processes, as well as all systems of perception, organization, interaction and memory that are semiotic yet non-conscious. Such semiotically organized (and organizing) processes cannot alone give rise to language[11] – but they constitute the biological and material relational conditions without which such linguistically-aided semiosis could not be. Indeed, although instantiated variously and on many different levels across species, such processes take place in all living beings as a consequence of the organism having to negotiate its internal relations with those of the environment, and vice versa: these are the processes, then, that make one meaningfully alive, or in other words, that make meaning a part of life.

---

[9] Instead of giving here a review of that tradition on which the contemporary biosemiotics stands, we refer the reader to two extensive anthologies (Favareau [ed.], 2010a; Maran et al. [ed.], 2011).

[10] More precisely, if pre-linguistic sign systems would be those which precede human linguistically-aided semiosis either in human ontogeny or in phylogeny, then, in addition, there certainly exist many more sign systems in other species that are not direct predecessors for human linguistically-aided semiosis, and these are also dealt with in biosemiotics.

[11] We follow here Thomas A. Sebeok's restricted usage of the term *language* as designating just one of many ways of instantiating semiosis – in this case, the one that is unique to the human species. And while we have also stressed here that the reader may substitute the more correct term "linguistically-aided semiosis" for the more common but conceptually problematic term "language" in each of its uses here, biosemioticians again follow Sebeok in insisting that what makes human semiosis distinctive is not the use of linguistic forms per se, but, rather, the understanding of thirdness relations *qua* thirdness relations, as also noted by John Deely (1990 and 2002). Similarly, if we accept Terrence Deacon's (Deacon 1997) understanding of the concept *symbol* as enabling this very ability, then we might redefine Sebeok's concept of "language" as the type of semiosis which necessarily includes such symbols among its signs. This is the sense in which we will be using the word henceforth, and therefore all future instances of the term in this article will heretofore go unremarked, in the security that the reader now understands what we are actually referring to when we use this term.

Biosemiotics is thus a more general field of investigation than is linguistics. For from the point of view of all existing semiotic systems used by living organisms, human language is just one type among many sign systems, although it is arguably the richest, as it is predicated upon and surrounded by the multiple component and supporting semiotic systems that, while not assuming the form of linguistic discourse, arise from the communication and meaning-making processes characteristic of all forms of life. To the biosemiotician, then, who studies the sign processes of any species of living beings (including the human being, both internally biologic and before they learn to speak), human language appears as a very particular case of semiosis that requires its own models and approaches, given its origin and deep commonality with other non-human semiotic systems, as well as the particular peculiarities of semiotic processing that human language both entails and requires.

When Ferdinand de Saussure formulated the field of a general science of signs, he postulated that linguistics would form but a part of it, noting that "if we are to discover the true nature of language, we must learn what it has in common with all other semiological systems".[12] "The task of the linguist," he specified, "is to find out what makes language a special system within the mass of semiological data".[13] Saussure's *semiology* was thus an attempt to develop a more general study of signs and meanings that would then contribute to the understanding of language itself. Because Saussure's semiology limited itself to an investigation of purely human, dyadic, and psychological sign relations, however, it could not account for the triadic, non-psychological and pre-linguistic sign relations found throughout the rest of the living world.[14]

It was precisely because of this limitation, in fact, that the contemporary discipline of *biosemiotics* was largely the brainchild of a linguist whose investigations into "the meaning of meaning" circled him outwardly from language to culture to animality to the biological organization of life itself – and, accordingly, from the human-focused *semiology* of Saussure to the more encompassing *semeiotic* (general sign logic, however instantiated) of the philosopher and logician Charles Sanders Peirce.

It behooves us to say a few words about this scholar as his project now, as the development of biosemiotic inquiry from linguistic inquiry is illustrative in its own right of the way in which these now somewhat segregated sites of inquiry may be able in the future to mutually and more productively inform one another.[15]

Thomas A. Sebeok: From Linguistics to Biosemiotics

Thomas Albert Sebeok (1920–2001) is now remembered primarily as a "semiotician" (and as a biosemiotician). However, his contributions to linguistics proper over the course of his lifetime are both lasting and vast. At the age of 16, Sebeok

[12] Saussure 1916 [2003, p. 17].

[13] *Ibid.*

[14] Hoffmeyer 1993 [1996]; Deely 1990; Short 1988.

[15] For more detailed accounts of the history of biosemiotics, cf. Favareau (ed.), 2010a, b; Kull 1999.

began his university studies at Magdalene College in Cambridge, where he studied under the philosopher and rhetorician Ivor Armstrong Richards, whose 1923 *The Meaning of Meaning* (co-authored with Charles Kay Ogden), was an early and lasting influence. Not long after, Sebeok's collaborator Marcel Danesi (together with Albert Valdman) notes:

> Sebeok immigrated to the United States and matriculated at the University of Chicago, where he enrolled in a semiotics course taught by Charles Morris. He earned a Ph.D. in oriental languages and civilizations at Princeton. But with a firm grounding in American structuralism acquired in Leonard Bloomfield's classes at Chicago, [Sebeok] commuted to Columbia to pursue his studies of linguistics under the tutelage of Roman Jakobson, his dissertation director, whose broader views on the place of language within the humanities and social sciences would shape Sebeok's intellectual development.
>
> In 1943 the twenty-four-year-old doctorandus arrived in Bloomington to assist the Amerindianist Carl Voegel in managing the country's largest Army Specialized Training Program in foreign languages. As enrollment swelled into the thousands and the number of less-taught tongues rose to fifteen, he took over the helm. During that period he also participated in the famous Broadway Project in which linguists were entrusted with the task of developing pedagogical materials to support instruction in the less-taught languages [...], as well as in the spoken varieties of commonly taught languages like French and German disdained by the philologists and literary scholars who staffed the foreign language programs of Academia in his era.[16]

By the age of 23, Sebeok had already published nine articles on linguistics, including an article in the prestigious journal *Language* in 1942, entitled "An examination of the Austro-Asiatic language family". Following his graduation from Princeton in 1945, Sebeok joined the faculty of Indiana University at Bloomington, which became his home university for the next 45 years. Ever a prolific researcher, writer, and academic impresario, the resume of Sebeok's first decade and a half at Indiana University – long before he got into his stride as world-renowned semiotician, much less biosemiotician – illuminates much about the man:

> A linguist studying Finno-Ugric languages, Sebeok's linguistic fieldwork took him to Central and Eastern Europe, including Lapland and the former Soviet Union. He also carried out linguistic studies in the former Mongolian People's Republic, Mexico and in the U.S. (among the Winnebago Indians of Wisconsin and the Laguna Indians of New Mexico). In addition to these studies in grammar and phonology, his interest in anthropology, folklore and literary studies led to publications dealing with folksongs, charms, games, poems and the supernatural. He published a ground-breaking volume on *Myth* in 1955, and in 1960, *Style in Language*. At the same time, he contributed to the creation of the new field of psycholinguistics, publishing with Charles Osgood, the famous classic text, *Psycholinguistics*, in 1954. He also made some of the first computer analyses of verbal texts. By 1960, Sebeok had established himself as a scholar known for crossing academic boundaries not only in his own research, but also in collaboration with scholars in adjacent fields. This carried over into his roles as an editor of books and journals, a founder and officer of several academic organizations, conference organizer, and mentor.[17]

[16] Danesi and Valdman 2004, pp. 312–313.

[17] Brier 2003, p. 103. This quotation, slightly edited by Søren Brier, initially comes from the press release of the Indiana University of January 3, 2002, distributed by George Vlahakis (http://ils.indiana.edu/news/story.php?story_id=364; website consulted on February 16, 2014).

The reader should take note here that as of this point, we have only recounted the first two decades of Sebeok's six decade career. Appointed to the directorship of the then newly created Research Center for Anthropology, Folklore, and Linguistics at Indiana University, Sebeok edited the influential book series *Current Trends in Linguistics* from 1963 to 1976, comprising 18 large-sized volumes devoted to Southeast Asian linguistics, Sub-Saharan African linguistics, Soviet and East European linguistics, Ibero-American and Caribbean linguistics, linguistics in Oceania, Western Europe and North America, as well as volumes devoted to the examination of diachronic, areal, and typological linguistics and the historiography of linguistics. In 1966, Sebeok authored the 600-plus page volume *Portraits of Linguists: A Biographical Source Book for the History of Western Linguistics*. Yet, as again Danesi and Valdman remind us:

> Sebeok's trailblazing in the various language sciences and arts – ethnolinguistics (*Studies in Cheremis Folklore* 1952), psycholinguistics (*Psycholinguistics* 1954, with the Illinois psycholinguist Charles E. Osgood), stylistics (*Style in Language* 1960), zoosemiotics (*Animal Communication* 1968) – must not be viewed as sallies outside of the traditional narrow purview of linguistics, but rather as laying the foundation for what might be viewed as his mature work in semiotics proper. As his new academic base at Indiana University progressively transformed itself into the Research Center for Language and Semiotic Studies, Thomas Sebeok, then a Distinguished Professor of Linguistics, established undergraduate and graduate programs in semiotics, coming to share with his friend Umberto Eco the signal honor of holding a designated chair in semiotics.[18]

Thus, as the 1970s began, while a growing number of scholars were all working away at various independent lines of inquiry into the problems of information processing, intercellular communication, behavioral psychology, neurobiology and animal ecology – and yet before the boom in popularity of such self-consciously "interdisciplinary fields" as "artificial intelligence", "dynamic systems research" or "cognitive neuroscience" – Sebeok, the academic polymath who once described himself as something akin to academic *Apis mellifera*, who "dart solitary from flower to flower, sipping nectar, gathering pollen [and] serendipitously fertilizing whatever they touch",[19] was to pioneer the practices that modern-day university refers to as "interdisciplinarity" in the course of founding the project that today bears the title of *biosemiotics*.

In short, Sebeok's search for the higher-order explanatory logic that could account for both human and non-human communicative practices – and, later, for the naturalistic emergence of meaning-bearing processes and relations of every kind in living systems – led him to the works of semiotician Juri Lotman, logician and philosopher Charles Sanders Peirce, and biologist Jakob von Uexküll whose pioneering volume *Theoretische Biologie* (1928) "unfolded a wholly unprecedented, innovative theory of signs, the scope of which was nothing less than semiosis in life processes in their entirety" enthused Sebeok.[20]

[18] Danesi and Valdman 2004, p. 313.

[19] Sebeok 1995, p. 121.

[20] Sebeok 1998, p. 32.

Sebeok's subsequent three decades worth of work in developing the circle of researchers who would aid him in the project of developing the contemporary inter-discipline of Biosemiotics have been well documented for those interested in following his trajectory from words to signs.[21] Two major bibliographical surveys[22] of books and journal articles authored by Sebeok at the time of his death list almost 600 single-authored or co-authored entries. Reviews, forewords, encyclopedia articles, and editorial work promoting other scholars raise the total number of publications to which Sebeok contributed to over 1200 scholarly works.[23]

Indeed, Sebeok's obituary states that of all of his 81 years worth of accomplishments, "he was most proud of having [brought into being] a group of theoretical biologists and semioticians to pursue this field of investigation".[24] This "bringing together" consisted not just in Sebeok's indefatigable efforts at creating publication venues for these authors, encouraging them in their often neglected efforts, and bringing their ideas to a larger audience by citing them insistently in his own widely-read work – all of which he did unceasingly – but, much more importantly, in his tireless efforts to realize an active and ongoing, cross-disciplinarily *community* of scholars who would work together on the puzzles of organismic sign-processing long after he was gone.

Yet, Sebeok's "parallel interest in animal communication eventually led to his 'transfiguration from a linguistic technician to a practicing semiotician' (Sebeok 2011: 455), but the implications that his later findings hold for linguistics proper have yet to be widely considered or appreciated".[25] For from his first book *Spoken Hungarian: The Basic Course*, published in 1944, when he was only 24 years old, to his last book that was published in his lifetime, *Forms of Meaning* (2000), 56 years later at the age of 80, Sebeok's organizing focus and passion – no matter how all-envelopingly it developed in an "ever-widening centrifugal" vision – never strayed from the central human puzzle and inquiry into "the meaning of meaning" that he was introduced to as a 16 year-old in the classes of I.A. Richards.

Indeed, throughout his development from linguist to anthropologist to biosemiotician, Sebeok came to refer to linguistics simply, but respectfully and warmly, as "the branch of semiotics that is devoted to [the study of] verbal signs"[26] – and this subset of biosemiosis, too, he would taxonomize as falling securely within the realm of human culture, i.e., "that realm of nature where the logosphere – Bakhtin's dialogic universe – impinges in infant lives [and] then comes to predominate in normal adult lives".[27]

It is with the view of signs *qua* signs that biosemiotics, via this volume, now returns to Sebeok's home discipline of Linguistics to see what new cross-fertilization can result.

---

[21] Cf. Cobley et al. (eds.), 2011.

[22] Deely (ed.), 1995; Umiker-Sebeok 2003.

[23] Cf. also Deely and Danesi (eds.), 2012.

[24] Brier 2003, p. 104.

[25] Pelkey 2012.

[26] Sebeok 1986, p. vii.

[27] Sebeok 2001, p. 69.

The Latency of Biosemiotics in the Study of Linguistics

Sebeok had already proposed the term *zoosemiotics* for the inquiry into animal semiosis as far back as 1963,[28] and since 1976 was well on his way to developing his notion of an even more encompassing *biosemiotics* after his re-discovery of the works of Jakob von Uexküll, whose *Theoretische Biologie* explicated the meaning-imprinting agent-object relations that comprise every organism's experiential world. Long before the development of human language, Sebeok realized from his reading of Uexküll, the relations whereby one thing "stands for" something other than itself – which is the canonical definition of a sign relation[29] have been underpinning and organizing the lives of all creatures.

Accordingly, Sebeok felt no need to *demote* the zoosemiosis of the animal world to the status of a "protolanguage" but rather, to investigate it as a robust network of interacting meaning-making interactions taking place within and across living systems – rather than as a formal, and possibly purely psychological, system of combinatory elements and rules. Indeed, with the application of Peirce's architectonic logic of irreducibly triadic and recursively interdependent, multi-level sign relations, Sebeok and his growing coterie of biologists and semioticians began to deeply challenge the prevailing Saussurean dyadic understanding of a sign as "the union of sense and sound pattern, both parts of the sign being psychological"[30] – along with the concomitant assumption that to understand human meaning-making, one must look first and foremost at human language, and its particular (and, as Deacon[31] will later argue, evolutionarily late and even semiotically anomalous) set of structural inter-relations.

Clearly, what Sebeok and his colleagues in the biosemiotic movement were doing was to attempt to understand language as a more biologically emergent (or exapted) system of sign relations – while, at the same time but from the other direction – an equally popular (if not more so) tendency was developing on the part of various other researchers attempting to understand biology as operating within the principles of a kind of language-based system in itself.

The discovery of the genetic code and the birth of molecular biology in the 1960s provided great impetus for this view, which is indeed, still taught widely in most Western universities today. "Science can now translate at least a few messages writ-

[28] Sebeok 1963, p. 465 and 1972, pp. 178–181. The 1960s through the 1980s saw a great deal of interest in the idea that the study of behaviour and communication in different species might be thought to reveal mechanisms that could be interpreted as antecedents to language. Charles Hockett's "design features" analysis, by which the evolutionary sequence from "animal communication" to "human language" is supposed to be derivable (e.g., Hockett and Altmann 1968) well exemplifies this approach. Sebeok himself was a central figure – and naysayer! – in the debates over claims that apes, parrots, and other animals were successfully being taught how to use human language by scientific researchers in the 1980s (e.g., Sebeok and Umiker-Sebeok [eds.], 1980).

[29] Deely 1990; Peirce 1865 [1982].

[30] Saussure 1916 [2013, p. 17].

[31] Deacon 1997.

ten in DNAese into the chemical language of blood and bone and nerves and muscle", wrote George and Muriel Beadle in 1966, "one might also say that the deciphering of the DNA code has revealed our possession of a language much older than hieroglyphics, a language as old as life itself, a language that is the most living language of all".[32]

This has led to the application of certain general linguistic concepts and models to the description and analysis of biological systems. Examples of these types of approaches include the application of the concepts of structuralist linguistics (cf. Saussurean semiology) in biology[33] and of Chomskyan "tree-structure" mappings (and later, the development of binary-based neural net architecture) in the quest to build "natural language processing" machines and programming. Because so many of these models assumed a semiological or purely computational approach, however, they were still unable to satisfactorily model the kind of complex, adaptive, self-organizing behaviors characteristic of actual biological systems embedded in information-consequential environments.

Biosemiotics, accordingly, is an attempt to re-frame biology in terms of the sign-processes that give rise to intelligent behavior, rather than vice versa. And it is precisely this perspective that we believe may be of use to those examining "language" phenomena, as well.

Language Is Not the Primary Creator or Locus of Meaning: Aspects of the Biosemiotic Perspective Relevant to the Advancement of Linguistic Inquiry

Biosemiotics' primary focus is on the underlying sign processes that accompany the formation and impact of each linguistic sign, on those levels of a language user's life that may not be consciously given – i.e., on the various biological-level processes of recognition, memory and emotion. These psychological and physiological processes include the meaning-making mechanisms without which no word can be understood or sensibly said. To the extent, then, that biosemiotics may be able to provide useful models for understanding and explaining such phenomena on the biological level, its adoption by linguists could both inform and extend the boundaries of current inquiry in both fields.

Indeed, a biosemiotically extended linguistics would differ from most traditional linguistics in the recognition that "signs" occur on many other levels of life processes, most of which do not take the form of – but yet themselves inform – the practices of linguistically-aided semiosis. Since the discovery of sign processes operative in nature that do not derive from human culture, and the corresponding

[32] Beadle and Beadle 1966, p. 207.

[33] Jakobson 1971 and 1988; Florkin 1974.

understanding regarding the placement of the lower semiotic threshold zones,[34] we can be certain that our life is informed by many sign processes that do not take the form of human language. Such pre-linguistic biological sign relations often co-occur with language in the lives of human beings, however, participating in producing and understanding linguistically-aided semiosis, yet are rarely studied in linguistics proper.

The role of biosemiotics that we envision for linguistics, then, is an extension of the current importance of general semiotics for linguistic inquiry. Accordingly, we can list briefly some theoretical aspects through which biosemiotic studies can have an importance for linguistics.

1. The recognition that levels of semiosis are not independent.[35] In an organism as a whole, or in the semiosphere as a whole, sign processes are interrelated and interconnected into complex, mutually reinforcing webs. Such an understanding includes several implications that are germane to future linguistic inquiry, including the following observations:
   (a) Semiotic systems are distributed systems. Signs are relations, and thus all semiotic systems take the form of self-organizing webs, or swarms. As Paul Cobley puts it, "the most obvious cultural implication of biosemiotics is its abolition of the individual/collectivity dyad".[36] This general feature has been well explained by biosemiotic mechanisms,[37] and it is not at all restricted to linguistics. The same is true for all general semiospheric features – e.g., heterogeny, distinctions between centre and periphery, border effects, etc.[38]
   (b) Symbolic relations presuppose indexical relations which presuppose iconic relations.[39] Thus, a careful consideration of the biosemiotic (e.g., pre-linguistic iconic and indexical) levels should be inescapable in any principled examination of the underlying working of symbolic systems such as human language.
   (c) The biological processes that underlie the accomplishment of linguistically aided semiosis (both in its comprehension and production) are obviously themselves not explainable on the level of language – rather, such semiosis requires the existence and participation of more fundamental sign processes which have access to and are responsible for the needs of the communicative and self-communicative (living) systems involved.
   (d) With the inclusion of a biosemiotic level of semiosis, it becomes possible to extend or generalise the concept of translation. Instead of describing translation as a process that takes place only between sign systems, we can see that it also – and perhaps more fundamentally – takes place between *umwelten*.[40]

[34] On the latter concept, cf. Kull et al. 2009.
[35] Deacon 1997; Hoffmeyer 1993 and 2008.
[36] Cobley 2010, p. 240.
[37] Hoffmeyer 1993 [1996] and 2008.
[38] Lotman 1990.
[39] Deely 1990; Deacon 1997.
[40] Cf. Kull and Torop 2003.

2. In addition to recognizing the distributed nature of semiosis, biosemiotics also recognizes that such distributed sign processes also function as modelling processes, the creators of *umwelt*. Sign systems (including language) function as *modelling systems* for the organism that employ them. Such cognitive and modelling functions are of central focus in biosemiotics, which includes *umwelt study* and *cognitive semiotics* as its subfields.

   Research in such fields have shown that different major sign types are related to different mechanisms of learning.[41] Such learning mechanisms include learning via trial and error, associative learning, and imitative learning – all of which are pre-linguistic mechanisms without which human language itself cannot function.

   Another important aspect of this kind of research is that while the iconic and indexical level sign relations that are available to non-human animals are acquired exclusively on the basis of some presently existing regularities – thus inevitably "modelling" the world in a functional (i.e., not necessarily "mentalistic") sense – symbolic relations such as are found in human language, in contrast, provide their users the possibility to semiotically detach from the deterministic limitations of mere physical actuality, and in the case of formal languages, even to model the world in counterfactual, conditional, decontextualized and/or future predictive modes of reasoning, as well. It is for this reason that the models attained through formal purely symbolic language require the processes of the biosemiotic level to make these very models testable.
3. On the level of organisms, it has become apparent that sign systems develop within an organism through ontogeny.[42] *Developmental semiotics* is the part of semiotics that studies the ontogeny of semiotic capacities, including the early stages of language acquisition. Developmental semiotics is also an approach to describe and explain pre-linguistic logic – i.e., the stages of the complexes of sign relations (for example, the semiotic logic that results in the spatial orientation of animals) that are necessary precursors to the development of inferential, syllogistic, and ultimately, linguistic semiosis.
4. *Evolutionary semiotics* studies the general mechanisms of the evolution of sign systems. The study of behaviour and communication in different species has revealed mechanisms that can be interpreted as antecedents to language. In this case, the phenomena discovered are often seen as the developmental or evolutionary stages of semiosis towards the development of language.[43] The contemporary challenge to neo-Darwinian theory posed by the semiotic theory of evolution[44] demonstrates the primary role of epigenetic and learning processes in the evolution of semiotic systems.

---

[41] E.g., Kull 2014.

[42] E.g., Krampen 1991, p. 11 *sq.*; Linask et al. 2015.

[43] Uexküll 1940; Sebeok 1997; Hoffmeyer 2012.

[44] E.g., Deacon 2011.

Conclusions

Covering dozens if not hundreds of different specialized sub-fields dedicated to the study of language, *linguistics* is a word used much like *biology* within its twin homes of academia and research. That is to say that it is an "umbrella term" for a huge number of different undertakings, across a number of quite radically different levels of organization, each of which examines a set of phenomena of quite a different nature than the others, and all of which are joined together only by some vague family resemblance to the study of "language" – which is itself, *in toto*, a concept upon which there is no settled definition.

Thus, the speech pathologist, the philologist, the syntactician, the linguistic anthropologist, and the neurolinguist all labor away – much like the virologist, the physiologist, the ecologist and the cardiologist – making profound breakthroughs in their specialized fields of endeavor… yet almost entirely speaking across each other and failing to attempt to synthesize the results of all these findings in the effort to discover the higher-order logic that joins together all these constituent phenomena in a satisfactorily explanatory way.

Also, like biology, linguistics has established some of its firmest claims – about phonetics, phonology, morphology, and formal syntax – through the application of experimental science upon combinatorial units held artificially very discrete for the purposes of the analysis. Analyses offered about higher-order levels of organization – e.g., semantics, pragmatics, discourse analysis and linguistic anthropology – are far less "settled" and convergent. As is the case in any other science, the success of breaking up the object of study into dedicated individual "disciplines" has come with a price for the effort to effectively understand the whole – which will not become reconstituted by a simplistic "adding together of all the parts".

The reason why such an "additive logic" will fail illuminates the higher-order emergent logic that we seek. For just as the most fine-grained explanation of the precise biophysics and biochemistry of neuronal activation in the brain can, by itself, tell us nothing at all about the images, thought or subjective experience being "represented" or "communicated" by such physio-chemistry, nor can an exclusive examination of the words used by a group of people tell us how those words are endowed with the power to organize activities, identities and lives.

In both cases, what is lost by a too-exclusive focus on the constituent components of a system is the realization that those components themselves arose as a result of a distributed dynamic of being-in-the-world that is both the origin and the prevailing ongoing force behind their current being and use as parts of a larger system. "Life is organized according to a semiotic dynamic", writes Jesper Hoffmeyer,[45] and this distributed dynamic of being-in-the-world (and of beings-in-their-worlds) is the focus of the interdisciplinary research agenda called biosemiotics.

[45] Hoffmeyer 2008.

Given that language, too, it is now well known, is an emergent, dynamic, distributed and semiotically complex adaptive system, the addition of the biosemiotic perspective to the future of language study seems a direction well worth pursuing.

References

Andresen, J. T., et al. (2014). *Linguistics and evolution: A developmental approach*. Cambridge: Cambridge University Press.

Atkinson, Q. D., & Gray, R. D. (2005). Curious parallels and curious connections – Phylogenetic thinking in biology and historical linguistics. *Systematic Biology, 54*(4), 513–526.

Beadle, G., & Beadle, M. (1966). *The language of life: An introduction to the science of genetics*. Garden City: Doubleday.

Brier, S. (2003). Thomas Sebeok: Mister (bio)semiotics. *Cybernetics and Human Knowing, 10*(1), 102–105.

Chomsky, N. (1965). *Aspects of the theory of syntax*. Cambridge: MIT Press.

Chomsky, N. (2007). Approaching UG from below. In U. Sauerland & H. M. Gärtner (Eds.), *Interfaces + recursion = language?* (pp. 1–29). New York: Mouton de Gruyter.

Cobley, P. (2010). The cultural implications of biosemiotics. *Biosemiotics, 3*(2), 225–244.

Cobley, P., Deely, J., Kull, K., & Petrilli, S. (Eds.). (2011). *Semiotics continues to astonish: Thomas A. Sebeok and the doctrine of signs* (Semiotics, communication and cognition, Vol. 7). Berlin: Mouton de Gruyter.

Danesi, M., & Valdman, A. (2004). Thomas A. Sebeok. *Language, 80*(2), 312–317.

Deacon, T. W. (1997). *The symbolic species: The co-evolution of language and the human brain*. London: Penguin.

Deacon, T. W. (2011). *Incomplete nature: How mind emerged from matter*. New York: Norton.

Deely, J. (1990). *Basics of semiotics*. Bloomington: Indiana University Press.

Deely, J. (Ed.). (1995). *Thomas A. Sebeok: Bibliography 1942–1995*. Bloomington: Eurolingua.

Deely, J. (2002). *What distinguishes human understanding*. South Bend: St. Augustine's Press.

Deely, J., & Danesi, M. (Eds.). (2012). *Semiotic prologues*. Ottawa: Legas.

Di Sciullo, A. M., & Boeckx, C. (Eds.). (2011). *The biolinguistic enterprise: New perspectives on the evolution and nature of the human language faculty*. Oxford: Oxford University Press.

Favareau, D. (Ed.). (2010a). *Essential readings in biosemiotics: Anthology and commentary* (Biosemiotics, Vol. 3). Dordrecht: Springer.

Favareau, D. (2010b). Introduction: An evolutionary history of biosemiotics. In Favareau (Ed.), 2010, pp. 1–77.

Florkin, M. (1974). Concepts of molecular biosemiotics and of molecular evolution. In M. Florkin & E. H. Stotz (Eds.), *Biochemistry* (Comprehensive biochemistry, Vol. 29A, pp. 1–124). Amsterdam: Elsevier.

Hockett, C. F., & Altmann, S. (1968). A note on design features. In T. A. Sebeok (Ed.), *Animal communication: Techniques of study and results of research* (pp. 61–72). Bloomington: Indiana University Press.

Hoffmeyer, J. (1993 [1996]). *Signs of meaning in the universe*. Bloomington: Indiana University Press.

Hoffmeyer, J. (2008). *Biosemiotics: An examination into the signs of life and the life of signs*. Scranton: Scranton University Press.

Hoffmeyer, J. (2012). The natural history of intentionality: A biosemiotic approach. In T. S. S. Schilhab, F. Stjernfelt, & T. Deacon (Eds.), *The symbolic species evolved* (Biosemiotics, Vol. 6, pp. 97–116). Dordrecht: Springer.

Jakobson, R. (1971). Linguistics in relation to other sciences. In R. Jakobson, *Selected writings, vol. II: Word and language* (pp. 655–696). The Hague: Mouton.

Jakobson, R. (1988). Die Biologie als Kommunikationswissenschaft. In R. Jakobson, *Semiotik: Ausgewählte Texte 1919–1982* (pp. 367–397). Frankfurt am Main: Suhrkamp.
Koerner, K. (Ed.). (1983). *Linguistics and evolutionary theory: Three essays by August Schleicher, Ernst Haeckel, and Wilhelm Bleek* (Amsterdam classics in linguistics, Vol. 6). Amsterdam: John Benjamins.
Krampen, M. (1991). *Children's drawings: Iconic coding of the environment.* New York: Plenum Press.
Kull, K. (1999). Biosemiotics in the twentieth century: A view from biology. *Semiotica, 127*(1–4), 385–414.
Kull, K. (2014). Catalysis and scaffolding in semiosis. In K. R. Cabell & J. Valsiner (Eds.), *The catalyzing mind: Beyond models of causality* (Annals of theoretical psychology, Vol. 11, pp. 111–121). New York: Springer.
Kull, K., & Torop, P. (2003). Biotranslation: Translation between umwelten. In S. Petrilli (Ed.), *Translation translation* (pp. 315–328). Amsterdam: Rodopi.
Kull, K., Deacon, T., Emmeche, C., Hoffmeyer, J., & Stjernfelt, F. (2009). Theses on biosemiotics: Prolegomena to a theoretical biology. *Biological Theory, 4*(2), 167–173.
Linask, L., Magnus, R., & Kull, K. (2015). Applying Jakob von Uexküll's concept of umwelt to human experience and development. In G. Mey & H. Günther (Eds.), *The life space of the urban child: Perspectives on Martha Muchow's classic study* (pp. 175–192). New Brunswick: Transaction Publishers.
Lotman, J. M. (1990). *Universe of the mind: A semiotic theory of culture.* London: I.B. Tauris.
Maran, T., Martinelli, D., & Turovski, A. (Eds.). (2011). *Readings in zoosemiotics* (Semiotics, communication, and cognition, Vol. 8). Berlin: Mouton de Gruyter.
Peirce, C. S. (1865 [1982]). Teleological logic. In E. C. Moore & M. H. Fisch (Eds.), *Writings of Charles S. Peirce: A chronological edition, vol. 1: 1857–1866* (pp. 303–304). Bloomington: Indiana University Press.
Pelkey, J. (2012). Semiotics continues to astonish [book review]. *Linguist List* 23.733 (http://permalink.gmane.org/gmane.science.linguistics.linguist-list/35554; website consulted on February, 16 2014).
Ritt, N. (2004). *Selfish sounds and linguistic evolution: A Darwinian approach to language change.* Cambridge: Cambridge University Press.
Saussure, F. de (1916 [2013]). *Course in general linguistics.* London: Bloomsbury Academic.
Sebeok, T. A. (1963). Communication among social bees; porpoises and sonar; man and dolphin. *Language, 39*, 448–466.
Sebeok, T. A. (1972). *Perspectives in zoosemiotics.* The Hague: Mouton.
Sebeok, T. A. (1986). Preface. In T. A. Sebeok (Ed.), *Encyclopedic dictionary of semiotics* (Vol. 1, pp. v–viii). Berlin: Mouton de Gruyter.
Sebeok, T. A. (1995). Into the rose-garden. In J. Deely (Ed.), *Thomas A. Sebeok: Bibliography 1942–1995* (pp. 116–125). Bloomington: Eurolingua.
Sebeok, T. A. (1997). The evolution of semiosis. In R. Posner, K. Robering, & T. A. Sebeok (Eds.), *Semiotics: A handbook on the sign-theoretic foundations of nature and culture* (Vol. 1, pp. 436–446). Berlin: Walter de Gruyter.
Sebeok, T. A. (1998). The Estonian connection. *Sign Systems Studies, 26*, 20–41.
Sebeok, T. A. (2001). Biosemiotics: Its roots, proliferation, and prospects. *Semiotica, 134*(1–4), 61–78.
Sebeok, T. A. (2011). Summing up: In lieu of an introduction. In Cobley, Deely, Kull, Petrilli (Eds.), 2011, pp. 451–459.
Sebeok, T. A., & Umiker-Sebeok, J. (Eds.). (1980). *Speaking of apes: A critical anthology of two-way communication with man.* New York: Plenum Press.
Short, T. L. (1988). Why we prefer Peirce to Saussure. In T. Prewitt, J. Deely, & K. Haworth (Eds.), *Semiotics 1988* (pp. 124–130). Lanham: University Press of America.
Tallis, R. (2011). *Aping mankind: Neuromania, darwinitis and the misrepresentation of humanity.* London: Acumen.

Uexküll, J. von (1928). *Theoretische Biologie* (2nd Rev. ed.). Berlin: Verlag von Julius Springer.
Uexküll, J. von (1940). *Bedeutungslehre* (BIOS. Abhandlungen zur theoretischen Biologie und ihrer Geschichte, sowie zur Philosophie der organischen Naturwissenschaften, Vol. 10). Leipzig: J.A. Barth.
Umiker-Sebeok, J. (2003). Thomas A. Sebeok: A bibliography of his writings. *Semiotica, 147*(1), 11–73.

Language and Biosphere: Blurry Contours as a Condition of Semiosis

Anton Markoš and Dan Faltýnek

Abstract The first part of this paper shows how our concepts (seemingly precise and clear-cut) in both linguistics and biology in fact represent very fuzzy and improperly defined entities that shimmer against the background of the world. The second part develops on such a knowledge: we argue that both language and evolution are distributed among many entities and processes on many time- and space- scales. As such, they can be grasped only by tools that allow extraction of meaning – like semiosis and/or narration.

Keywords Discrete and digital • Evolution • Culture

> *In a collective world, individual experience acquires implicit content based on acting in a cultural landscape. […] Older structures shape impressions that prompt us to action. […] Intuitive dealings with the world are increasingly shaped by the shared content. With mimesis, hominids make bodily displays based in traditions as they become players on the stage of life. […] Indeed, the language is so embedded in action that transcription leaves the event opaque.*
>
> *(Cowley 2012, pp. 20–21, 30)*

> *Dispersers, for instance, bring not only their genes into their new population, but also their phenotype, which brings key information on the conditions that prevail outside of the population. They also bring their cultural habits (e.g., dialects),*

This research was supported by the Grant Agency of Czech Republic 13-24275S (AM), and by the grant CZ.1.07/2.3.00/20.0161, financed by the European Social Fund and the National Budget of the Czech Republic. We are grateful to Stephen Cowley who helped to comb the blurry contours of the manuscript and to Fatima Cvrčková (Figures ©).

A. Markoš (✉)
Charles University, Praha, Czechia
e-mail: markos@natur.cuni.cz

D. Faltýnek
Palacky University, Olomouc, Czechia
e-mail: dan.faltynek@upol.cz

E. Velmezova et al. (eds.), *Biosemiotic Perspectives on Language and Linguistics*, Biosemiotics 13, DOI 10.1007/978-3-319-20663-9_3

> *so that high immigration rates can lead to cultural meltdown in a single generation, which is equivalent to the loss of a genetic structuring. Such cultural meltdown should affect the inclusive heritability of a local population and, thus, its evolutionary dynamics.*
>
> (Danchin 2013, p. 356)

Life as a Text: What Is the Metaphor?

The metaphor of life – or even of the world – as a text has been prolific in our civilization for many centuries. *Books of Life* abound in religious and scientific literature, as well as in fiction.[1] What is meant implicitly or explicitly by "book" is a text written by an alphabetic string of "digits". As shown by Roy Harris[2] the trend can be traced to Aristotle, who tends not to differentiate between letters (*grammata*) and sounds (*stoicheia*). The introduction of writing was, according to Harris, an epoch-making task that has led to the "scriptism", a way of thinking fundamentally different from "primitive" thinking. The difference resides in the fact that the alphabet (and later also special sets of numerals) establishes invariants of the "digital" world (note that all such digital inscriptions can be, using a simple algorithm, transformed into a string in binary code). A letter of the alphabet dwelling in the virtual realm has no shape (form); its defining properties are (1) its position (succession) in a given alphabet, and (2) the fact that it is unambiguously different from other digits (i.e. no transient digits are allowed, e.g., something "between", say, A and B). Moreover, (3) the script contains *only* digits belonging to a given alphabet, and those digits are immutable, do not alter into other digits of the set. The script also does not accept anything not belonging to the given alphabet (of course, this precondition can be somewhat mollified by introducing, e.g., capitals, spaces, commas, italics, numerals, etc.).

Scriptism resides in projecting the properties of reality into a digital, virtual world whose elements can be distinguished, or calculated, absolutely (i.e. attainable with whatever accuracy chosen). Such digital, accurate representations were, later, considered to be somehow "better" compared to the floating "real" world: elements of language, like speech sounds, words, meanings, even the truth of proposals, are *stipulated by letters* (*digits*), elements of living beings are stipulated by genes. Illustrative in this respect is the view of Marcello Barbieri[3] who assumes such (absolute) correspondence between the written code and elements of natural language, and coins the idea of "code biology" drawn from an analogy. Similarly, Jesper Hoffmeyer and Claus Emmeche[4] proposed a view of life based on code duality, in

[1] Cf., e.g., Auerbach 1946; Blumenberg 1986.

[2] Harris 2009.

[3] Barbieri 2003, p. 94 *sq*.

[4] Hoffmeyer and Emmeche 1991.

interplay of digital and analogue information. However, the very idea of correspondence and similarity between two "worlds" is dubious.[5] Elsewhere, we argue[6] that the very possibility of attaining digitality in the real, bodily world is an impossible task: digitality is a quality of the artificial world of objective reality. What has been considered "digital" by many authors is, in fact, discreteness characterized by more or less blurred contours of basic units. Hence, biosemiotics (and semiotics in general) cannot even be conceivable outside the realm of foggy appearance of things dwelling in the real world: semiosis works with concepts dependent on contexts, not with clear-cut, immutable scientific terms belonging to the realm of formal language and objective reality.

It follows that distinguishing discrete units in the real world is always the result of an effort developed by individuals or communities of living beings (e.g., users of a language). Even if they distinguish such quasi-digitality of elements intuitively, it is impossible to do so with absolute accuracy as in case of digital units from the virtual realm: instead, distinguishing is a result of historically established, and/or individually recognized, differences. Hence, "discretion" (B is B, and not V or something in between), is never absolute, it is given by the context of the epoch, community, or the topic. Yet, in our culture, there is a tendency to overlook such a principal difference between the virtual (digital) and real worlds; natural sciences offer a good example.[7]

Not surprisingly, the discovery of the DNA structure in the 1950s fits this picture. Taking a scriptist view, Roman Jakobson asserts that "the deciphering of the DNA code has revealed our possession of a language much older than hieroglyphics [*sic. – A.M., D.F.*], a language as old as life itself, a language that is the most living language of all".[8] He believes that "the genetic code, the primary manifestation of life, and, on the other hand, language, the universal endowment of humanity and its momentous leap from genetics to civilization, are the two fundamental stores of information transmissible from the ancestry to the progeny".[9] In such an atmosphere, DNA was perceived (in tune with the scriptic tradition) as an inscription of the "language of life". Oddly, it was assumed that in both DNA and languaging, the very inscription (not acts of articulation and interpretation) was *the* "representation" of language. For the same reason, human apprehension of language was reduced to the use of a "code" that specified properties of the living. Even more remarkably, no effort was made to find out how this strange language might be "spoken". Indeed, for many the existence of a code was seen not only just necessary, but also a sufficient condition for a language. This taking code and language for synonyms was rooted in, and supported by, the expansion of formal languages developed in computer sciences. Since formal languages *are* codes, there occurred a loss of cultural sensitivity to how they differ from their natural counterparts. Rather, there is a very

[5] Cvrčková and Markoš 2005.

[6] Markoš and Faltýnek 2011; Markoš et al. 2009.

[7] Markoš and Faltýnek 2011.

[8] Jakobson 1971, p. 678.

[9] *Ibid.*, p. 681. The text does not seem to differentiate between language and script.

general tendency to conflate the formal appearance of the structure with its "content". This is possible because, in fact, many aspects of language *can* be modeled by formal instruments. In words of Umberto Eco: "As long as these simplifying models succeed in explaining many phenomena, they may well reproduce some 'natural' order or reflect some 'universal' functioning of the human mind. [It is important to avoid] the *ultimate* assumption that, when succeeding in explaining some phenomena by unified structural models, one has grasped the format of the world as an ontological *datum*".[10]

In the light of simplifying approaches the metaphor of DNA being a code *and*, at the same time, a (formal) language has been taken for granted.

Beyond College Textbooks

Indeed, textbooks abound with statements like the following: "[A] living system must be able to replicate itself. To do so, an organism must possess a *complete description of itself*. This description [...] specifies *every step* required for a cell to construct an exact replica. As each generation begets the next, a copy of the instruction set is given to each descendent. [...] [T]he information required for self-replication resides in the genetic material as a molecule called DNA".[11] This leads to a conclusion: the habitus, or appearance of every living being (seen as an analogy of articulate speech) can, and should be explained as an "utterance" of a sort prescribed by its genetic material.[12]

We are far from ridiculing such deterministic axioms – after all, they have begotten a piece of solid science. In this context, our modest goal is to stress that the inherent scriptism gives rise to simplifications. A world that is modelled in line with such axioms leaves no space for (bio)semiotics. In the present work, we disregard the fact that it undoubtedly *is* in *our* powers to inscribe, i.e. reduce quite precisely, ultimately even in a digital way, both language and life, and to create their scientific, i.e. formal, models. In what follows, we reject the scriptist view, to bring home the parallels between spoken language and other appearances of the living. By so doing we aim to assess whether, and how, the natural-language metaphor can help the understanding of both the natural and the artificial.

We start with examples demonstrating how seemingly digital elements of our world have blurred contours upon closer inspection; later we justify the language metaphor of life, i.e., the analogy between the structure of language and life. In both cases the basic process entering the fluffy (non-digital) flow of events is semiosis: deciphering the signs based on experience, written records, and contexts, and weaving a narrative that allows an understanding history that gives the world its sense.

[10] Eco 1976 [1979, p. 47].

[11] Rawn 1989, p. 665; italics ours. – *A.M., D.F.*

[12] For criticism of such a free interchange between the virtual/digital, and natural world/language, cf. Love 2004 and 2007; Port 2010; Markoš and Faltýnek 2011; Markoš et al. 2009.

All features are distributed in cultures whose members negotiate meanings of their life affairs.

Elements

Speech Sounds and Molecules

Speech sounds are recognized intuitively; this led many to assume that speech depends on underlying features or a code that allows even a child to differentiate the elements of its mother tongue. There is, however, an alternative view: learning to hear speech sounds may result from a process of sense-making. It depends on coming to attend to utterances as utterances of something.[13] The community of speakers will facilitate the learning.

Even an illiterate speaker (e.g., a little child) should be able to differentiate speech sounds of his/her mother language. Hence, the ability to understand speech sounds as *units* may be secondary, only after the child learns the alphabet and the rules of writing, counting, and creating alphabetical lists.[14] In short, there is no reason to assume that to grasp the complex features of a given sound one needs a descriptive apparatus involving some meta-language, or better, some of the methods of experimental linguistics (cf. Dr. Higgins in *My Fair Lady*). Efforts to classify articulate sounds of a given language and to describe their difference are motivated by the determination to recognize an order lying *behind* practical speaking. They are very old: Indian or Latin textbooks from archaic era do successfully classify speech sounds in a way acceptable even today.[15] As a result of such a long tradition, the average literate person of our civilization will take it that sounds and letters are commensurable *via* a simple code. What follows is the conviction that speech sounds (as representatives of sounds) are as clearly distinguishable as are the letters (digits) of a written text. Even if spoken utterances may look quite disorganized and fuzzy,[16] it is believed that – in principle at least – they can be "articulated" into a sequence of digital, clearly delimited units. In fact, the situation is not that clear, and it is not easy to find physical invariants for phonemes: example in Fig. 1 illustrates the fact.

To speak a language, however, does not require such knowledge at all. To describe differences between sounds means to orient the language, the tool of our inquiry, to itself.

[13] E.g., Cowley and Vallée-Tourangeau 2013.

[14] Harris 2009.

[15] Priscianus circa 500 [2001].

[16] E.g., Cowley 2012; Steffensen et al. 2010; Favareau 2008.

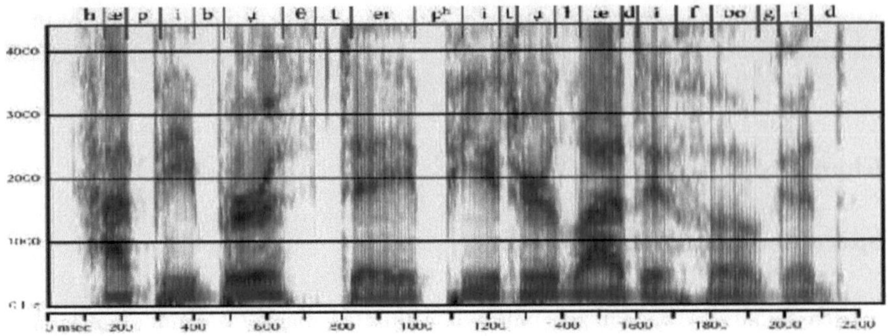

Fig. 1 A spectrogram of the utterance "Happy birthday, Peter Ladefoged!" (cf. http://home.cc.umanitoba.ca/~robh/archives/arc0512.html; website consulted in July 2014). Note from the transcription into the universal phonetic alphabet, that discreetness (but not digitality) can be established on the spectrogram by an experienced linguist. A skilled reader of spectrograms will be able to distinguish vowels and types of consonants. However (s)he cannot possibly know from the spectrogram signatures, what language is represented, or, of course, what is being said. *Horizontal* axis: time (ms); *vertical*: frequencies (Hz)

Jumping to Chemistry

A similar pattern arises when considering chemical literacy as it is used in contemporary biology. Chemistry adopted the "letter" notation for atoms almost two centuries ago: as words and higher structures of speech emerge from structured strings of letters, so molecules (and higher structures; even the world itself) emerge from structured composites of atoms.

The water molecule H_2O (or HOH) offers a simple example of such a linear notation. Even a more sophisticated structural formula of the molecule (Fig. 2a), showing the 105° angle of OH bonds, suggests that reality depends on a stable and well-determined structural relations (as a scriptist analogy of written words). In fact, the angle represents the statistical average of a very dynamic structure: it would be more "realistic" to view a molecule as a "cloud" without sharp boundaries. Thus, just as with the inscription that reduces the spectrogram (Fig. 1), such a linear, blackboard modeling of the molecule is a plausible, but still reduction.

The contrast between the linear inscription of the formula and the "real" shape of the molecule sticks out even more in another simple, 5-atom **non-linear** molecule of formic acid (Fig. 2b), linearly inscribed as HCOOH.

The unequal status of both hydrogen atoms is emphasized by the script. However, the *equal* status of both oxygen atoms is neither apparent, nor is it evident from the figure. The second hydrogen atom "belongs" equally to both, it can even dissociate away; but the two oxygen atoms are indistinguishable, forming with the adjacent C, a simple "cloud", the *carboxyl* group. (In fact, the whole molecule is such a cloud.)

If we now turn to a more complex molecule – amino acid methionine (a constitutive component of proteins, Fig. 2c), in addition to previous features, we notice the parts of the molecule can *rotate around* each bond (symbolized by sticks connecting individual atoms).

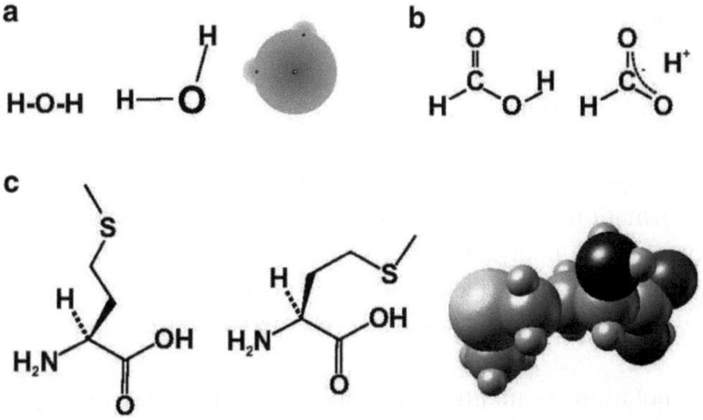

Fig. 2 "Digitality" of chemical notation. (**a**). Water molecule. *Left*: a scheme of an average emplacement of the three atoms (HOH) in the molecule. *Right*: the "electron cloud" model. Note that (i) the model is not to the scale: the atomic nuclei are very small and should be invisible at this magnification; (ii) the electron cloud is really an ever-changing field, it does not have an appearance of a soap *bubble*. (**b**). Formic acid. The carboxyl group —COOH defies any linear description; the most correct entry should be a kind of a pictogram e.g. ♋. With a perspective to memorize hundreds of pictograms, reduction to a string (plus a simple instruction on how to read it) is much more feasible, of course. (**c**). Methionine. *Left*: demonstration of two out of many conformations available to a relatively small molecule. *Right*: another way of depicting the contours of the molecule, in one of its many conformations (http://fineartamerica.com/featured/methionine-molecular-model-dr-mark-j-winter.html; website consulted in July 2014)

As said above in the example with formic acid, the first – carboxylic group is symmetrical (in spite of the notation), in that it does not matter if it rotates or not. The groups attached to the second carbon (—NH2,—H,—COOH, and—CH2—CH2—S—CH3) vary as to their mutual positions, and so do groups attached to 3rd, 4th carbon, and to sulfur atom. Obviously, there is no stable molecular shape, seemingly indicated by static formulas and structures in Fig. 2a–c. In summary, the "cloudy" appearance of the molecule may assume a plethora of shapes, and the number of possibilities increases with the size and complexity of a given molecule. All conformations come with some probability, the set of conformations "in use" depending, however, on the environmental context. As to the size ("weight"): if that of hydrogen atom is posed as 1, then water is 18, formic acid 46, and methionine 149. Compare to biological polymers: proteins and sugars have molecules in the range from tens of thousands up to millions, and DNA string can be even bigger: this fact points towards an enormous conformational *potential*.

Now, let us draw an analogy to much larger molecules "weighing" hundreds of thousands units; their cloudy nature comes to the fore even more clearly.

Cloudy Strings

Proteins are synthesized as long *linear* and *nonrandom* polymers of amino acids. The sequence in which amino acids are inserted into the chain will to a great extent determine the range of possible conformations of the given macromolecule. Not all possible conformations are allowed in the natural environment of a given protein; yet there exists a great many of them.[17]

As shown in Fig. 3, the sequence of amino acids in a particular protein chain can be favorably written down as a sequence of letters, each assigned to one particular amino acid.

The letter notation, as mentioned above, is a reduction from a great variety of shapes. By deriving the letter notation from a great variety of shapes, the researcher gains advantages enabling him analyzing the string, looking for motifs, repetitions, etc., by employing methods similar to those used in linguistic analysis. Even chemical modification of amino acid residues[18] can be easily expressed in such a letter form; however, one should not forget that any of such modifications may – even drastically – change the appearance of the molecular "cloud". Hence, the *shape* of a protein molecule (the "cloud") is a function not only of the sequence of "letters", but also of the context (internal environment of the cell, external cues, etc.) into which the protein is embedded. Here, the analogy with the situation in language (i.e. letter string versus utterance) may prove to be useful. On one side, there is a simple way of "digitalization" as a string of letters, on the other, both amino acids

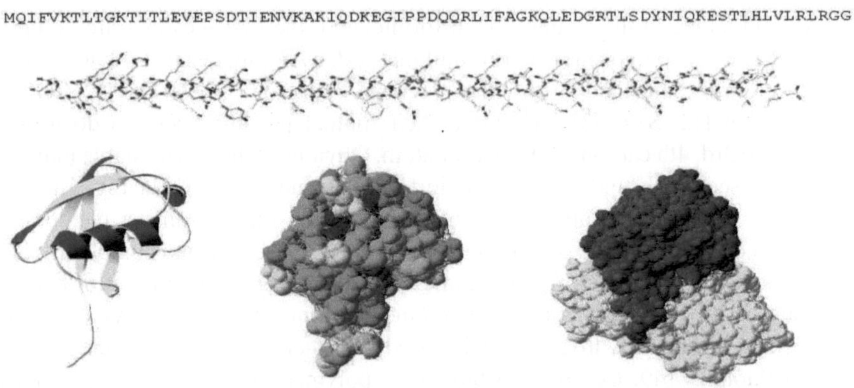

Fig. 3 Protein structure (example of small protein ubiquitine). *First line*: linear sequence of amino acids in a standard "letter" notation; below part of the stretched sequence of amino acids when letters are replaced by amino acid formulas as in Fig. 2c. *Lower line left*: a ribbon model of the spatial conformation (one of many possible functional structures) of the protein; middle: the contour model of the same structure; *right*: a complex of protein ubiquitine with another protein, ubiquitin ligase (source: Protein Data Bank: 1UBQ, 1FXT)

[17] More on the ecosystem of protein shaping the behavior of its members, cf. Markoš et al. 2013.

[18] As if introducing a diacritic of a sort, cf. Markoš and Švorcová 2009.

```
5´ACTTCTCCCTTTCAGTTTTCCTATGTCATTT 3´
3´TGAAGAGGGAAAGTCAAAAGGATACAGTAAA 5´
```

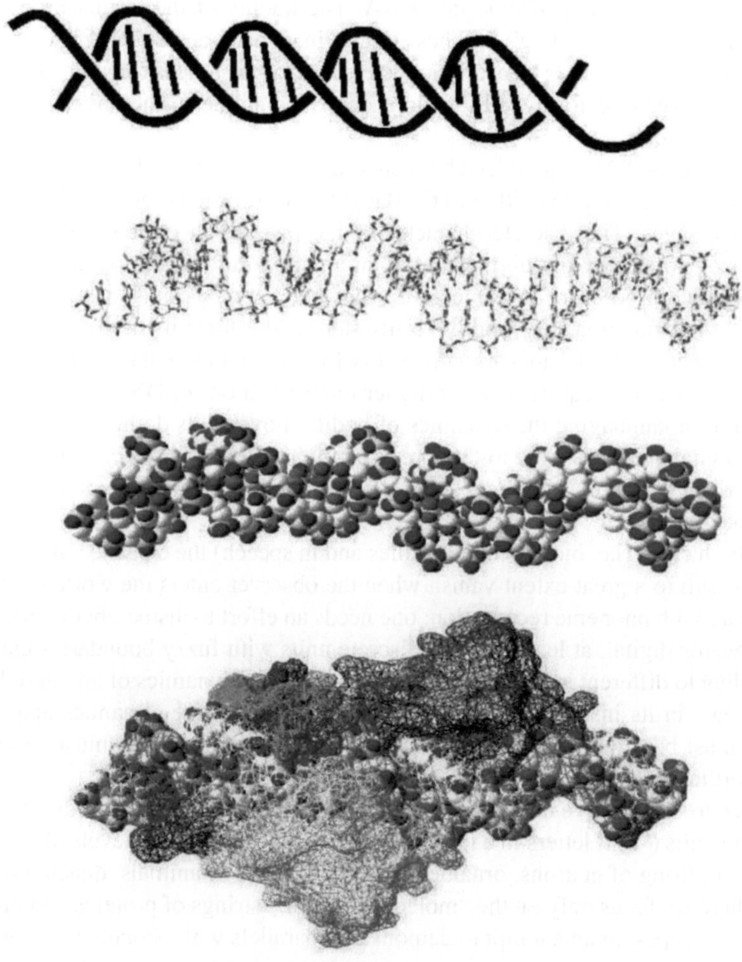

Fig. 4 DNA sequence and structure. The letter notation of the double helix is followed by the spatial ribbon sketch and its more detailed representations showing molecular residues, and the contour model. *Lowest part*: a complex of DNA with several DNA-binding proteins (IRF-3, ATF-2, Jun) (source: Protein Data Bank: 1T2K)

and speech sounds often assume far from typical "conformation" tailored according to the context. Once again there is a close analogy with languaging in that the letter string that describes an act of utterance entirely obliterates its dynamic complexity. Not all conformations of the "packed" string are allowed, yet the set may be quite big.

What was said for proteins holds for DNA and RNA as well, with their four-digit "alphabet" A, C, G, T (Fig. 4).

As shown above, the *bodily* character of the genuine molecule may not be a net digitalization. DNA, however, is a special case shaped by evolution towards the maximal fidelity of two key processes involving the copying of the string, i.e. DNA *replication*, or its *transcription* into RNA. The fidelity of the former process that includes also "proofreading", reaches extraordinary accuracy by making one "typo" per billion (10^9) "digits"; transcription is more error-prone, making one error in ten thousand. (Needless to say, all functions described are performed by sophisticated teams of proteins.)

At this level, DNA can indeed be seen as a medium store for linear, digital information that is transmitted either to the daughter copy, or to (shorter and redundant) RNA messages. The discrete character of the inscription is preserved *up to the bounds* of possibility of "digitalization" in a real world (similar as in human artifacts as microprocessors or CD records). However, at other levels where DNA also plays a vital function, its "bodily" traits step to the fore: if the reduction to the "string of beads" model appears satisfactory for interpretation of copying processes, it will not meet the requirements of higher-order functions of DNA. Note the parallel again: in languaging the dynamics of bodily movements dominate, whereas its quasi-digital qualities come to the fore in script and records. The case affords clear comparison with how a phonetician ascribes phonological structure to the read out of a spectrogram.

In both cases (i.e. biological molecules and in speech) the elegance of the digital writing will to a great extent vanish when the observer enters the world of shapes. Compare with phoneme recognition: one needs an effort to dismember utterances if not towards digital, at least towards discrete units with fuzzy boundaries that vary according to different speakers, and contexts.[19] Such a dynamics of an uttered word disappears in its inscription; the relation between the voice dynamics and digital script must be established by conventions, and such conventions must be learned (e.g., orthography).

Back to the putative counterparts of words in living beings: aperiodic strings of discrete units ("as if letters in a text") can be recognized at many levels of organization (e.g., firing of neurons, ornaments, vocal displays in animals, dancing figures, etc.); here we focus only on the "molecular" level – strings of proteins and/or DNA and their shapes, in an attempt to demonstrate parallels with phonemes and words.

Shaping the Message

Parallels abound also when we compare external and internal forces influencing the shape of 1D strings. The sketches in Fig. 4 aim to depict the floating external shapes of the DNA molecule. What looks, on the "digitalized" model, as a perfectly symmetrical helix of double rosary, is in fact a dynamical bodily formation, comparable

[19] Similar knowledge can be gained from other types of analogue recordings, e.g., cuts in the vinyl of the gramophone records.

to that in proteins. First, the nucleotide building blocks are not the same: in the real world, each "letter" has a characteristic shape with – as we have seen above – blurred contours; the lineup of nucleotides in the DNA string necessarily leads to a complicated bodily structure. Second, the dynamics of the molecule is modulated by factors, like (1) thermal motion inducing internal vibrations (as described on small molecules above), and (2) continuous bombardment by zillions of small molecules like water, ions, small organic molecules, etc. Such physical forces blur the image (shape of the string), introduce noise, but do not *contribute* to information content of the string. This is provided by two processes: (3) *specific* interaction with other molecules present (mostly proteins), and (4) introducing chemical "tags" on both DNA and DNA-binding molecules.

Hundreds of proteins interact with DNA in specific ways, recognizing specific shapes on the macromolecule. The binding region on DNA for a given protein is about six nucleotides ("letters") long, however, it is not "read" letter by letter; the protein binds to the unique *shape* created by interaction of that sequence, *and* by the surrounding milieu. The interaction is thus influenced not only by the presence or absence of partners involved (i.e. motifs on the protein and DNA), but also by ones distal to the site. They contribute to fine "tuning" (stretching, bending), causing the shape of the very long DNA molecule to be in an endless (but not random) dynamical transfiguration. By this way, the "reading" DNA "text" (in other words – gene expression) is being put into a proper context.[20]

The last factor discussed here concerns the memory of the system by attaching chemical tags to both DNA and DNA-binding proteins (by adding or removing functional groups, like, phosphate, methyl, sugar); such tasks are performed by another cohort of special proteins. No need to add that such tags ("diacritics") may again dramatically contribute to the overall shape of the DNA molecule – locally but also to a considerable distance; thus creating often long-lasting contexts (like keeping the cell type in a differentiated state). Again, a situation may be taken as a parallel to longer-lasting (but not eternal) protocols, rituals, polite phrases, etc. in a language.

Similar factors are at work in proteins that are also assembled as linear polymers of 20 constituting units – amino acids. Here, however, the linear factor comes to word only at the "assembly line": proteins never act as strings, but as pure and specific shapes that interact with other shapes to produce the resulting "utterances". The analogy with language is straightway: as in speech it matters little that the results can be analyzed as words, speech sounds, or written down with letters. Our sentences are not immediately parsed into words and word types: the utterance is intended, and also received, as a whole. The same holds also for the "living matter". In both cases we also notice another property: mutations in forms of "copy error", improper understanding, shift in the word usage, weird metaphors, etc., will have gotten mostly buffered by the overall context; on some occasion, however, a single

[20] More about this, cf. Markoš and Faltýnek 2011; Markoš and Švorcová 2009; Markoš et al. 2013.

error may cause a dramatic shift of context resulting in a new developmental pathway into the "adjacent possible".[21]

The principle is recurrent for all four biological macromolecules contributing, in tight cooperation, to biological form: DNA, RNA, proteins, and polysaccharides. At the cellular level, this flexible "ecosystem" of molecular "words" may be, perhaps, compared with utterances, and even narratives in linguistics. Moreover, depending on the context, the system will decide which part of DNA is to be exposed in a "digital" form, thus allowing differential synthesis of selected proteins; in the next step, newly appearing (or disappearing) proteins change the state, i.e. the context, of the very ecosystem.

Distributed in the World

Avatars: Critters or Zombies?

"Understandably, we have been so fascinated by the fantastic capacity of genes to encode and transfer information across generations that we have become oblivious to evidence of other mechanisms of inheritance", claims Étienne Danchin[22] in his review devoted to different forms of inheritance and heritability of traits (i.e. bodily appearances).

Danchin refers to genome-wide association studies based on a very high number of genetic markers in the genome (i.e. total DNA) of a given species (more than 500,000) that were compared to the heritability of (bodily) traits. It seems that heritability and variation correlated with genetic markers only in several per cent. It follows that the rest is a function of epigenetic factors like environment, learning, cultural habits and rituals, imprinting, cell (and body) structures, extracellular matrix, etc. (Fig. 5). The principal idea is that critters are *born* into a world with zillions such factors being negotiated *before* the fact of this birth[23]; this, again applies also to languaging.

Danchin introduces a metaphor of the avatar, i.e. "a material form taken by an abstract entity",[24] just as a CD is an avatar of a film or music, or DNA is an avatar for heritable information. The motif *Life as its own designer* giving a title and pervading our book[25] may easily replace the notion: we can work not with the concept of embodiment, but directly with bodies – of an individual, of a community, etc.

One is tempted to continue to consider any living being (or even a community) molded by avatars belonging to the experience of the lineage. It is expressed

[21] *Sensu* Kauffman 2000.

[22] Danchin 2013, p. 351.

[23] More on the topic cf. Kauffman 2000.

[24] Danchin 2013, p. 352. Reflecting the origins of the term (meaning 'incarnation' or 'embodiment') but in conflict with the nowadays common usage *sensu* 'virtual form taken by a material (human) entity' known e.g. from virtual worlds such as *Second Life*.

[25] Markoš et al. 2009.

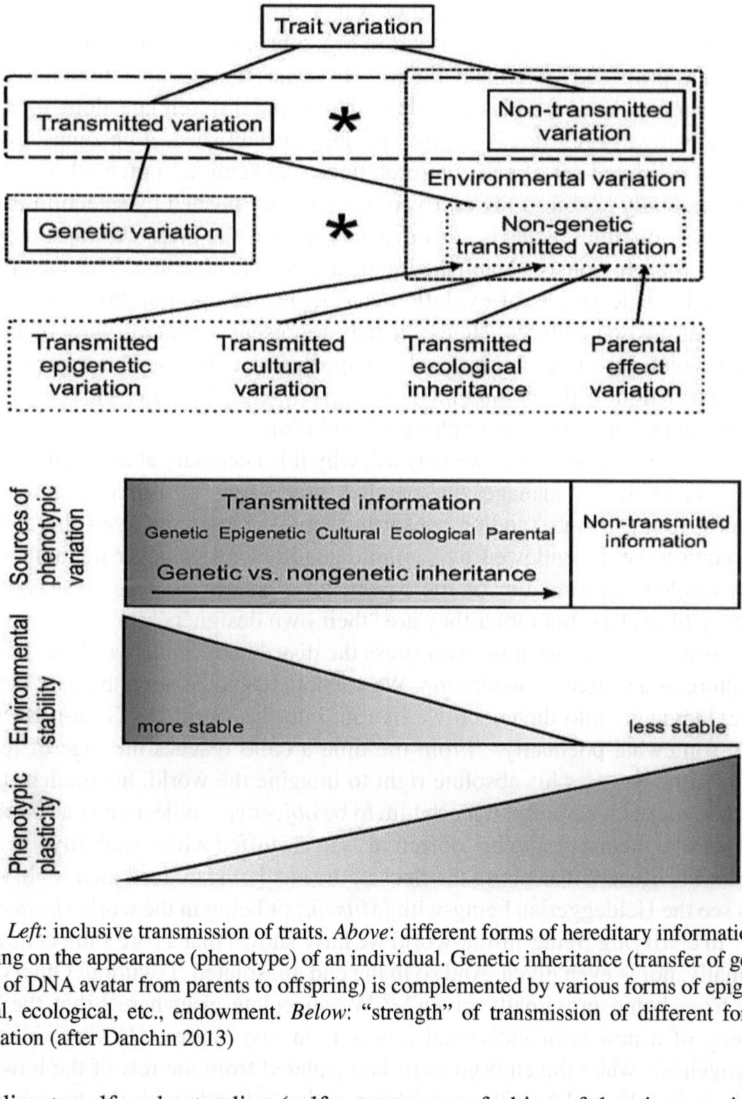

Fig. 5 *Left*: inclusive transmission of traits. *Above*: different forms of hereditary information participating on the appearance (phenotype) of an individual. Genetic inheritance (transfer of genes in a form of DNA avatar from parents to offspring) is complemented by various forms of epigenetic, cultural, ecological, etc., endowment. *Below*: "strength" of transmission of different forms of information (after Danchin 2013)

according to self-understanding (self-esteem, even fashion of the given period) of a given individual. Trans-generational transfer of any trait is thus mediated by a plethora of mutually overlapping avatars; of these DNA is but one in many – albeit most reliable as concerns the copying fidelity. The bias when preferring avatars hidden in DNA as a single avatar is due to the fact that information in DNA is encoded in a quasi-digital form. As a result, given contemporary techniques, it is easily readable, copy-prone, and easy to manipulate. If, however, we model living beings as creations of avatars coming from a single information source (DNA), we end up with Dawkinsian programmable automata; acting as zombies, without reflection, as perpetuators, copiers of information into the next generation. To get a full-fledged view, it is necessary to accept the idea that information put to use by living beings is *distributed* and not always in a digital, easy-to-grasp form as is the case in DNA. Moreover, the lower part in Fig. 5 brings us to the concept of frozen evolution as proposed by Jaroslav Flegr.[26] He argues that young communities (populations, species) are very plastic, able to invent and differentiate along many evolutionary pathways (cf. dogs). As the species is established, it becomes "frozen", relying on a limited set of self-interpretations: the semiotic potential of the community gradually becomes lower. Even when hardly pushed by selection in a particular direction, the species is *elastic* and as soon as the pressure stops, it quickly returns to the previously established appearance (cf. pigeons). The metaphor of plastic and elastic phases of evolution can be, in our opinion, broadened to languages, nations, or cultures. Hence, it may represent a general view of the life-course (evolution) of communities, be it species or cultures. The trajectory may end by extinction of the community, or – driven, e.g., by some shocking event – may rejuvenate and enter a new phase of evolution.

In the context of our quest, we may ask why it is necessary at all to introduce the avatar metaphor. To manage the parallel processing of different information resources, living beings (and/or consortia thereof) must command the ability of interpretation, i.e. be endowed by semiotic qualities, of whatever memory source, be they labeled "information" or, more poetically, "avatar". Living beings are not at the mercy of avatars, but rather they are "their own designers".[27]

This brings us back to those who stress the distributed character of our language and culture as a source of inspiration. We are not created de novo, but *born* into our culture, language, into the epoch we live in, into the world. As Gaston Bachelard puts it, somewhat poetically: "From the time a child reaches the 'age of reason', from the time he loses his absolute right to imagine the world, his mother, like all educators, makes it her duty to teach him to be *objective* – objective in a simple way adults believe themselves to be 'objective'. He is stuffed with sociability. He is prepared for his life as a man along the lines of the ideal of stabilized men".[28] In stricter words see the Heideggerian being-with (*Mitsein*) or being in the world (*in-der-Welt-sein*): "In clarifying Being-in-the-world we have shown that a bare subject never 'is' proximally, nor is even given. And so in the end an isolated 'I' without Others is just as far from being proximally given".[29] We speculate elsewhere[30] that the whole ontogeny of a new-born individual proceeds in two phases: (1) very short early embryogenesis when the embryo must be insulated from the rest of the biosphere; this phase is followed by (2) establishing multiple bonds with the biosphere and active participation in the business of the world.

---

[26] Flegr 2008.

[27] Kull 2000; Markoš et al. 2009.

[28] Bachelard 1960 [1971, p. 107].

[29] Heidegger 1927 [1962, §25].

[30] Pátková et al. 2012.

Mimesis

In a seminal paper "Mimesis and language" Stephen J. Cowley argues as if he were complementing what we learned above from Danchin's paper.[31] The work is a representative for a rich body of literature that pursues the idea of *distributed language*: "As we concert activity, we re-enact the phonetic gestures and visible expression of our fellows. We come to hear phenomenological patterns that can be described as languages and their parts".[32]

As all living beings are born into structures established before their birth, they must be instructed – forced, led by instructing or mimesis, etc. – how to live (they do not necessarily need to follow such instructions). Mimetic, imitative skills, then, belong to very powerful tools of lived experience.[33] Their prerequisites constitute mutual understanding (or more modern a wording: mutual embodiment), negotiation, narrative. Alas, all rumination on the topic always concerns human society and humans.[34]

We therefore conclude that sign production and sign relations exist in parallel in both natural languages and living systems (from cells up to biosphere/semiosphere). It is tempting to extend the analogy to hermeneutics and the narrative. The hermeneutic nature of the living is discussed elsewhere,[35] here we shall focus on the analogy between the narrative as known from literature studies, and self-shaping of the living as a narrative act of a kind. We leave aside futile disputes among literary scientists concerning fiction/non-fiction borderlines[36]; as narrative, we shall take any text that is self-contained and supplies *some* version of the world (hence, not manuals or special texts like legal treatises or scientific papers and books). A narrative becomes a solution when dealing with fuzzy contours of world appearances, memory, tradition, or texts; it gives a given individual, a given culture, a given species an Ariadne's thread that helps them in orientation in the labyrinth of possible interpretations.

We are aware that all examples about being-in-the-world are understood as solely the doings of humans. Yet, in the light of biological evolution the analogies are more than tempting. When taking into account the impossibility – in this world – of attaining fully digital working regime of functioning, we learn that the fuzzy parallel processing of world appearances needs semiosis, negotiation, and storytelling to be introduced into the affairs of evolutionary biology.

---

[31] Cowley 2012.

[32] *Ibid.*, p. 17; cf. also the epigraph.

[33] Cf., e.g., Moore 2013.

[34] As examples of an attempt to broaden the scope to non-human animals cf., e.g., Kleisner and Markoš 2005 and 2009; Markoš et al. 2009.

[35] Markoš 2002; Markoš et al. 2009.

[36] E.g., Cohn 1999.

Epilogue

Elsewhere, Anton Markoš speculates about the evolution of horses, in a way that deserves a brief retelling here. Contemporary horses (*Equus*) are separated from their three-toed ancestor *Hyracotherium* by some 55 millions of years. The interval had been punctuated by hundreds of species coming and becoming extinct, creating a plethora of variants, in different ages, of species-cultures encompassing "horse-kind" (the word echoing *humankind*). It is tempting to see their evolution not only as governed by a Blind Watchmaker (Richard Dawkins) who draws on two endlessly tinkering engineers – Mutation and Selection (Konrad Lorenz). Perhaps the contemporary state mirrors also an intrinsic "effort" (or even effort) of individuals, populations, species to display their "horse-ness" in order to represent best – here and now – their self-awareness in an ever-changing world.

References

Auerbach, E. (1946). *Mimesis. Dargestellte Wirklichkeit in der abendländischen Literatur*. Bern: A. Francke AG.

Bachelard, G. (1960 [1971]). *The poetics of reverie. Childhood, language, and the cosmos*. Boston: Beacon Press.

Barbieri, M. (2003). *The organic codes*. Cambridge: Cambridge University Press.

Blumenberg, H. (1986). *Die Lesbarkeit der Welt*. Frankfurt am Main: Suhrkampf.

Cohn, D. (1999). *The distinction of fiction*. Baltimore: Johns Hopkuns University Press.

Cowley, S. J. (2012). Mimesis and language: A distributed view. *Interaction Studies, 13*(1), 17–40.

Cowley, S. J., & Vallée-Tourangeau, F. (2013). Systemic condition: Human artifice in life and language. In S. J. Cowley & F. Vallée-Tourangeau (Eds.), *Cognition beyond the brain* (pp. 255–273). Dordrecht: Springer.

Cvrčková, F., & Markoš, A. (2005). Beyond bioinformatics: Can similarity be measured in the digital world? *Journal of Biosemiotics, 1*, 87–105.

Danchin, É. (2013). Avatars of information: Towards an inclusive evolutionary synthesis. *Trends in Ecology & Evolution, 28*(6), 351–358.

Eco, U. (1976 [1979]). *A theory of semiotics*. Bloomington: Indiana University Press.

Favareau, D. (2008). Collapsing the wave function of meaning: Collapsing the epistemological function of talk-in-interaction. In J. Hoffmeyer (Ed.), *A legacy for living systems: Gregory Bateson as precursor to biosemiotics* (pp. 169–211). Berlin: Springer.

Flegr, J. (2008). *Frozen evolution, or, that's not the way it is, Mr. Darwin*. Prague: Charles University.

Harris, R. (2009). *Rationality and the literate mind*. New York: Routlege.

Heidegger, M. (1927 [1962]). *Being and time*. London: SCM Press Ltd.

Hoffmeyer, J., & Emmeche, C. (1991 [2005]). Code-duality and the semiotics of nature. *Journal of Biosemiotics, 1*, 37–91.

Jakobson, R. (1971). Linguistics in relation to other sciences. In R. Jakobson, *Selected writings, vol. II: Word and language* (pp. 655–696). The Hague: Mouton.

Kauffman, S. A. (2000). *Investigations*. New York: Oxford University Press.

Kleisner, K., & Markoš, A. (2005). Semetic rings: Towards the new concept of mimetic resemblances. *Theory in Biosciences, 123*(3), 209–222.

Kleisner, K., & Markoš, A. (2009). Mutual understanding and misunderstanding in biological systems mediated by self-representational meaning of organisms. *Sign Systems Studies, 37*(1–2), 299–310.

Kull, K. (2000). Organisms can be proud to have been their own designers. *Cybernetics and Human Knowing, 7*(1), 45–55.
Love, N. (2004). Cognition and the language myth. *Language Sciences, 26*(6), 525–544.
Love, N. (2007). Are languages digital codes? *Language Sciences, 29*(5), 690–709.
Markoš, A. (2002). *Readers of the book of life*. New York: Oxford University Press.
Markoš, A., & Faltýnek, D. (2011). Language metaphors of life. *Biosemiotics, 4*(2), 171–200.
Markoš, A., & Švorcová, J. (2009). Recorded vs. organic memory. *Biosemiotics, 2*(2), 131–149.
Markoš, A., Grygar, F., Hajnal, L., Kleisner, K., Kratochvíl, Z., & Neubauer, Z. (2009). *Life as its own designer: Darwin's origin and western thought*. Dordrecht: Springer.
Markoš, A., Švorcová, J., & Lhotský, J. (2013). Living as languaging: Distributed knowledge in living beings. In S. J. Cowley & F. Vallée-Tourangeau (Eds.), *Cognition beyond the brain* (pp. 71–92). Dordrecht: Springer.
Moore, R. (2013). Imitation and conventional communication. *Biology & Philosophy, 28*(3), 481–500.
Pátková, I., Čepl, J. J., Blahůšková, A., Neubauer, Z., & Markoš, A. (2012). Developmental plasticity of bacterial colonies and consortia in germ-free and gnotobiotic settings. *BMC Microbiology, 12* (http://www.biomedcentral.com/1471-2180/12/178; website consulted in July 2014).
Port, R. F. (2010). Rich memory and distributed phonology. *Language Sciences, 32*(1), 43–55.
Priscianus, C., circa 500 (2001). *Prisciani institutionum grammaticalium librorum I-XVI, indices et concordantiae*. Curantibus C.G. Roman, M.A. Gutierrez Galindo. Hildesheim/New York: Olms – Weidmann.
Rawn, J. D. (1989). *Biochemistry*. Burlington: Neil Patterson Publishers.
Steffensen, S. V., Thibault, P. J., & Cowley, S. J. (2010). Living in the social meshwork: The case of health interaction. In S. J. Cowley, J. C. Major, S. V. Steffensen, & A. Dinis (Eds.), *Signifying bodies: Biosemiosis, interaction and health* (pp. 207–243). Braga: Portuguese Catholic University.

Language as Primary Modeling and Natural Languages: A Biosemiotic Perspective

Susan Petrilli and Augusto Ponzio

Abstract Our paper concerns general linguistics and discusses standpoints in both taxonomic and generative-transformational structuralism. The question that linguistics most often fails to address is "why so many languages?"; this is the enigma of Babel. We attempt an answer in a biosemiotic key, with special reference to Sebeok's global semiotics. What is implied is the problem not only of the plurality of natural languages (Fr. *langue*/It. *lingua*), but also of the different "languages" (Fr. *langage*/ It. *linguaggio*) of different discourse genres, as well as the infinite differentiation in individual speech. Babel does not only concern difference among languages (Fr. *langue*/It. *lingua*), but also the different ways in which single individuals use the word. Far from acting as an obstacle to communication, the otherness relation among the word of single individuals is the condition for communication to obtain, for expression and understanding.

Keywords Communication • Enigma of Babel • General linguistics • Language • Modeling • Natural language • Otherness • Plurilingualism • Understanding • Word

General Linguistics and Global Semiotics

General linguistics presupposes general semiotics simply because definition of the *verbal sign* presupposes definition of the *sign in general*. The sign model, in turn, is relative to the *vastness* and *extension* of the horizon of semiotics. Very often this model has been constructed neglecting a whole series of different types of signs – either because they are not considered as signs or because they are not considered to be semiotically relevant. Consequently, it is important that general semiotics should not be constructed on the basis of a limited survey of signs passed off as complete. In other words, the general science of signs must be careful not to elect a *part* and describe it as the *totality*.

This essay develops a series of problematics presented in Ponzio 2002; Petrilli and Ponzio 2002a; Petrilli 2014.

S. Petrilli (✉) • A. Ponzio
The University of Bari "Aldo Moro", Bari, Italy
e-mail: susan.petrilli@gmail.com; augustoponzio@libero.it

E. Velmezova et al. (eds.), *Biosemiotic Perspectives on Language and Linguistics*, Biosemiotics 13, DOI 10.1007/978-3-319-20663-9_4

As claimed in the entry "Thomas A. Sebeok" (by John Deely) in *Encyclopedia of Semiotics*,[1] a turning point in the history of semiotics occurred during the first half of the 1960s, when Thomas A. Sebeok (1920–2001) extended the boundaries of the science of signs well beyond the limits of what then was commonly recognized as "semiology".

The latter, semiology, is based on the verbal paradigm and suffers from the *pars pro toto* fallacy. That is, it exchanges the part for the whole. Sebeok calls this tendency in the study of signs the "minor tradition". He opposes it to what he calls the "major" tradition, considering the temporal and thematic extension of the latter. The major tradition is represented by John Locke (1632–1704) and Charles Sanders Peirce (1839–1914) and goes back to early studies on signs and symptoms (ancient medical *semeiotics* or symptomatology) with Hippocrates (460 BC–377 BC) and Galen (circa 130 AD–circa 210 AD).

Thanks to Sebeok semiotics today emerges as "global semiotics".[2] In fact, through numerous publications he promotes a new vision of semiotics where sign sciences converge with life sciences. The underlying assumption is that *living matter and sign matter converge*. As a result of its "global" or "holistic" approach, semiotic research today on the "life of signs" is directly interested in the "signs of life". Therefore, from the perspective of *global semiotics*, *semiosis* (that is, the relation, or process or situation in which something is a sign) and *life* converge given that semiosis is the criterial attribute of life. After Sebeok's work – amply inspired by Peirce, but also Charles Morris (1901–1979) and Roman Jakobson (1896–1982), Sebeok's immediate masters – our conception of both the semiotic field and of the history of semiotics has changed significantly.

It follows that global semiotics also presents itself as a *critique* of semiotic theory and practice vitiated by oversimplifying anthropocentric and glottocentric tendencies.

Global semiotics extends its gaze well beyond the signs that human beings use to communicate – the subject matter of semiology as formulated by Ferdinand de Saussure (1857–1913) – and includes not only *zoosemiotics* (term introduced by Sebeok[3]), comprehensive of *anthroposemiotics* and the study of the signs of the other great kingdoms (*phytosemiotics* and *mycosemiotics*), but also *microsemiotics* and *endosemiotics*. As such global semiotics converges with *biosemiotics*.

The subject matter of global semiotics or *semiotics of life*[4] is the *semiosphere* conceived as converging with the *biosphere*. The term *semiosphere* is taken from the work of Juri Lotman (1922–1993),[5] but is understood by Sebeok in a far broader sense. In fact, Lotman limits the field of reference of the term *semiosphere* to human culture and states that outside the semiosphere thus described there is no communication.[6] On the contrary, from the perspective of global semiotics which maintains that *semiosis* converges with *life*, the semiosphere is identified with the *biosphere*, a term used by Vladimir Vernadsky (1863–1945) in 1926,[7] and therefore is understood as a

[1] Deely 1998.

[2] Sebeok 2001.

[3] Sebeok 1963.

[4] Petrilli and Ponzio 2001 and 2002a.

[5] Lotman 1991.

[6] *Ibid.*, pp. 123–124.

[7] Vernadskij 1926.

'semiobiosphere'. The semiosphere in Lotman's sense is limited to human culture, that is, to *anthroposemiosis*, consequently to the verbal and nonverbal signs forming its languages (Fr. *langages*/It. *linguaggi*). As such Lotman's semiosphere only accounts for a limited portion of the semiobiosphere. Instead, taken in its totality, the semiosphere extends across the whole sign network that goes to form the living world.

Considered in the context of global semiotics, general linguistics is part of *anthroposemiotics*. General linguistics studies verbal language, oral and written. However, it neither focuses on a given natural language (Fr. *langue*/It. *lingua*), nor even on a given discourse genre or literary genre. Instead, general linguistics focuses on certain general aspects (at times with claims to universality) as they characteristically present themselves in a given natural language, as the condition itself of its being a language.

To contextualize linguistics in global semiotics is not only functional to a classification of the sciences, but it also guarantees that the general *sign* concept used by linguistics is drawn from general semiotics and, therefore, is truly general and not partial.

Modeling, Communication and Dialogue

Now we shall explain two notions which are interconnected and fundamental in semiotics: *modeling* and *dialogism*. Without them it is not possible to understand a third notion: *communication*. This notion is generally privileged in the study of signs over the other two.

The concept of modeling comes from the so-called Tartu-Moscow school (A.A. Zaliznjak, V.V. Ivanov, V.N. Toporov. Ju.M. Lotman[8]). It is applied to natural language (Fr. *langue*/It. *lingua*), which it describes as a "primary modeling system",[9] and to the other human cultural systems described as "secondary modeling systems".

On our part, instead, we implement the term *modeling* in Sebeok's sense. Sebeok extends the concept beyond the sphere of anthroposemiosis and connects it to the biologist Jakob von Uexküll and his concept of *Umwelt* ('surrounding world').[10] In Sebeok's interpretation, *Umwelt* means 'external world model'. On the basis of research in biosemiotics, we know that the modeling capacity can be observed in all life-forms.[11] "Modeling systems theory" has recently been reformulated by Sebeok in collaboration with Marcel Danesi.[12] They study semiotic phenomena as modeling processes. In light of semiotics oriented in the sense of modeling systems theory, semiosis can be defined as a capacity with which all life-forms are endowed to produce and understand signs according to specific models, organizing perceptive input as established by each species.[13]

[8] Cf. Lucid (ed.), 1977; Rudy 1986.

[9] Cf. Deely 2007.

[10] Cf. Kull 2010.

[11] Cf. Sebeok 1979, pp. 49–58, 68, 82 and 1991, pp. 117–127.

[12] Sebeok and Danesi 2000.

[13] *Ibid.*, p. 5.

The applied study of modeling systems theory is called "systems analysis". It distinguishes between *primary*, *secondary* and *tertiary* modeling. The primary modeling system is the innate capacity for simulative modeling, that is, a system that enables all organisms to simulate the world in species-specific ways.[14]

Sebeok introduces the term *language* for the primary modeling system specific to the genus *Homo*. The *primary modeling system* is not natural language (Fr. *langue*/It. *lingua*), as instead the Tartu-Moscow school maintains, but rather language in the sense of the French *langage* and Italian *linguaggio*. Instead, natural language (Fr. *langue*/It. *lingua*) appears quite late in human evolution and is a *secondary modeling system*. Consequently, cultural sign systems that presuppose natural languages are *tertiary modeling systems*.

Secondary modeling subtends modeling processes of both the indicational and extensional types. Indicational modeling has been registered in various living species. Instead, extensional modeling is a uniquely human capacity insofar as it presupposes *language* (primary modeling system specific to human beings alone) which Sebeok distinguishes from speech, from natural language, *lingua-langue*, a secondary modeling system.[15]

Tertiary modeling subtends highly abstract modeling processes of the symbolical type[16] which in addition to language understood as *linguaggio-langage* also presuppose natural language, *lingua-langue*.

Communication presupposes modeling, given that communication occurs internally to a world produced by the modeling processes it presupposes. It is precisely by considering the communication/modeling relation and the fact that the communicative relation is impossible if not on the basis of modeling able to engender an *Umwelt*, as understood by J. von Uexküll, that we can formulate a response to Winfried Nöth's question "Is communication possible?" and thus escape the paradoxes produced by reflecting on the notion of communication taken in isolation.[17] Modeling systems, in turn, also evolve from communication as it occurs in the species, and from the environment – being the context of modeling produced by adaptation. But communication always occurs on the basis of the type of modeling that characterizes a species. For example, as a system specific to the genus *Homo*, therefore already present in hominids, language regulates communication with the environment. Evolution of the species in the genus *Homo* to *Homo sapiens sapiens* occurs through adaptation, but necessarily according to its species-specific modeling system (which from the very moment of its appearance assigns it to a special niche with respect to other species, as close as they may be homologically).

By *dialogue*[18] is understood the way in which an organism in its specific *Umwelt* relates to the intraspecific and extraspecific organic, and to the inorganic. Semiosis

[14] *Ibid.*, pp. 44–48.

[15] *Ibid.*, pp. 82–95.

[16] *Ibid.*, pp. 120–129.

[17] Nöth 2013.

[18] We obviously cannot dwell now upon Paul Cobley's reconstruction of the relation between our conception of "dialogue" and that of Emmanuel Levinas and of Mikhail Bakhtin whose position is

is generally *dialogic* (cf. below). The notion of dialogism does not contradict, but rather supplements and confirms those notions that insist on the autonomy of the living organism, for example, J. von Uexküll's *functional cycle* and Humberto Maturana and Francisco Varela's *autopoiesis*. Furthermore, *dialogue* must be distinguished from *communication*. Communication is only one aspect of semiosis. The other two are *modeling* and *dialogism*, as we have already stated.[19]

Dialogism, modeling and communication – which in the human being are characterized species-specifically – belong to semiosis in general and for this reason can be traced, in different forms, degrees and modalities, in all living beings. The dialogic character of verbal semiosis, its modeling and communicative functions, are specific characterizations of the human species of capacities that can be traced in semiosis generally in any living being. We will describe this condition more closely in the sections that follow. In them we present a series of considerations we must necessarily keep account of in the study of the semiosis of language understood as '*langage/linguaggio*' (primary modeling), and as '*langue/lingua*' (secondary modeling), and of other cultural sign systems that presuppose language understood as '*langue/lingua*' (tertiary modeling).

Language and Endosemiotic Systems

In his essay "The evolution of semiosis", Sebeok begins from Peirce's definition of semiosis as an irreducible teleonomic process, consisting in the relation between a sign, its object and its actual or potential interpretant.[20] On the basis of this triadic model, Sebeok takes his distances from semiotic theories that claim to explain semiosis through such notions as *information*, *code*, *message*, all of which express a dichotomic vision of the sign. All the same, Sebeok uses such notions to explain the evolution of semiosis on the planet Earth. He resorts to them to explain the crucial difference between non-semiosic, quasi-semiosic or proto-semiosic phenomena relating to non-biological atomic interactions and inorganic molecules, on the one hand, and semiosis as the criterial attribute of life, on the other.

---

particularly interesting in the present context given his focus on corporeity and the biological sciences. In any case, it is above all owing to the relation Cobley establishes with Th.A. Sebeok that we wish to signal his "brief note" of 2007 (Cobley 2007). For a very effective synthesis of Sebeok's contribution to semiotics and to biosemiotics in its current configuration, cf. also Deely 1998. The implications of the relation between dialogue and alterity (or otherness) from a biosemiotic perspective and what Peirce calls "agapasm" and "evolutionary love" are evidenced, passing through Levinas, by Donald Favareau (Favareau 2013).

[19] A relation comes to be established among authors who have enquired into the "origins" of life and its different worlds from different perspectives. These authors include Bakhtin, Driesch, J. von Uexküll (cf. the essay on "contemporary vitalism," in *Bachtin e il suo circolo* 2014, presented in a bilingual – Russian-Italian – edition, originally published by Ivan Kanaev, in a specialized journal of biology, in Russia, in 1926, but in reality written by Bakhtin). On the relation among these authors, taken into consideration as part of a dialogue with ourselves (lasting several years now), cf. Kull 2007 and 2013.

[20] Sebeok 1997, p. 436.

As regards the evolutionary process of semiosis, Sebeok implements *information* and *semiosis* to indicate two different evolutionary phases. Semiosis is what distinguishes the animate from the inanimate. Before semiosis there was information. The essence of information is change; the prerequisite of semiosis is life. Information is possible without semiosis. But semiosis is not possible without information. Semiosis and life include information, they imply it. "Cosmic expansion is accompanied by a departure from a state of maximum entropy, and information (as a measure of the nonuniform, orderly properties of physical systems) evolved out of that initial state of utter chaos".[21]

That the terms *information*, *code*, *message* characterize so-called "codification semiotics" does not stop them from being implemented again by trends in so-called "semiotics of interpretation", as in the case of global semiotics or semiotics of life. In his explanation of the functional cycle, Thure von Uexküll (1908–2004) implements the terms *code* and *context* connecting them to the Peircean triad, representamen, interpretant and object or referent.[22] Any term whatsoever can be used in semiotics so long as it is defined rigorously by other terms.

The term *code* has been employed to characterize both properly human sign systems as well as human and non-human endosemiosic sign systems; for example, to characterize verbal language (*langue*) as much as the genotypical system, or "genetic code". Influenced by the predominance of linguistics in the study of signs, initially the terms *code* and *language* were used indifferently for both verbal and nonverbal sign systems, including the genotypical. But this led to what Sebeok describes as much "fruitless debate"[23] about whether the genetic code is (like) a language or not.

Once the modeling procedure specific to mankind – which subtends "speech" or the so-called "*langue/lingua*" – is named *language*, it is legitimate to ask whether language (*langage/linguaggio*) (including verbal language) and the genetic code device are homologous. It would seem so. As Sebeok observes, this is determined by the principle of articulation traceable in both language and the genetic code, that is, by the fact that both function on the basis of what he calls *syntax*, but which is better denominated *syntactics*.[24] The fact that language, a secondary modeling system, incorporates a syntactic component (articulation), as Sebeok says, is singular: this feature is not present in other zoosemiotic systems, although it abounds in endosemiotic systems, such as the genetic code, the immune code, the metabolic code, and the neural code.[25]

This way, semiosis and information, the genetic code, just like other endosemiotic systems, and language, including verbal and nonverbal language, are connected by a genetic structure. Beginning from this, each system is then characterized in terms of its own specific quality. In the information-semiosic–semiotic and non-

[21] *Ibid.*, pp. 436–437; cf. also Sebeok 1986, pp. 15–16.

[22] Uexküll 1998, art. 110, pp. 2187–2188.

[23] Sebeok 1997, pp. 437–438.

[24] Petrilli and Ponzio 2002b and 2007.

[25] Sebeok 1991, pp. 57–58.

life–life *continuum* likenesses like differences are qualitative and structural. In this sense, in the case of likeness, it is not a question of *analogy* (casual and superficial likeness) but of *homology* (profound, genetic and structural likeness), to use terminology from genetic biology. This confirms the conception introduced by Ferruccio Rossi-Landi (1921–1985) when he maintains that to determine the specificity of verbal language, it will be necessary to study any homologies with other sign systems, therefore to proceed according to the homological method.[26]

Binarism, Triadism and Dialogism

From what we have stated so far, it results that the dichotomies *code/message*, *information/redundancy*, *first/second articulation*, etc., can be applied to both semiosis and information. What counts is that these notions be functional to explaining the different aspects of information and of the semiosic and semiotic universe. For example, the concept of redundancy from information theory is valid both in linguistic studies of the utterance or text and in biosemiotic studies of the genetic code.

Binarism helps explain certain endosemiosis related phenomena (the term *endosemiotics* was coined by Sebeok in 1976[27]) as much as certain aspects of properly human semiosis. From an endosemiotic point of view, the fundamental binary opposition in the ontogenesis of an organism is that between the *ego* and *alter* concepts, studied by Sebeok in his research on the "semiotic self".[28] On the other hand, we know that phonology avails itself of binary opposition to identify pairs of distinctive traits.

From the point of view of global semiotics which aims not to neglect any sign phenomenon in the planetary biosphere, binarism cannot be excluded. Implementing the expression "ecumenicalism in semiotics", introduced by Sebeok,[29] we can claim that global semiotics is ecumenical because it elaborates on terms taken from information theory and code semiotics (semiology) and applies them to the vast range of semiosical phenomena, from verbal languages and cultural systems to the genetic code, the immune system, the metabolic code, and the neural code, etc., as listed above.[30]

Instead, what should be rejected is the orientation that establishes binarism as the only feature of semiosis or that restricts it to the cultural world. These are the fundamental limits of traditional binarism as results from the well-documented entry "Binarism"[31] (by Paul J. Thibault) in the *Encyclopedia of Semiotics*.[32] Such limits

[26] Rossi-Landi 1968 and 1972.

[27] Sebeok 1976 [1985].

[28] Sebeok et al. 2001.

[29] Sebeok 1979, pp. 61–83.

[30] Sebeok 1997, pp. 438–440; cf. also Bouissac (ed.), 1998.

[31] Thibault 1998.

[32] Bouissac (ed.), 1998.

are also determined by the fact that the research interests of major exponents of semiotic binarism (Ferdinand de Saussure, Nikolai Trubetzkoy, Noam Chomsky, Morris Halle, Roman Jakobson, Claude Lévi-Strauss) were restricted to the field of verbal and cultural phenomena.

As regards binarism, the vision of semiotic research as it emerges in *Semiotik/Semiotics*,[33] in *Encyclopedia of Semiotics*[34] and in Sebeok's global semiotics no doubt transcends any opposition between semioticians with a Saussurean/Hjelmslevian/Greimasian orientation[35] and semioticians of Peircean inspiration. These two trends in semiotics would seem to converge with the opposition between *binarism* and *triadism*, respectively. However, we believe that the central question in semiotics considered on a theoretical level as well as from the point of view of the history of these two different trends, is not the opposition between binarism and triadism.[36]

Instead, the opposition is between a sign model that tends to oversimplify the complex process of semiosis, on the one hand, and a sign model, like Peirce's, that would seem to account for the different aspects of a process thanks to which something is a sign, on the other.

The validity of the latter is not determined by its triadic configuration, but rather by given aspects of Peircean triadism: its categories, sign typologies, dynamism according to a model that describes signs as regulated by deferral from one interpretant to another. The categories of *firstness*, *secondness* and *thirdness*, the triad *representamen*, *object* and *interpretant*, the triadic tendency of signs in the direction of symbolicity, indexciality, and iconicity all contribute to delineating and supporting a conception of semiosis featuring otherness and dialogism.

Peircean logic is dialogic and polylogic. However, its merit does not lay in its triadic formula. Proof is Hegelian triadism which abstracts from the constitutive dialogism of life and gives rise to unilinear and monologic dialectics. Under the entry "Binarism" in *Encyclopedia of Semiotics*, Hegelian philosophy is strangely described as superseding the theory of binary opposition featured by structuralism with Claude Lévi-Strauss (1908–2009).[37] In his 1970–1971 notes, Mikhail Bakhtin (1895–1975) describes the formation process of Hegelian dialectics.[38] It has its roots in the live dialogic context of semiosis, but transforms dialogical relations into abstract concepts, judgements and standpoints of the single and solitary conscious. Peirce himself took a stand against the constitutive sclerosis of Hegelian dialectics which rather than remain open and contradictory presents itself as the expression of a hypochondriac search for the conclusion, oriented unilaterally towards a synthesis.[39]

---

[33] Posner et al. (eds.), 1997–2004.

[34] Bouissac (ed.), 1998.

[35] Johansen 1998; Parret 1998.

[36] Cf. Petrilli 2013.

[37] Thibault 1998, p. 81.

[38] Bakhtin 1970–1971 [1986].

[39] On the relation between *dialogue* and *dialectics* in Peirce and Bakhtin, cf. Ponzio 1984 and 1990; Ponzio et al. 2006.

The alternative in semiotics is not between binarism and triadism, but between *monologism* and *polylogism*. The limit of the sign model proposed by Saussurean semiology is not determined by binarism as such, as claimed instead by Thibault.[40] Rather, it is determined by the fact that binarism finds expression in the concept of *equal exchange between sign and meaning* and reduces complex linguistic life to the dichotomic scheme represented by code and message.[41]

Language and the Origin of the Word

The question of the origin of speech, verbal language, is generally dismissed by the scientific community as unworthy of discussion, having given rise to unfounded discussions (an exception is the book by Giorgio Fano [1885–1963] *Origini e natura del linguaggio*[42]).

On the basis of more recent studies, the problem of the origin of verbal language has been reexamined and evidenced in all its complexity. One of the most systematic proposals comes from Sebeok who explains the species-specific character of *speech* (verbal language) in terms of the human primary modeling system, *language*. Moreover, he describes speech as arising at a certain point in evolution through adaptation, as a function of communication, where adaptive processes are regulated by *language* understood as 'modeling'. Consequently, Sebeok intervened polemically and ironically on various occasions to cool down hot enthusiasm towards theories and training practices (particularly fashionable at the time in the United States of America), which aimed to demonstrate that animals can speak.

According to Sebeok's modeling theory, *language* (understood as the 'primary modeling system specific to *Homo*') appeared and developed through adaptation much earlier than speech in the course of human evolution through to *Homo sapiens*. Originally, language was not a communicative device. Chomsky also maintained that language is not essentially communicative, but by *language* he understands 'verbal language', what Sebeok calls "speech".[43] Instead, according to Sebeok, verbal language has a specific communication function from the very moment it appears. Chomsky's theory of verbal language does not keep account of the difference between language (*langage/linguaggio*) and verbal language, and without this difference it is not possible to explain the origin, nor the functioning of verbal language.

In short, language is a *modeling device* with which the first hominid was endowed and thanks to which, from an evolutionary point of view, development was possible from the first species of *Homo* through to *Homo sapiens sapiens*.[44] Other animals

[40] Thibault 1998. For an analysis of binarism in Saussure, cf. §222 "Binarität" in *ibid.*

[41] Ponzio 1990, pp. 279–280.

[42] Fano 1972. This book is now also available in English translation (1992).

[43] Ponzio 2012b.

[44] Sebeok 1994, pp. 117–128.

are also endowed with a modeling system through which they produce their worlds; language is that which belongs to mankind. But man's modeling system is completely different from other primary modeling systems. Its specific characteristic is what Peirce called "the play of musement" (and expression used by Sebeok as the title of one of his books[45]) and what Giambattista Vico (1668–1744) called "poetic logics".[46] These expressions refer to the human capacity, unlike other animal species, to produce multiple models, therefore, to use an expression from Gottfried Wilhelm Leibniz (1646–1716), to invent and simulate an infinite number of "possible worlds".

Speech, like *language*, understood as 'modeling' also appeared through adaptation, but with a communicative function, and much later with respect to language, precisely with the appearance of *Homo sapiens*. As the human species evolved, language also took on a communicative function through the process of *exaptation* (an expression introduced by Stephen Jay Gould and Elisabeth S. Vrba[47]), thereby empowering the communicative function of speech; and speech also took on a modeling function thereby enhancing the modeling function of language, as it materialized in each of the multiple natural languages: *language evolved as an adaptation*; whereas *speech developed out of language as a derivative exaptation* over a succeeding period of approximately two million years.[48] Language is the primary evolutionary adaptation that characterizes the hominid. Speech developed from language-as-modeling as a result of the evolution of physical and neurological capacities, about 300,000 years ago.

Exapted for communication first in the form of *speech* and later as *script*, language-as-modeling also enhanced the human capacity for nonverbal communication, giving rise to the development of a broad and complex range of nonverbal languages. Through a process of exaptation speech took on a modeling function in turn, thereby acting as a secondary modeling system. Such transformation favored development of the human semiotic capacity on the cognitive, organizational, inventive levels, etc. Beyond increasing the capacity for communication through speech itself as much as through nonverbal languages, speech-as-modeling favoured the proliferation and specialization of *languages* understood now not only in the sense of 'natural languages', but also of 'sectorial languages', etc.

The relation between language-as-modeling and speech has involved mutual adjustment of the encoding with the decoding capacity, of language "exapted" for communication, first for the sake of speech, for "ear and mouth work" and subsequently for script and other forms of communication, with speech for (secondary) modeling, "for mind work". All the same, absolute mutual comprehension remains a distant goal, so that the whole system still remains to be perfected.[49] As Sebeok observes:

---

[45] Sebeok 1981.

[46] Cf. Danesi 1993.

[47] Gould and Vrba 1982.

[48] Sebeok 1986, pp. 14–16; italics ours. – *S.P., A.P.*

[49] Sebeok 1991, p. 56.

> As to why this process of exaptation took several million years to accomplish, the answer seems to be that the adjustment of a species-specific mechanism for encoding language into speech, i.e. producing signs vocally, with a matching mechanism for decoding it, i.e. receiving and interpreting a stream of incoming verbal/vocal signs (sentences), must have taken that long to fine tune, a process which is far from complete (since humans have great difficulties in understanding each other's spoken messages).[50]

At this point, another process of exaptation in the evolution of anthroposemiosis we should note is the distinction between "manual work" and "intellectual work". In Rossi-Landi's terminology this is the distinction between "nonlinguistic (nonverbal) work" and "linguistic (verbal) work".[51] These two different types of work have only just come together,[52] and this as a result of developments in technology and communication. Insofar as it unites *hardware* and *software* the computer is the most obvious expression of the type of adjustment that leads towards the development of an ever more efficient communication system.

Syntactics and Writing in Language

Plurilingualism (including "internal plurilingualism"), the multiplicity of languages, internal and external, results from the human modeling capacity to invent multiple worlds. This is the capacity for the "play of musement" or, as Vico says, for "poetic logic" proper to the human being. As much as Chomskyan linguistics insists on the creative character of (verbal) language, which presupposes an innate universal grammar (*à la* Descartes), it does not explain the proliferation of multiple natural languages (*langue/lingua*).

Before presenting itself as speech with communicative functions which subsequently renew and enhance nonverbal sign behaviors (nonverbal languages), language is a *modeling "procedure"*, that is, a construction model of the world. We prefer the term *procedure* over *system*, recovered by Sebeok from the Tartu-Moscow school.[53] The specific function of language-as-modeling is to signify, interpret and confer sense.

All animals have construction models of the world and following Sebeok that belonging to the human animal is denominated language. However, language differs totally from modeling procedures in other animals. What does not differ is the *type* of sign implemented (icon, index, symbol, etc.). The specific characteristic of human modeling is articulation, or as Sebeok says, *syntax*, which enables us to produce different signifying itineraries with the same objects that function as interpreted signs and interpretant signs. The term *articulation* recalls decomposition into elements. *Syntax* projects the idea of the temporal-spatial distribution of these

[50] Sebeok 1997, pp. 443–444.

[51] Rossi-Landi 1968 and 1975.

[52] Rossi-Landi 1985 [2006]; Petrilli and Ponzio 2005, pp. 232–296.

[53] Cf. Sebeok 1991, p. 49.

objects. However, *syntactics*, a term introduced by Morris to denominate one of three branches of semiotics (the other two being *semantics* and *pragmatics*), avoids confusing syntax in the linguistic-verbal sense with syntax in the sense of neopositivist logic. The term *syntactics* avoids the ambiguity connected with the word *syntax*, a term proper to linguists and neopositivists (*logical syntax* in the formulation of Rudolf Carnap [1891–1970]). The syntactics of language determines the possibility of using a finite number of elements in different combinations to produce an infinity of different meanings.

On our part, we prefer the term *writing* to *syntactics*. Writing alludes to the combinatorial procedure through which a finite number of elements produces an infinite number of senses and meanings. Writing thus described is antecedent to speech, the condition of possibility for speech. The phonetic sign itself is writing given that it only functions on the basis of combination; writing belongs to language before the stylet or pen impresses letters on tablets or on pergamen or on paper, as Emmanuel Levinas (1906–1995) says.[54] Therefore, language-as-modeling is writing, it subsists *avant la lettre*, before the invention of *writing* understood as 'transcription', that is, as a system for the transcription of vocal semiosis, before the connection with phonation and formation of natural languages.

Language today is influenced by phonetic material, while maintaining the features that characterized it antecedently to transcription. These are evidenced in the articulation of verbal language and its iconic character (signification through position, extension, as when the adjective in the superlative or the verb in the plural become longer, as pointed out by Jakobson[55]). When writing emerges subsequently as a secondary covering to fix vocalism, it uses space to preserve the oral word, giving it a spatial configuration.[56]

Articulation in verbal language (André Martinet's double articulation) is an aspect of language-as-modeling which articulates the world on the basis of differentiation and deferral – *difference/différance.*[57] Articulation is firstly distancing, spacing out by language-as-modeling insofar as it is writing. To signify by positioning the same things differently is already writing in itself. Articulation *of* verbal language and *through* verbal language (secondary modeling) is achieved on the basis of signification by position.

Insofar as it is syntax, or *syntactics*, or more precisely *writing* antecedent to phonation and independent from the communicative function of transcription, language-as-modeling implements pieces that can be put together in an infinite number of different ways, thereby giving rise to an indeterminate number of models that can be dismounted to construct different models with the same pieces. So, as Sebeok says,[58] by virtue of language, human beings not only produce their own world, like other animals, but they also produce an infinite number of possible worlds: this is

[54] Lévinas 1982.

[55] Jakobson 1965.

[56] Kristeva 1969 [1981].

[57] Derrida 1967.

[58] Sebeok 1986.

the "play of musement". The "play of musement" is fundamental in scientific research and all forms of investigation, in simulation, from lying to fiction, and in all forms of artistic creation. So "creativity" is proper to *language* understood as 'writing', as a 'primary modeling device' and a 'derivative in verbal language' (though mistakenly described by Chomsky as specific to the latter).

The formation itself of speech and of relative verbal systems, natural languages, presupposes *writing* such as we have defined it in this paper (in contrast to *transcription*). Without the capacity for writing, humankind would not be in a position to articulate sounds and identify a limited number of distinctive traits, *phonemes*, to reproduce phonetically, nor to arrange phonemes in different ways to form words (*monemes*), nor words syntactically to form an infinity of different *utterances* expressing different meanings and senses, nor to produce texts, those complex signs whose meaning is qualitatively superior and irreducible to the sum of its parts.

To recapitulate: writing is inherent in language-as-modeling, given that it confers different meanings to the same elements by repositioning them chronotopically. In other words, writing is inherent in language as a signifying procedure insofar as it is characterized by *syntactics*. The phonetic sign itself is writing. Language was already writing, even before the invention of writing as transcription.

The a priori is not speech. The a priori is language and its writing mechanism. The language of music articulates space-time thanks to language-as-modeling. Musical scores, like verbal language, are an expression of the human capacity for language, writing, articulation, ultimately for the properly human.

Language and Communication

To maintain that communication is not the specific function of language can be confusing, as in Chomsky's case. When Chomsky claims that communication is not specific to language, he is not referring to what Sebeok understands by *language* distinguishing it from *speech*, in spite of the fact that Sebeok cites him in support of his own position. But by *language* Chomsky understands 'verbal language', 'speech', and speech arises specifically for communication as Sebeok maintains.

Natural language is a (secondary) modeling system (whereas original language-as-modeling is a primary modeling system), and communication through natural language presupposes a particular modeling of the world. But Chomsky lacks the concept of modeling. On the contrary, *modeling* is present in the "theory of linguistic relativity" as formulated by Edward Sapir (1884–1939) and Benjamin Lee Whorf (1897–1941). However, given that it does not trace (secondary) modeling in natural languages (*langue/lingua*) back to language as (primary) modeling, the theory of linguistic relativity (like the Chomskyan approach) does not explain the multiplicity of natural languages which it presents as closed universes.

When explaining the specific grammars of natural languages, Chomsky proceeds from natural language (*langue/lingua*) to language (*langage/linguaggio*); he describes *language* as an innate "faculty" of speaking, rather than as a modeling system, as a species-specific representation of the world, through verbal and nonverbal signs. He uses the term *grammar* to refer to verbal language, when instead *language* is a term which underlines the latter's characteristic capacity for modeling, verbal and nonverbal. Therefore, by *grammar* Chomsky understands a 'device that generates the sentences of different natural languages'. As such it is endowed with a phonological component, a syntactical component and a semantical component. But this grammar – unlike that of the natural languages – as described by Chomsky claims to be universal. In this sense, it resembles an *Ursprache*, an original verbal language (*langage/linguaggio*), a universal natural language (*langue/lingua*). The claim is that despite multiplicity and diversity all natural languages can be traced back to the innate structures of universal grammar. This is conceived in terms of "Cartesian" innatism, updated in biologistic terms, moreover on the basis of opposition (now outdated) between rationalism and empiricism, as though philosophers such as Immanuel Kant (1724–1804), Ernst Cassirer (1874–1945) or Edmund Husserl (1859–1938) never existed.

Chomsky denies verbal language its communicative function. He isolates natural languages from their historical-social context (nor is it incidental that he should deny sociolinguistics the status of science). Furthermore, he considers them independently from nonverbal languages, as though interpretation were possible uniquely through verbal signs, through *renvoi* from one verbal interpretant to another (surface and deep structures).

Not making a distinction between *language* (as primary modeling) and *verbal language* (*natural language*) gives rise to forms of psychological reductionism as in the case of Philip Lieberman.[59] He attempts to explain the origin of language with concepts from Chomskyan linguistic theory. On this account, complex anthropogenetic processes are explained in terms of the linear development of given cognitive capacities. Moreover, all this is described in the language of traditional syntactics.[60]

As far as the communicative aspect of human languages is concerned, to describe *communication* simply as the 'exchange of information between emitters and receivers', as though they were preconstituted and external to the communication process, is reductive. "Communication" is a far vaster phenomenon than that described by semiology of Saussurean derivation. This is also true of communication as conceived by information theory. We must also add that this notion of communication as understood precisely by information theory found its most intelligent and perhaps most renowned formulation in an essay by Jakobson, "Linguistics and poetics", of 1960.[61] Here we trace the main concepts of communication semiotics taken from information theory: code, message, emitter, receiver, channel and context. Jakobson adds the important concept of function (Prague linguistic circle). "Communication" must be recognized in its effective historical-social consistency.

---

59 Lieberman 1975.

60 Rossi-Landi 1985 [2006, p. 229].

61 Jakobson 1960; subsequently Jakobson 1971.

Its development and functioning must be contextualized in the global sign network of human semiosis. Global semiosis is the condition of possibility for communication in the restricted sense, that is, the exchange of messages. Thus described, communication in the human world converges with *social reproduction*, of which communicative exchange, that is, the exchange of messages and goods, constitutes only one aspect. Identification of the object of communication-transmission, formation of "personal experiences" to communicate, coming to awareness, taking standpoints, interindividual relationships and intentional communication are all developed in the communication process thus understood.

Even needs, including "communicative needs" are formed in the communicative process. Needs, as demonstrated by Karl Marx (1818–1883) in his critique of "bourgeois" economy, develop as part of the process of social reproduction and are inconceivable outside communication. Consequently, to explain the origin of language *à la* Lamarck affirming, as does Friedrich Engels (1820–1895) in *Dialektik der Natur* (1883), that it arises when human beings have something to say to each other ("einander etwas zu sagen haben"[62]) is at the very least an oversimplification (on this aspect, cf. Rossi-Landi's critique of Engels[63]).

Communication is the place where meanings, messages and experience, intentional acts including information transmission, are all formed. Reality, the way we perceive it, is organized and developed in the social processes of communication. As Rossi-Landi claims, from an evolutionary perspective verbal language does not emerge from an abstract need to communicate, but rather from specific communicative needs determined in the social. At the basis of communication, including in its primitive nonverbal forms, is the human species-specific modeling (and not communicative) procedure of language (in the sense described above as understood by Sebeok). Moreover, as Rossi-Landi claims, "language cannot be reduced to mere communication, otherwise the linguistic capacity could not be placed in a coherent phylogenetic framework of nervous structures and psychical functions".[64]

At this point, it is clear that to establish that nonverbal languages precede verbal languages or vice versa is a mistake. Today's nonverbal languages, insofar as they are languages, do not precede verbal languages. Instead, nonverbal sign behaviours do. As much as nonverbal sign behaviour can be traced in the animal kingdom at large, development in the human world is conditioned by the species-specific procedures of *language* understood as 'primary modeling'. If such sign behaviours become "languages" and in turn (tertiary) modeling procedures, this is thanks to the mediation of natural languages (secondary modeling). As such, these sign behaviours are posterior to verbal language (speech), though they increase the interpretive and communicative possibilities of the latter.

[62] Engels 1883 [1962].

[63] Rossi-Landi 1985 [2006, pp. 225–226].

[64] *Ibid.*, pp. 233–234.

The Enigma of Babel

Chomsky's linguistic theory does not succeed in explaining the multiplicity of different languages. Moreover, this situation of multiplicity contradicts the uniqueness of the innate universal grammar hypothesis.

Plurilingualism does not only consist of different natural languages, but also of a multiplicity of different languages (*langage/linguaggio*) internally to the same natural language (*lingua*). Chomskyan linguistics neglects this type of multiplicity as well, given that it considers natural language (*lingua*) as a unique and unitary code. Chomsky's linguistics does not explain the plurality of natural languages nor the plurilingualism internal to each natural language. Though he insists on the "creative character of language", by *language*, as anticipated, he understands 'verbal language'. Furthermore, he remains anchored to the assumption that verbal language is endowed with a universal grammar, whatever the specific natural language and the specific grammar we are dealing with. This universal grammar has the same features and components (phonological, syntactic, semantic) of the specific grammars whose rules are reconducted to those of universal grammar. Given these premises, Chomskyan linguistics is not able to address the "enigma of Babel".

Plurilingualism does not only consist in the fact that verbal languages are multiple and cannot be reconducted to a single univocal and omnicomprehensive system that can supplant them all, or that functions as a model to study, understand, characterize them in theoretical terms. Plurilingualism is also given by the fact that all languages flourish in direct or indirect, implicit or explicit relationships with other languages acting as possible interpretants through which meaning is constituted, developed and transformed. Plurilingualism involves relations of translation, but also of derivation and mutual completion. Moreover, the multiplicity of different languages shares in the common language of a given culture, etc.

Natural languages form and develop through mutual relations of interaction and exchange; each one of them originates in the life of another natural language (*langue/lingua*), in its internal subdivisions and stratifications, in the internal dialectics of its languages (*langage/linguaggio*) and in the external dialectics of relations with other natural languages (*langue/lingua*), etc. The more complex a natural language becomes in terms of expressive capacity, terminological specification and specialization, of semantic-ideological extension, in terms of enhancement of its languages (*langage/linguaggio*) and discourse genres, the more it participates in the linguistic life of other verbal systems.

Work on internal and external plurilingualism in natural languages and on the relation between verbal and nonverbal signs is relatively recent. This is because linguistics has often underestimated, even ignored constitutive interlingualism among languages (*langage/linguaggio*) and discourse genres that go to form the different verbal sign systems – and linguistics has often played a leading role among language sciences.

Among those who have contributed most to underlining the importance of plurilingualism in the life of a natural language (*langue/lingua*) and of all cultural sign systems generally, we wish to recall Bakhtin and Peirce. Bakhtin addressed the question of polylogism and plurilingualism at a time in political-cultural history when a mechanistic and monolinguistic view of the world prevailed, the Stalinist. Peirce has indirectly helped us understand the vital importance of internal and external plurilingualism for natural language (*langue/lingua*) with his theory of "the infinite deferral of interpretants". Signs as such must necessarily relate to other signs that interpret them and determine their meaning at each occurrence in dynamical and open relations, of the endolingual and interlingual orders. In Italy, Giacomo Leopardi (1798–1837) was aware early-on of the essential nature of plurilingualism, at the time perhaps him alone, and not only in Italy, with respect to his time.[65] Leopardi thematizes plurilingualism, external and internal to the same natural language (*langue/lingua*), as a necessary, indeed constitutive factor of natural language. He returns to this issue on several occasions, though the question of plurilingualism is central to his reflections on language. Leopardi takes his distance from those philosophical-linguistic tendencies that, to echo Bakhtin,[66] only know two poles in linguistic life between which all linguistic phenomena are forcefully organized: the unitary system of a given language (*langue/lingua*) and individual use of this language by the speaker.

If plurilingualism is given naturally so to say – though it may be stronger or weaker depending on the historical-cultural situation –, this means that it is a feature of linguistic life that cannot be refrained, one we can actively intervene upon to favour transformation of plurilingualism into *dialogized pluridiscursivity*.[67]

Dialogized pluridiscursivity describes a situation that overcomes mere cohabitation among multiple languages (*langage/linguaggio*), and in certain cases among natural languages (*langue/lingue*), not only in the same culture and the same language (*langue/lingue*), but also within the same person, in the direction of a relation of communication, confrontation, and mutual interpretation. This is the condition of possibility for speech to be able to take its distances from a given language and achieve a metalinguistic and critical awareness of them.

Among common ideals wrongly indicated as favouring the quality of life, we find monolingualism and univocality (but this tendency can also be traced in those philosophical orientations that refer to logical-formal languages as the criterion to evaluate natural languages, that is, historical-social languages): a single language (*langue/lingua*), a single meaning for each signifier, and an unchanging verbal system devoid of internal languages that provoke semantic gaps from one language to another. This situation of monolingualism is expected to guarantee perfect communication, the exact expression of reality and of one's own personal experiences. "New Speak" as hypothesized by George Orwell in his novel *1984* is a heavy satire of the myth of the "perfect language (*langue/lingua*)".

[65] Cf. Ponzio 2001.

[66] Bakhtin 1952–1953, pp. 67–75.

[67] *Ibid.*, p. 75 *sq.*

Currently there exist about six thousand natural languages, while we know of about eight thousand different languages, dead or living[68]; the difficulty in establishing the precise number[69] is linked, above all, with the possibility of distinguishing between languages (*langue/lingua*) and dialects. Just over a hundred or so of the total are languages accompanied by writing (transcription) systems.[70]

The biblical myth of the Tower of Babel describes the passage from an original situation of happy monolingualism to the "confusion of languages", to "the chaos of plurilingualism". According to this myth, the happy original world, a world that human beings slowly lost featured uniqueness and linguistic univocality.

All the same God punishes by raising the bet (what sort of a God would he be otherwise?). God humiliates by giving. Plurilingualism is a gift, even if often misunderstood. In the situation of Pentecost understanding the language (*langue/lingua*) of others consists in hearing it resound in one's own language (*langue/lingua*). This means that encounter among different languages does not effectively occur. Each language only knows itself and remains closed and satisfied in its own identity. Instead, in the Babel of languages (*langue/lingua*) different languages effectively encounter each other and mutually experiment each other's irreducible alterity. Nostalgia of "original monolingualism" can even be traced beyond myth and the popular imaginary in certain philosophical and linguistic conceptions. On their account, the multiplicity of languages (*langue/lingue*) can be traced back to a single original language, an *Ursprache*, universal linguistic structures subtending all languages (*langue/lingua*), so that divergences only concern surface structure. This describes Chomsky's position. In reality, monolingualism, which is also monologism, is but one aspect of a totalitarian attitude towards pluralism and differences, made to pass as a necessary condition for living together.

Plurilingualism and polylogism – like plurivocality, ambiguity, vagueness –, rather than a punishment, a malediction, a fall from a condition of original happiness, are fundamental conditions, indeed irrevocable for communication, expression and understanding.

With reference to Chomsky's linguistic theory, Dell Hymes in his essay "Speech and language"[71] observes that the more we insist on hypothetical universals and their relationship to a "faculty of language", the more existing languages become mysterious. Why many languages and not one only? Differences are not eliminated and resemblances are far from being universals *à la* Chomsky. True language often begins where abstract universals finish.

To study verbal language (*langage/linguaggio*), as Chomsky does, in terms of biologistic innatism, and to judge socio-cultural, historical forces in linguistic development as marginal, does not explain the fact that the supposedly universal biological structures of verbal language do not produce a single language, but many, nor that social conditioning and social differences produce the condition of internal plurilingualism.

---

[68] Mauro 1994.

[69] Michel Malherbe counts three thousand (Malherbe 2010).

[70] *Ibid.*

[71] Hymes 1973.

An explanation is possible if we acknowledge that *language* understood as a 'human species-specific modeling procedure' distinct from *verbal language* is capable of producing multiple worlds and using the same "material", as this term is understood by Louis Hjelmslev (1899–1965), to achieve multiple linguistic universes: in this case the great multiplicity of languages and expressions (on the phonological, syntactic, semantic and pragmatic levels) of "reality" depends on the propensity that language (*langage/linguaggio*) has for plurilinguism and polylogism, for the "play of musement".

A clear sign of limits in Chomsky's conception is the opposition between the *essential properties* of language (*langage/linguaggio*) which are determined biologically and expressed by a "universal grammar", on the one hand, and "incidental facts" that distinguish among different languages (*langue/lingua*), on the other.[72] Monological reductivism is always connected to a monolingual vision. As observed by Jakobson, Chomsky's followers most often only know but one language (*langue/lingua*), English, and from the English language they draw their examples.[73] What Chomskyan theory does not succeed in explaining is not only the existence of the multiplicity of natural languages (*langue/lingua*). The concept of innate grammatical structures also prevents an adequate understanding of the creative character of language. Under this latter aspect, George Steiner[74] agrees that critical readings of Chomsky's approach have demonstrated that his "mentalism" is as naively deterministic as the behavioural theories of language, such as Skinner's.[75]

Steiner advances the hypothesis that the proliferation of different languages (*langue/lingua*) derives from the fundamental need for "distancing" characteristic of language (*langage/linguaggio*), for developing the "otherness" dimension inherent in the "identity" of "lived individuality". Human language (*langage/linguaggio*) is the process of signification that is forever renewing itself, in such a way that a language (*langue/lingua)* is never fixed, nor is it absolutely unitary: as it presents itself through a given language (*langue/lingua)* the world is never univocal and definitive; a given language develops points of view that are *other*, possibilities of saying the world that are *other* by comparison to another language (*langue/lingua*). Indeed, a given natural language is constituted and develops as a function of this possibility. In this sense, Steiner states that language (*langage/linguaggio*) is the main instrument through which man refuses the world as it is. He maintains that to move across languages (*langue/lingue*), to translate, even when we cannot move altogether freely, leads to discovering the human spirit's almost disconcerting taste for freedom.[76]

We are on the way towards unraveling the enigma of Babel where such characteristics as ambiguity, semantic ductility, polysemy, hermetism, simulation, fiction, allusion, reticence, the implicit, otherness are all considered as essential aspects of verbal language, rather than as secondary, weak points, surface traits. Instead of

---

72 Chomsky 1975.

73 Jakobson, quoted in *New Yorker*, 8 May 1971, pp. 79–80 (Steiner 1975, p. 245 *sq.*).

74 Steiner 1975.

75 *Ibid.*, p. 288.

76 *Ibid*, p. 473.

uttering *the same* reality, verbal language tends to take its distances from it by producing other meanings, other modalities of saying, by uttering *another* reality. Languages (*lingue/langues*) do not simply renew, as understood by generative-transformational grammars, they literally create.[77]

As Tullio De Mauro observes, "*variation* is not something that hits languages (*lingue*) from the outside: it installs itself in all points of the reality of a language (*lingua*) as a necessary consequence of its semantics and pragmatics, both of which, in turn, necessarily draw the characters of extensibility and flexibility from the functional needs of each language (*lingua*) in itself".[78] This is what Leopardi had already maintained when he stated the need for internal and external plurilingualism, for semantic vagueness. He asserts that it is absolutely, materially impossible to impose a single language (*langue/lingua*), without giving rise to internal transformations and to other languages (*langue/lingua*), precisely as a way of spreading and imposing itself to a maximum degree.[79]

Language (*Langage/Linguaggio*) as Primary Modeling Species-Specific to Man and Natural Language (*Langue/Lingua*)

Language *(langage/linguaggio)* as the capacity to construct multiple possible worlds finds form and expression through its materialization in a given language (*langue/lingua*).

The "play of musement", no doubt founded on the capacity for language-as-modeling (*langage/linguaggio*), is enhanced by natural language (*langue/lingua*), the more it uses the instruments provided by the latter and fully exploits its resources and potential. On the other hand, languages (*langue/lingua*), themselves the historical result of this "play of musement", are founded on the capacity for language (*langage/linguaggio*), each testifying to its capacity to construct multiple worlds.

But the capacity for language (*langage/linguaggio*) and the "play of musement" also find in a given language (*langue/lingua*), as it has been constructed historically, a limit on their possibilities. The restriction of language (*langage/linguaggio*) by a natural language (*langue/lingua*) can be superceded in the relation with another natural language. To know another natural language, in fact, *does not only serve to supercede barriers of a communicative order*, but also of the *cognitive*, *critical*, *ideological*, *inventive*, *emotional orders*, etc. Knowledge of one or more languages in addition to one's own constitutes an obvious advantage in terms of deconstruction and reconstruction, given that such a capacity is not limited to or conditioned unilaterally by the mother-tongue (*lingua*).

[77] *Ibid.*, p. 228.

[78] Mauro 1994, p. 80.

[79] Ponzio 2001.

Consciousness towards one's own natural language, which is favoured by the gaze of another language, promotes the possibility of experiences that do not converge with one's own language and that not only enrich speaker linguistic consciousness, but also the linguistic consciousness of the language itself. Already in its lexicon, every language (*langue/lingua*) contains instruments and materials with which it presents itself as a metalinguistic device capable of self-reflection, making of itself the object of reflection. One language (*langue/lingua*) empowers the linguistic consciousness of another language, providing not only instruments and materials that enhance and refine its self-awareness, but also an external point of view beginning from which it can improve the way it sees, describes and evaluates itself.

The relation between a language (*langue/lingua*) and experience of one's own body by the speaker of that language deserves attention in itself. Here, we can only address the issue briefly. Learning a mother tongue involves losing many sounds and with them the relative capacity to produce those sounds which, instead, the child who is only beginning to learn how to speak possesses, as testified by infant lallation. To learn a foreign language means to recover (at least in part, relatively to a given language [*langue/lingua*]) the phonatory capacity and sounds debarred by the mother tongue. To articulate the phonemes of another natural language the learner must reactivate physiological capacities that have atrophied because they were not foreseen by the first language, but which could have developed in a different linguistic community.

The implication is that knowledge of natural languages different from one's own offers the possibility of recovering capacities long-abandoned, therefore of renewing the relationship between word and body, speaker and one's own body. This last aspect should not be underestimated when it is a question of motivating foreign language learning. In fact, the search for new experiences, the desire to perceive new sensations, to experiment the body and savour the exotic are certainly more attractive than the drudgery of training to use a given means to satisfy given ends, in this case, the need to communicate which is the motivation generally proposed to promote the study of foreign languages.

The proliferation of natural languages and the concept of linguistic creativity (Chomsky) both testify to the "capacity of language", understood as a 'primary modeling device capable of producing an indeterminate number of possible worlds'. Both derive from the human modeling capacity to invent multiple worlds, that is, from the propensity for the "play of musement".

Modeling works on what Hjelmslev calls "purport",[80] an amorphous *continuum* both on the acoustic level and the semantic. Every natural language gives a particular form to this purport, like sand, as Hjelmslev says, which takes the shape of its container. Every natural language (*langue/lingua*) articulates the indistinct material of expression and content in different ways. This is what Sebeok calls secondary modeling. The phonic material of the *continuum*-purport is organized into "distinctive features", known as phonemes, in the different natural languages (*langue/lingua*),

[80] Hjelmslev 1943 [1961, pp. 32–33].

just as the continuum of the colours of the solar spectrum is divided differently, for example, in English and Welsh.[81] All this can be explained on the basis of creativity as it characterizes *language* understood as a 'human species-specific modeling procedure' (primary modeling).

To use Rossi-Landi's terminology, "linguistic work" produces different paradigms that correspond to the different worlds of different natural languages. The same thing occurs with articulation and organization of the social continuum in different cultures, for example, in the systems of family relations analyzed by Lévi-Strauss.[82]

Human language-as-modeling, writing, produces interpreted signs and interpretant signs on purport as understood by Hjelmslev, on the levels of content and expression. Purport in Hjelmslev's sense is similar to Hamlet's cloud (Shakespeare): it changes aspect from one moment to the next. Signs shape purport differently in different natural languages, each tracing their own specific subdivisions upon it. Purport is physical, acoustic, for what concerns the form of expression, but it is also the amorphous "mass of thought", for what concerns the form of content. Thanks to linguistic work as deposited in different historical-languages, the same material can be formed or restructured differently in different languages like sand put into different shapes or clouds taking different forms, as Hjelmslev claims.

Purport is always other with respect to a given configuration. All the same, however, it always gives itself as *signified*; it *obeys* a form and presents itself as *substance*.

Language (*Langage/Linguaggio*) and Cognitive Processes

Chomsky limits linguistic creativity to verbal language, moreover separating the latter from its communicative function. Instead, creativity is proper to *language* understood as a 'human species-specific modeling device'. Creativity in verbal language and the capacity to be freed of the communicative function is determined by the fact that verbal language is grounded in language-as-modeling, which has no limits on the capacity for innovation and inventiveness. Similarly, that writing can get free of its (mnemotechnic) function (which consists in transcribing verbal oral language) and present itself as creative writing is possible for the same reason.

Reflection on language and speech throws light on what it means to be "*sapiens*", or rather "*sapiens sapiens*", an expression used to characterize mankind in the most advanced phase of development. While the human being shares in semiosis like all other living beings, it is the only animal capable of "semiotics", that is, of contemplating semiosis. *Semiotics* thus understood alludes to the universal propensity of the human mind, as Sebeok claims, for reverie focused on its long-term cognitive strategies and daily maneuverings.[83]

[81] Cf. Johansen 1998, pp. 2275–2282.

[82] Cf. Lévi-Strauss 1958; Ponzio et al. 1994 [1999, pp. 50–53].

[83] Sebeok 1991, p. 97.

Verbal language plays a fundamental role in all this. It provides the form through which consciousness and thought exist and take shape, but it is not the origin.[84] We have already commented on the relation between verbal language and language-as-modeling. As the possibility of constructing different worlds, language-as-modeling is enhanced by verbal language, by other auxiliary artificial stimuli or "stimuli-means" [*stimul-sredstvo*] – different number and calculation systems, different mnemotechnic devices, different writing systems, schemes, diagrams,[85] and by the manipulative, productive activity of artifacts (like the former string these are specified historically and socially and as such relate to them dialectically).

At the same time, however, language as pre-verbal modeling subtends the manipulative activity of verbal and nonverbal languages.[86] The production of artifacts and transformation of material objects into signs proceed at the same pace (on the phylogenetical level as well, that is, in the process of homination). And while they presuppose language as primary modeling, the central element of such transformation is the human body.

The human body is the primary material of manipulative material and sign material: this involves the primacy of gesture and voice, even before the latter becomes an articulate phonic language. Moreover, the instruments used for work represent an extension on the human body [*Leib*]. With respect to one's own body, external material reality, both in its sign function and in its instrumental function, can be considered as secondary material that presupposes reference to the human body.[87]

The relation between semiosis and thought also emerges as the connection between *meaning* and *concept*. In any case, meaning is distinguished from concept. The interpretive itinerary that goes to form meaning converges in part with the *class* that forms the concept. For example, interpreteds-interpretants, that is, meaning in the botanical sense for the phonia "tree" only enter a part of the class that forms the concept *tree* (understood in the same sense). In fact, if, in this interpretive itinerary, we have interpretants that are trees (the olive tree is an interpretant of the sign "tree"), we also have interpreteds-interpretants that are not trees, beginning from the same phonia itself "tree", which expresses the concept *tree* through its meaning, but is not a tree and therefore does *not belong* to the logical class *tree*. Knocking at the door is generally interpreted as "someone is behind the door and wants to enter". The two things, like the interpretant formed by the action of opening the door, are on the same interpretive route, but they do not enter the same logical class and do not form a concept. Smoke signifies fire, that is, it has fire as an interpretant – just like the word *fire* – but *smoke* and *fire* do not enter the same concept.

Therefore, meaning and concept are closely connected. Every meaning expresses a concept and, vice versa, every concept requires a meaning, that is, an interpretive route. All the same, however, meaning and concept must be kept distinct. *The concept is a class of objects which may or may not be grouped together in subclasses, and the*

[84] Rossi-Landi 1985 [2006, p. 252].

[85] Vygotskij 1934 [1990].

[86] Cf. Rossi-Landi 1985 [2006, pp. 217–269].

[87] Voloshinov's essays of 1926–1930 cf. in Ponzio (ed.), 2014, pp. 271–333, 1461–2069.

class may eventually enter a larger class. Meaning is an interpretive route formed of connections among signs, of deferrals from interpretant to interpretant. The meaning 'tree' and the concept *tree* are two different things even if one implies the other.

Utterance and Answering Comprehension

Until it deals exclusively with the elements of natural language and the sentence, linguistics cannot account for answering comprehension. Instead, answering comprehension (or if we prefer, responsive understanding) is connected with the utterance, intertextuality and dialogue. The objects of linguistics are limited to interpretation in terms of identification, that is, interpretation understood in terms of identification rather than of answering comprehension. Consequently, in linguistics *quietude* is the condition for interpretation-identification. Quietude is the condition for perceiving sounds and identifying verbal signs. Following Bakhtin in "From notes made in 1970–71",[88] a distinction can be made between *quietude* and *silence*, which corresponds to the distinction between the conditions for *perceiving a sound*, the conditions for *identifying a sign* and the conditions for *responding to the sense of a sign*. Quietude is associated to the first two cases, silence to the third, i.e. to the conditions for responding to the sign and understanding sense. Quietude is the condition for perceiving sound and the distinguishing features of language; for identifying the repeatable elements of language, those belonging to the system of language on the phonological, syntactical and semantical levels. Instead, silence is the condition for understanding the sense of the utterance, sense in its unrepeatability; silence is the condition for response to the utterance in its singularity. Quietude is associated with *language* understood as '*langue*' and with its physical (acoustic and physiological) substratum. Silence is associated with the utterance and with sense, with the social-historical materiality of the sign. Whilst quietude is an expression of the logic of identity, silence is associated with high degrees of alterity and as such is an expression of the properly human.[89] It ensues that silence can reach high degrees of critique and creativity. In terms of interpretive capacity it is associated with *responsive understanding* and *responsible engagement*. According to this analysis quietude is associated with signality and silence with semioticity.[90]

Both taxonomical linguistics and transformational generative linguistics – which shifts its attention from the elements of natural language and the sentence to the relations that generate them – belong to the same orientation. We are alluding here to the tendency to neglect the relation of answering comprehension (or responsive understanding) among utterances, their sense. Unlike *meaning* understood in terms of identification, answering comprehension, *signifying processes* that develop in terms of sense and significance require *silence* as the condition of their production. Nor does silence represent a limit on sense and significance.

[88] Bakhtin 1970–1971 [1986].

[89] Ponzio 1993, pp. 138–154; Petrilli 2014, pp. xx, 112–114.

[90] Bakhtin 1970–1971 [1986, pp. 133–134]; Petrilli 2014, Chapter 6.

It ensues that neither taxonomical linguistics nor generative linguistics have anything interesting to say about the utterance as the live cell of discourse, about its dialogical character, its essential vocation for answering comprehension. Nor do they have anything to say about the different forms of silence, about the indirect, deferred, allusive, parodic, ironical utterance, about its ambiguity and polysemy, its implied sense, implicit meaning, potential for disengagement, capacity for "shift".[91] Neither taxonomical linguistics nor generative linguistics have anything to say about literary writing which is made of different forms of silence.[92] In his 1959–1961 essay, "The problem of the text", Bakhtin says that the "writer" does not use language (*lingua*) directly, but "has the gift of indirect speaking".[93]

Insofar as it is based on the notion of the system of rules, on the code, insofar as it can only move in the space that extends from sound to the verbal sign identified in phonological, syntactical and semantical terms, that is, the space of quietude, this type of linguistics, code linguistics, can also be named "linguistics of quietude".

Encounter, mutual methodological and terminological exchange between linguistics of the sentence, on the one hand, and mathematical information theory, on the other, is not incidental. The denomination itself of *code linguistics* derives from this exchange. As for information theory, this type of linguistics, code linguistics, is only familiar with noise as an obstacle to interpretation, that is, to interpretation reduced to de-codification, recognition and identification. Once the utterance is reduced to the relation between code and message, proper to the signal, noise is connected to some imperfection in the channel, to interference from the external context, or to lack of rules that restrict the relation between message and code and consequently allow for ambiguity. In any case, noise thus described is connected with quietude, the condition for perception of the signal.

The problem of sense and significance goes beyond the limits of code linguistics or "linguistics of quietude". It concerns linguistic reflection that is not limited to *historical natural language* (*langue/lingua*) understood in terms of code, to linguistic relations among elements in the system of language (*langue/lingua*), to relations among sentences, or to transformational processes (from "deep structures" to "surface structures"). Rather, the question of sense and significance concerns dialogical relations among verbal signs insofar as they are utterances, on the one hand, and interpretants of answering comprehension, on the other.

The background from which dialogical relations emerge is silence. Quietude and the absence of noise constitute the physical condition for the utterance, the minimal condition that concerns it in the signality dimension, that of recognition and identification, but they will not suffice for the utterance to subsist as a sign and have sense.

Silence is both the situation or position the utterance begins from and the situation or position it is received in. The condition of possibility of the word's freedom is silence, a choice made by the speaker, a position chosen by the speaker; freedom

[91] Barthes 1982.

[92] Ponzio 2010 and 2012a.

[93] Bakhtin 1959–1961 [1986, pp. 110–115].

involves the violation of silence and not simply violation of quietude; at the same time, it presupposes silence as *a listening position*.

From silence, the utterance's freely chosen starting point, to the silence it calls for, to which it turns, to which it gives itself and which receiving silence welcomes in listening: this is the movement of the utterance. Between emitter silence and receiver silence there is no substantial difference: silence as the utterance's starting point is in turn a listening position; the utterance is effectively a response, an answering comprehension response. Vice versa, silence as a listening position is the starting point for interpretation of the answering comprehension order, the beginning of a response in the form of an utterance when the responsive interpretant is of the verbal type. The utterance turns to the silence of responsive listening. Once the silence of responsive listening is eliminated, what remains is quietude. Obviously the utterance does not address quietude, on the contrary it withdraws from it. Quietude as we are describing it here belongs to the system of *language* (*langue/ lingua*) understood as 'repetition', 'iteration', as reproduction of the "order of discourse" (Michel Foucault[94]). Instead, silence belongs to the sphere of the non repeatable utterance; it participates in the open unfinalized totality of the logosphere, as Bakhtin says in his 1970–1971 notes quoted above.[95]

Silence allows the utterance to withdraw from investigative, coercive quietude, quietude of the linguistic system. Roland Barthes (1915–1980) speaks of the "fascist" character of the system of language.[96] This does not consist in stopping a person from speaking, but in obliging that person to speak, to reiterate fixed meanings, sanctioned by the order of discourse. Quietude imposes speaking, but not listening. Silence is listening. Insofar as it is responsive listening, silence is a pause in the unrepeatable utterance.

The "linguistics of quietude" corresponds to a communication system dominated by quietude. Code linguistics is the expression of the centripetal forces of the social. Monologism, the tendency towards univocality and the lowering of the sign to the level of signality, as established by the equal exchange relation between signifier and signified, only belongs to the linguistics of quietude secondarily: in the first place, they belong to the social form that has chosen quietude as the background for speaking. The linguistics of quietude is simply an expression of this state of affairs.

Homologation of the communicative universe reduces listening to wanting to hear. It limits the spaces of silence where freedom to listen is as necessary as freedom of the word. Consequently, due to such homologation processes the communicative universe ends by investing the verbal sign solely with the conventional characteristics of the signal or the natural characteristics of sound.

From necessity of the natural to repetition of the conventional, or to say it with Peirce, from indexicality to symbolicity: this is the sphere reserved to the sign when it loses its ambivalence, ductility, and possibility of attracting an interpretant characterized by originality, autonomy, absolute alterity. Peirce attributes such characteristics to *iconicity*.

[94] Foucault 1971.

[95] Bakhtin 1970–1971.

[96] Barthes 1979.

Enclosed in the universe of quietude and the obligation to speak according to laws, conventions, habits, the sign loses its character as a challenge, a provocation with respect to identity, the closed totality; it loses the possibility of questioning that which seems firm and definite, as though it were natural. Instead, such an attitude is possible through silence, which means not to collaborate with the closed universe of discourse, to withdraw from monologism, to supercede the logic of equal exchange between *signifiant* (signifier) and *signifié* (signified), between the interpreted sign and the interpretant sign.

Constriction of the sign to the space of quietude, separation from silence and the freedom of listening (listening open to polysemy) deprives the sign of its species-specific *human* character, of its capacity for *language* (understood in Sebeok's sense as 'modeling', 'syntactics'). Quietude renders the sign mechanical and natural, making it oscillate between the conventional character of the signal and the natural character of sound, the natural character of that which does not make claims to sense.

A linguistic theory capable of accounting for the universe of language, expression and communication must be explicative and critical, well beyond the limits of a descriptive and taxonomic approach to language analysis. A global approach to communication in the human world must account for the social processes of linguistic production in relation to a critical theory of ideology. What we are describing here as the "linguistics of silence"[97] is turned to the live word, to the utterance as it develops out of the dialogic interaction among interpreted signs and interpretant signs, among voices in the real context of social relations. The linguistics of silence is oriented as listening, therefore it focuses on language oriented in the direction of dialogic heteroglossia, plurilingualism internal and external to the same natural language and answering comprehension, which also account for the human capacity for critique and creativity.

References

Bakhtin, M. M. (1952–1953). The problem of speech genres. In Bakhtin 1986, pp. 59–102.

Bakhtin, M. M. (1959–1961 [1986]). The problem of the text. In Bakhtin 1986, pp. 103–131.

Bakhtin, M. M. (1970–1971 [1986]). From notes made in 1970–71. In Bakhtin 1986, pp. 132–172.

Bakhtin, M. M. (1986). *Speech genres and other late essays*. Austin: Austin University of Texas Press.

Barthes, R. (1979). *Leçon*. Paris: Seuil.

Barthes, R. (1982). *L'obvie et l'obtus. Essais critiques III*. Paris: Seuil.

Bouissac, P. (Ed.). (1998). *Encyclopedia of semiotics*. New York/Oxford: Oxford University Press.

Chomsky, N. (1975). *Reflections on language*. New York: Pantheon.

Cobley, P. (2007). A brief note on dialogue. In Petrilli (Ed.), 2007, pp. 13–20.

Danesi, M. (1993). *Vico, metaphor and the origin of language*. Bloomington/Indianapolis: Indiana University Press.

De Mauro, T. (1994). Capire le parole. Roma/Bari: Laterza.

[97] Cf. Petrilli (ed.), 2007 and 2013.

Deely, J. (1998). Thomas A. Sebeok. In Bouissac (Ed.), 1998, pp. 557–559.
Deely, J. (2007). The primary modeling system in animals. In Petrilli (Ed.), 2007, pp. 161–180.
Derrida, J. (1967). *L'écriture et la différence*. Paris: Seuil.
Engels, F. (1883 [1962]). Dialektik der Natur. In K. Marx & F. Engels, *Werke* (Vol. 20, pp. 305–570). Berlin: Karl Dietz Verlag.
Fano, G. (1972). *Origini e natura del linguaggio*. Turin: Einaudi [English translation and introduction by Petrilli, S. (1992). *Origins and nature of language*. Bloomington: Indiana University Press].
Favareau, D. (2013). Agapastic exploration of the biosphere: Alterity and biosemiosis. In Petrilli (Ed.), 2013, pp. 143–150.
Foucault, M. (1971). *L'ordre du discours*. Paris: Gallimard.
Gould, J., & Vrba, E. S. (1982). Exaptation – A missing term in the science of form. *Paleobiology, 8*(1), 4–15.
Hjelmslev, L. (1943 [1961]). *Prolegomena to a theory of language*. Madison: University of Wiscontin Press.
Hymes, D. (1973). Speech and language. On the origins and foundations of inequalities among speakers. *Daedalus, 102*(3), 59–85.
Jakobson, R. (1960). Linguistics and poetics. In T. A. Sebeok (Ed.), *Style in language* (pp. 350–377). New York: Wiley.
Jakobson, R. (1965). Quest for the essence of language. *Diogenes, 13*(51), 21–37.
Jakobson, R. (1971). *Selected writings, vol. II: Word and language*. The Hague: Mouton.
Johansen, J. D. (1998). Hjelmslev and glossematics. In Posner, Robering, Sebeok (Eds.), 1997–2004, Vol. 2, pp. 2272–2289.
Kristeva, J. (1969 [1981]). *Le langage, cet inconnu*. Paris: Seuil.
Kull, K. (2007). Biosemiotic conversations: Ponzio, Bakhtin, Kanaev, Driesch, Uexküll, Lotman. In Petrilli (Ed.), 2007, pp. 79–89.
Kull, K. (2010). Umwelt and modeling. In P. Cobley (Ed.), *The Routledge companion to semiotics* (pp. 43–56). London/New York: Routledge.
Kull, K. (2013). On the sphere of understanding: A note on Driesch. In Petrilli (Ed.), 2013, pp. 151–161.
Levinas, E. (1982). *L'au-delà du verset*. Paris: Éditions de Minuit.
Lévi-Strauss, C. (1958). *Anthropologie structurale*. Paris: Plon.
Lieberman, P. (1975). *On the origins of language*. New York: Macmillan.
Lotman, J. M. (1991). *Universe of the mind. A semiotic theory of culture*. Bloomington: Indiana University Press.
Lucid, D. P. (Ed.). (1977). *Soviet semiotics*. Baltimore: Johns Hopkins University Press.
Malherbe, M. (2010). *Les langages de l'humanité*. Paris: Robert Laffont.
Nöth, W. (2013). Is communication possible? In Petrilli (Ed.), 2013, pp. 66–71.
Parret, H. (1998). Greimas and his school. In Posner, Robering, Sebeok (Eds.), 1997–2004, Vol. 2, pp. 2300–2311.
Petrilli, S. (Ed.). (2007). *La filosofia del linguaggio come arte dell'ascolto/Philosophy of language as the art of listening. On Augusto Ponzio's scientific research*. Bari: Edizioni dal Sud.
Petrilli, S. (2013). Reinterpreting the notions of 'triadism', 'dialectics' and 'automatism' with A. Ponzio. In Petrilli (Ed.), 2013, pp. 302–308.
Petrilli, S. (Ed.). (2013). *Writing, voice, undertaking*. New York/Ottawa/Toronto: Legas.
Petrilli, S. (2014). *Sign systems and semioethics. Communication, translation and values*. Berlin: Mouton de Gruyter.
Petrilli, S., & Ponzio, A. (2001). *Thomas Sebeok and the signs of life*. Duxford, Cambridge: Icon Books.
Petrilli, S., & Ponzio, A. (2002a). *I segni e la vita. La semiotica globale di Thomas A. Sebeok*. Milan: Spirali.
Petrilli, S., & Ponzio, A. (2002b). Sign vehicles for semiotic travels: Two new handbooks. *Semiotica, 141*(1–4), 203–350.

Petrilli, S., & Ponzio, A. (2005). *Semiotics unbounded. Interpretive routes through the open network of signs*. Toronto: Toronto University Press.
Petrilli, S., & Ponzio, A. (2007). *Semiotics today. From global semiotics to semioethics, a dialogic response*. New York/Ottawa/Toronto: Legas.
Ponzio, A. (1984). Semiotics between Peirce and Bakhtin. *Recherches Sémiotiques/Semiotic Inquiry, 4*(3–4), 306–326.
Ponzio, A. (1990). *Man as a sign*. Berlin: Mouton de Gruyter.
Ponzio, A. (1993). *Signs, dialogue and ideology*. Amsterdam/Philadelphia: John Benjamins [edited and translated from Italian by S. Petrilli].
Ponzio, A. (2001). *Enunciazione e testo letterario nell'insegnamento dell'italiano come LS*. Perugia: Guerra.
Ponzio, A. (2002). *Il linguaggio e le lingue*. Bari: Graphis.
Ponzio, A. (2010). *Encontres de palavras. O outro no discurso*. San Carlo: Pedro e João Editores.
Ponzio, A. (2012a). *In altre parole*. Milan: Mimesis.
Ponzio, A. (2012b). *Linguística Chomskyana e ideologia social*. Curitiba: Editora UFPR.
Ponzio, A. (Ed.). (2014). *Michail Bachtin e il suo circolo. Opere 1919–1930*. Milan: Bompiani.
Ponzio, A., Calefato, P., & Petrilli, S. (1994 [1999]). *Fondamenti di filosofia del linguaggio*. Roma/Bari: Laterza.
Ponzio, A., Bonfantini, M. A., & Petrilli, S. (2006). *I dialoghi semiotici. Sul dialogo, sulla menzogna e la verità, sui nuovi mass-media, sulla retorica e l'argomentazione, sulla testualità e la discorsività, sull'ideologia e l'utopia 1982–2006*. Napoli: Edizioni Scientifiche Italiane.
Posner, R., Robering, K., & Sebeok, T. A. (Eds.) (1997–2004). *Semiotik/semiotics. A handbook on the sign-theoretic foundations of nature and culture* (Vol. 4). Berlin/New York: Walter de Gruyter.
Rossi-Landi, F. (1968). *Il linguaggio come lavoro e come mercato*. Milan: Bompiani.
Rossi-Landi, F. (1972). *Semiotica e ideologia*. Milan: Bompiani.
Rossi-Landi, F. (1975). *Linguistics and economics*. The Hague: Mouton.
Rossi-Landi, F. (1985 [2006]). *Metodica filosofica e scienza dei segni*. Milan: Bompiani.
Rudy, S. (1986). Semiotics in the USSR. In T. A. Sebeok & J. Umiker-Sebeok (Eds.), *The semiotic sphere* (pp. 34–67). New York: Plenum.
Sebeok, T. A. (1963). Untitled review (subsequently sometimes referred to under the *ex post factum* title "Communication in animals and men") of three books: 1. Lindauer, M. (1961). *Communication among social bees*. Cambridge, MA: Harvard University Press; 2. Kellogg, W. N. (1961). *Porpoises and sonar*. Chicago: University of Chicago Press; 3. Lilly, J. C. (1961) *Man and dolphin*. Garden City: Doubleday & Co. Inc. *Language*, *39*(3) (July–September 1963), 448–466.
Sebeok, T. A. (1976 [1985]). *Contributions to the doctrine of signs*. Lanham: University Press of America.
Sebeok, T. A. (1979). *The sign & its masters*. Texas: The University of Texas Press.
Sebeok, T. A. (1981). *The play of musement*. Bloomington: Indiana University Press.
Sebeok, T. A. (1986). *I think I am a verb. More contributions to the doctrine of signs*. New York: Plenum.
Sebeok, T. A. (1991). *A sign is just a sign*. Bloomington: Indiana University Press.
Sebeok, T. A. (1994). *Signs. An introduction to semiotics*. Toronto: Toronto University Press.
Sebeok, T. A. (1997). The evolution of semiosis. In Posner, Robering, Sebeok (Eds.), 1997–2004, Vol. 1, pp. 436–446.
Sebeok, T. A. (2001). *Global semiotics*. Bloomington: Indiana University Press.
Sebeok, T. A., & Danesi, M. (2000). *The forms of meaning. Modelling systems theory and semiotics*. Berlin: Mouton de Gruyter.
Sebeok, T. A., Petrilli, S., & Ponzio, A. (2001). *Semiotica dell'io*. Rome: Meltemi.
Steiner, G. (1975). *After Babel. Aspects of language and translation*. London: Oxford University Press.
Thibault, P. J. (1998). Binarism. In Bouissac (Ed.), 1998, pp. 76–82.

Uexküll, T. von (1998). Jakob von Uexkülls Umweltlehre. In Posner, Robering, Sebeok (Eds.), 1997–2004, Vol. 2, pp. 2183–2190.
Vernadskij, V. I. (1926). *Biosfera*. Leningrad: Nauka. [The Biosphere].
Vygotskij, L. S. (1934 [1990]). *Pensiero e linguaggio*. Roma/Bari: Laterza.

Umwelt and Language

Morten Tønnessen

Abstract It is often asserted that the existence of human language sets us apart from non-humans, and makes us incomparably special. And indeed human language does make our Umwelt (Jakob von Uexküll), our lifeworld, uniquely open-ended. However, by committing what I term *the anthropocentric mistake*, i.e. falsely assuming that all true reality is linguistic, we close in on ourselves and our language-derived practices, and as a result we lose sight of much that truly matters (including a proper understanding of our human nature). Like Sebeok and Hoffmeyer I hold that language is a modeling system, but unlike them I argue that language is not external to the Umwelt, but internal to it. Language changes the human Umwelt not by escaping or sidelining it, but by fundamentally transforming it. In consequence supra-linguistic phenomena as well are modeled as internal to the human Umwelt. The Umwelt model presented is termed *the tripartite Umwelt model*, and includes three aspects of Umwelt: the *core* Umwelt, the *mediated* Umwelt and the *conceptual* Umwelt. Linguistic practices are placed within the latter, but it is furthermore claimed that a number of animals too have conceptual Umwelten, which are said to be characterized by predicative reasoning, the habitual, mental attribution of specific features to someone or something. The activity of languaging is presented as more-than-linguistic, with reference to the distributed language perspective. Given all the dark matter underpinning and surrounding verbal practices, a foray into the hinterland of language is called for. A section on the genesis and modalities of language addresses the origin and evolution of language, acquisition of language in childhood and a simple typology of the various linguistic modalities of the human Umwelt. The concluding section treats Ivar Puura's notion *semiocide*, and the question: how can we *language* as if nature mattered?

Keywords The anthropocentric mistake • Dark cognitive matter • Dark cultural matter • Distributed language • First-order languaging • Genesis of language • Perception • Semiocide • Umwelt theory

M. Tønnessen (✉)
University of Stavanger, Stavanger, Norway
e-mail: mortentoennessen@gmail.com

E. Velmezova et al. (eds.), *Biosemiotic Perspectives on Language and Linguistics*, Biosemiotics 13, DOI 10.1007/978-3-319-20663-9_5

> *[Man] knows that there are in the soul tints more bewildering, more numberless, and more nameless than the colours of an autumn forest [...] Yet he seriously believes that these things can every one of them, in all their tones and semi-tones, in all their blends and unions, be accurately represented by an arbitrary system of grunts and squeals. He believes that an ordinary civilized stockbroker can really produce out of his own inside noises which denote all the mysteries of memory and all the agonies of desire.*
>
> (Chesterton 1904, p. 88)

Introduction

> *[W]e move in science into an unknown language with unknown grammar and try, with a dictionary in our hands, to compose grammatically correct sentences.*
>
> (Markoš 2002, p. 180)

Language, writes Marcello Barbieri, is "the quintessential example of semiosis".[1] According to Martin Heidegger, Man is not simply a living creature who possesses language along with other capacities – no, "language is the house of Being in which man ek-sists [**sic** – 'stands out'] by dwelling, in that he belongs to the truth of Being, guarding it".[2] In Jesper Hoffmeyer's words, our species' evolutionary acquisition of language implied a "switch from an *umwelt* containing very few transformation rules to a grammatical *umwelt*".[3] We are fundamentally linguistic creatures. "Humans", says Thomas A. Sebeok, "have evolved a way of modeling *their* universe in a way that not only echoes 'what is out there' but which can, additionally, dream up a potentially infinite number of *possible worlds*".[4] Edmund Husserl was of a similar opinion: "Clearly it is only through language and its far-reaching documentations, as possible communications, that the horizon of civilization can be an open and endless one, as it always is for men".[5]

And so the stage is set. To most people, language largely constitutes reality. And yet language is free to evolve at the inkling of an eye or by the hunch of a confused mind. Without a doubt, language does in many senses open the world up to us – but it also conditions and constrains us. As David Abram writes, "[e]very attempt to definitively say *what language is* is subject to a curious limitation. For the only medium with which we can define language is language itself. We are therefore unable to circumscribe the whole of language within our definition".[6]

[1] Barbieri 2012b, p. 450.

[2] Heidegger 1977, p. 213.

[3] Hoffmeyer 1993 [1996, p. 102].

[4] Sebeok 1987, p. 347.

[5] Husserl 1936–1939 [1970, p. 358].

[6] Abram 1997, p. 73.

A second caveat is also required: there are phenomena that cannot (best) be described in a scientific language. The academic genre is given to objectification and generalization, and might thus not be capable of capturing all phenomena which are not easily objectifiable. This reminder is no less important given the implicit topic matter of this text, *subjective experience*. The reader should therefore keep in mind warnings à la those of Gabriel Marcel[7] with regard to the pitfalls of methods of objectification. By objectifying subjective phenomena, and describing them in scholarly language, we convert them into another genre, and consequently a different mode of being – and this scholarly mode of being is not in all respects true and faithful to the phenomena. In particular, the detachment necessitated by abstract analysis is (if it were to become our *only* mode of being) irreconcilable with full-fledged participation as incarnated, engaged beings on par with other creatures.

"I *Language*, Therefore I Model"

> *Sultan knows: Now one is supposed to think. That is what the bananas up there are about. The bananas are there to make one think, to spur one to the limits of one's thinking. But what must one think? One thinks: Why is he starving me? One thinks: What have I done? Why has he stopped liking me? One thinks: Why does he not want these crates any more? But none of these is the right thought. Even a more complicated thought – for instance: What is wrong with him, what misconception does he have of me, that leads him to believe it is easier for me to reach a banana hanging from a wire than to pick up a banana from the floor? – is wrong. The right thought to think is: How does one use the crates to reach the bananas?*
>
> (Coetzee 1999, p. 28)

Uexküll and Language

Jakob von Uexküll (1864–1944), notes Han-Liang Chang, "rarely referred to language communication".[8] On one of the rare occasions where he did refer to language, in a letter to Heinrich Junker dated 29th March 1937, Uexküll said that "[l]inguistics itself is rather remote from my area", though he complimented Junker for being "on the right path by making it into a biological science".[9] The German-Baltic biologist further noted: "Language interests me mainly as a means of communication between man and animals, and as a means of communication between animals themselves".[10]

[7] Marcel 1962.

[8] Chang 2009, p. 170.

[9] *Ibid.*, referring to Uexküll 1981 [1987, p. 176].

[10] *Ibid.*

In a 1917 article entitled "Darwin and the English Morality",[11] Uexküll comments on the difference that human language makes in our studies of animals and human beings.[12]

> It is clear that the mechanical effects of the physical and chemical forces alone do not lead us to insight about animal life, and, furthermore, that knowledge of these factors requires that the builder or operation manager affects the body machine.
>
> These factors have been known since ancient times – they were called "drives" or urges and distinctions were made between food drive and sex drive, one spoke about self-preservation drive, and in animal communities or animal states [*Tierstaaten*] the social drives were detected.
>
> As long as the topic is processes in the animal world, one must be satisfied with the identification of such drives, which one treats as given factors of nature and seeks to investigate objectively.
>
> But if the topic is humans, whose language we understand and whose utterances resemble our own – then we are capable of providing part of the drives with sensory content that makes psychological understanding possible.[13]

Is Language External or Internal to the Umwelt?

The reality of signs, and of Umwelten, entails that living beings are enmeshed in worlds of meaningful, significant phenomena and occurrences. Barbieri[14] and several other biosemioticians have suggested that even though there are examples of symbolic activity in animals, "[a] systematic use of symbols at the basis of our behaviour is indeed what divides human language from animal communication". As Sebeok believed and Hoffmeyer thinks, I too think of language as being a species-specific human capability that has tremendous impact on the character of human affairs and of the human being. However, as we shall see, I think about language in terms of *the conceptual Umwelt* – an "outer" yet, as a rule, thoroughly integrated layer of the Umwelt. Language, then, is intimately tied to perception – language *frames* perception, and simultaneously language is *grounded in* (core) perception – and, indeed, in a sense language *is* perception (as scholars within ecological linguistics freely admit, language is a perception system).

Sebeok and Hoffmeyer both see language as transcending the human Umwelt. Particularly relevant here is Sebeok's view on language as a secondary modelling system, whereas the Umwelt is the primary modelling system.[15] The distinction between primary and secondary modelling systems derives from the Tartu-Moscow

---

[11] Uexküll 2013, p. 454; cf. Uexküll 1917.

[12] Cf. Uexküll 1917, pp. 219–220.

[13] Cf. also the passage corresponding to *ibid.*, p. 236, where Uexküll addresses the difference, in his eyes, between English language and German language with regard to propagation of influence: "Every English word comes from an English heart".

[14] Barbieri 2012b, p. 449.

[15] Or more specifically, as Barbieri points out: "The primary modelling system consists [...] of two types of models, one that represents the environment [the Umwelt] and one that carries information about the body [the Innenwelt]" (Barbieri 2012a, p. 40).

school[16]; however, in that tradition Juri Lotman[17] and others regarded language as the primary modeling system. This was because it had a central position in culture, and so, any secondary modeling system was supra-linguistic or, in other terms, language-derived. While Sebeok positions the Umwelt as fundamental, he simultaneously positions language as external to it. Admittedly, he saw "organism-environment interaction (i.e. species-specific *Umwelt*) as a crucial component of the growth of language in the individual"[18] – but he nevertheless asserted that language ultimately escapes the Umwelt, a view also adopted by Hoffmeyer.

The claim that language is a modeling system has an important implication, namely that language is not first and foremost (and was not originally) a verbal communication system. "Language", wrote Thomas Sebeok and Marcel Danesi, "is, by definition, a secondary cohesive modelling system providing humans with the resources for extending primary forms ad infinitum".[19] In Prisca Augustyn's words, the Umwelt, "in Sebeok's working definition, 'is a model generated by the organism' […] to which language adds a secondary, cognitive dimension".[20] While language transcends the Umwelt, it also gives it depth or detail. Sebeok thought that language initially above all had served "the cognitive function of modeling, and, as the philosopher Popper as well as the linguist Chomsky have likewise insisted […], not at all for the message swapping function of communication. The latter was routinely carried on by nonverbal means, as in all animals, and as it continues to be in the context of most human interactions today".[21]

The Tripartite Umwelt Model

Figure 1 shows the tripartite model of the human Umwelt.[22] In addition to the three aspects of Umwelt, the illustration displays Uexküll's four main categories of functional cycles,[23] two of them in generalized form.

By *core Umwelt*, I mean the aspect of Umwelt within which one interacts directly and immediately with other creatures or Umwelt objects, in (to use a figure of

[16] Zaliznjak et al. 1977; cf. Chang 2009, p. 172.

[17] Lotman 1991.

[18] Augustyn 2013, p. 98.

[19] Sebeok and Danesi 2000, p. 108.

[20] Augustyn 2015, p. 180.

[21] Sebeok 1991, p. 334.

[22] A precursor to this model, which is the invention of the author, is the notions *conceptual world* and *conceptualized Umwelt experience* (cf. Tønnessen 2003, p. 290), representing two of seven distinctive human features. "The conceptual world", it is stated, "has its roots in sensory perception, and its concepts are meaningful only by reference – direct or indirect – to concrete objects of perception (cf. Uexküll 1928, pp. 334–340)" (*Ibid.*).

[23] Cf. Uexküll 1928, p. 101.

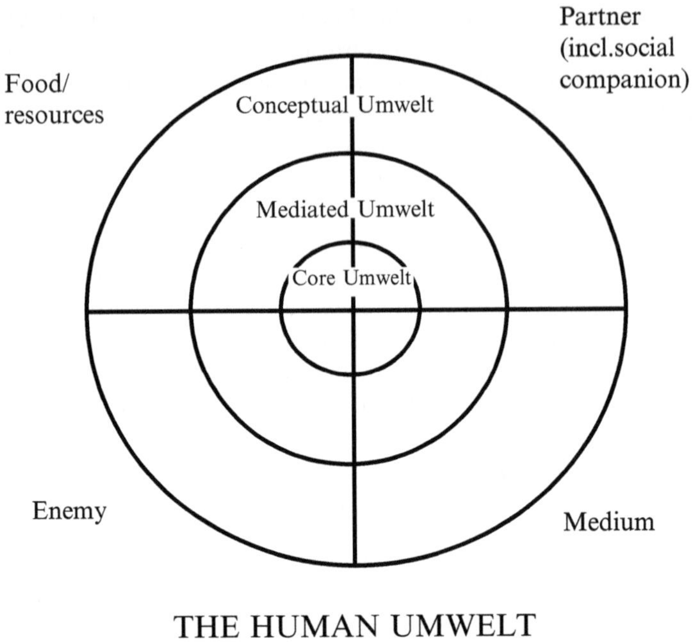

Fig. 1 The tripartite model of the human Umwelt

speech) "face-to-face" encounters.[24] By *mediated Umwelt*, I mean the aspect of Umwelt in which Umwelt objects are encountered indirectly by way of some mediation (memory, fantasy, anticipation, modern media, etc.). I suggest that this particular aspect of Umwelt can generally be associated with Uexküll's notion of the *Suchbild*, the *search image.*[25] By *conceptual Umwelt*, I mean the aspect of Umwelt in which one navigates among Umwelt objects in terms of predicative reasoning in general or human language in particular. Conceptual Umwelt objects are in the latter case Umwelt objects whose functional meaning is imprinted linguistically. Though the conceptual Umwelt is particularly central in the human case (to the point where we confuse linguistic reality with reality as such), a number of "higher" animals qualify for being attributed conceptual Umwelten as well, in so far as they are capable of conducting predicative reasoning.[26] I theorise that these three layers

[24] However, in all normal instances, i.e. whenever the perceiver is capable of having memories or at least is capable of anticipating events, our actual encounters with others involve mediation, and thus the mediated Umwelt, as well. Only in exceptional cases, in consequence, are "face-to-face" encounters *solely* located within the core Umwelt.

[25] Cf. Uexküll 2010, pp. 113–118. In the human context, the mediated aspect of Umwelt arguably dominates in modern culture, as reflected in cultural practices including day-long interaction with screens.

[26] Note that by attributing a conceptual Umwelt to an animal one does not attribute language to it. The question "Do animals have language?" is as controversial as the related question "Is Man an animal, yes or no?" The answers given often appear to be derived from emotion and identity rather

interact dynamically so that one or two of the layers are occasionally temporarily suspended (in other words, human perception is subsequently focused – more or less exclusively – on different Umwelt layers).

The conceptual Umwelt is the most novel in evolutionary terms and, thus, corresponds broadly to what Sebeok characterised as humans' secondary modelling system. But as we have seen, both Sebeok and Hoffmeyer think of human language as being external to the human Umwelt. For both of them the Umwelt represents the "animal" side of the human creature, whereas human culture can only be understood in terms of something (particularly language) that escapes the Umwelt. In my perspective, human language is a special case of more widespread systems of predicative reasoning, and enmeshed in the Umwelt that is our lifeworld, our phenomenal world. Language is *internal* to the Umwelt, i.e. *part* of the Umwelt, and there is a dynamic relationship between the conceptual side of Umwelt and the other aspects of Umwelt. This situates the Umwelt as a rich notion capable of serving as theoretical and methodological foundation for studies of the world of the living and the world of human affairs alike (for example, the tripartite model of the Umwelt may be applied as an ethogram in ethology, or for similar mapping purposes in ethnographic work).

The Role of Language and Predicative Reasoning in the Umwelt

While Sebeok held that supra-linguistic phenomena were constitutive of a tertiary modeling system, my assertion is rather that the impact of language on the human Umwelt is "thrown back in" and saturates other aspects of Umwelt. This concerns language-derived practices and far more. In short: the practice of languaging changes the human Umwelt not by escaping or sidelining it, but by fundamentally transforming it. In this process of recalibration, the core Umwelt may become "background" or otherwise loose in meaning.

Taking one step back, I will now explain what I mean by predicative reasoning, or the criterion for being endowed with a conceptual Umwelt. By *predicative reasoning*, I mean the mental act of ascribing a specific feature to someone or something. Animals that ascribe specific features to other living beings or objects in this manner are arguably capable of carrying out a fundamental form of logical reasoning. They thereby exercise a capacity which is indicative of rational judgment, and thus proto-linguistic capacities. An animal's capacity for predicative reasoning can be more or less advanced and complex. And as we see, we can define the conceptual Umwelt as related to any kind of reasoning.

---

than fact. At any rate the disputed terms (*animal, language*) have to be precisely defined, and a definition agreed on by all discussants, before such discourses take on the character of being meaningful. This is no small task, since the "ayes" and the "nays" both tend to operate with tailormade definitions that make their stands highly meaningful.

The inclusion into the tripartite Umwelt model of this notion makes it clear that I too conceive of the ability to *language* as a modelling system, or as an important aspect of the modelling system that is the Umwelt. This holds true even for proto-language in form of predicative reasoning, which must be assumed to be quite wide-spread among animals. These animals, too, have cognitive modelling capabilities that go well beyond the work performed by the core Umwelt, which is based in automated perceptual acts.

But some animals participate in human language. Animals that recognize, understand and act on a number of human words arguably have conceptual Umwelten that envelop elements of language (this rests on the assumption that they actually understand words *as* words). For example, sheep herding dogs respond to verbal commands such as "Come by", "Lie down", "Stop", "Stand", "Walk up", "Steady", "Right there", "There, now", "Look back" and "That will do".[27]

As Stephen J. Cowley points out, "[l]anguage and perception use bidirectional coupling that links experience with wordings; you thus anticipate what is (un)likely to come next".[28] In the human context this implies, for one thing, that "our world is encultured".[29] "We see and hear and otherwise experience very largely as we do because the language habits of our community predispose certain choices of interpretation", as Edward Sapir writes.[30] As we become a part of a particular culture or language, notes Abram, "we implicitly begin to structure our sensory contact with the earth around us in a particular manner, paying attention to certain phenomena while ignoring others, differentiating textures, tastes, and tones in accordance with the verbal contrasts contained in the language".[31] Bert H. Hodges strikingly observes that languaging binds us together and empowers us: "Humans may find their identity, partly at least, within the interactions we call linguistic. Perhaps language is metaphorically a kind of weak force that binds humans in ways that make them effective causal agents in the physical world".[32]

In order to make a more convincing case for the phenomenon of predicative reasoning, I will now, as background for this notion, outline the workings of the tripartite Umwelt in more detail. Specifically, my claim is that we can generally conceive of six types, or categories, of acts, and that these can be located within the three different aspects of the Umwelt:

Core Umwelt

Automated acts of perception
Automated mental acts

[27] Westling 2014, pp. 49–50.

[28] Cowley 2013.

[29] Cowley 2006.

[30] Sapir 1949, p. 162.

[31] Abram 1997, p. 255.

[32] Hodges 2007, p. 601.

Mediated Umwelt

Wilful acts of perception
Wilful mental acts

Conceptual Umwelt

Habitual acts of perception
Habitual mental acts

The elements involved are quite few: perceptual acts and mental acts which are each either automated (by which I mean the exact and physiologically based matching of something with something else), wilful (by which I mean the agenda- and interest-driven matching of something with something else) or habitual (by which I mean the learned matching of something with something else). But the distinctions implied are crucial: whereas *conscious animals* (with a brain, mind, and mental activity) carry out all six types of acts, non-conscious creatures, in so far as they perceive (in a broad sense), only carry out two, namely automated acts of perception and wilful acts of perception. These do not have any conceptual Umwelt, their Umwelten consist only of two aspects, the core aspect and the mediated aspect. Habitual, i.e. conceptual acts are reserved for conscious creatures (but even bacteria can carry out wilful acts of perception, i.e. make choices based on interpretation).

Here, language is implicitly said to be habitual. This is not to be associated with behaviourist language acquisition theories based on the work of, for instance, Burrhus Frederic Skinner.[33] According to this approach language is learned by way of simple stimulus-response mechanisms, and habit formation occurs as imitations of correct associations are encouraged via a sort of positive response. Within an Uexküllian framework, it does of course make sense to say that associative learning occurs, but language acquisition is more meaningfully looked upon as happening in the context of the individual Umwelt, or more specifically by way of the contextualization in (or integration into) the *Umwelttunnel* (i.e., the personally experienced chain-of-events throughout someone's life) of the learner. Language acquisition, therefore, is extensively based on interpretation (as well as on social expectations). Moreover, the characterization of language as habitual is not only relevant for language acquisition, but just as much for adult, mature language practices at large.

Previously I defined predicative reasoning as the mental act of ascribing a specific feature to someone or something, and contrasted it with automated acts of perception. We now see, for one thing, that it must also be distinguished from wilful acts of perception. In general terms automated acts can be said to be code-based, whereas both wilful acts and habitual acts are interpretation-based.[34] Simple crea-

[33] Skinner 1953.

[34] An implication of this claim is that the core Umwelt is generally code-based, and that the mediated Umwelt and the conceptual Umwelt are interpretation-based. If this is correct, the interpretive threshold is not located where animals *with* a nervous system meet creatures *without* a nervous system, as Barbieri holds, nor where the biotic meets the abiotic, as Hoffmeyer holds. Instead, it is,

tures such as bacteria are capable of interpretation, and thus of making choices, but they are not capable of predicative reason, which is a capacity that is displayed only by conscious (brained, mindful) creatures.

Languaging as Perception, Action and Self-Deception

> *I began to wonder if my culture's assumptions regarding the lack of awareness in other animals […] was less a product of careful and judicious reasoning than of a strange inability to clearly perceive other animals – a real inability to clearly see, or focus upon, anything outside the realm of human technology, or to hear as meaningful anything other than human speech.*
>
> (Abram 1997, p. 27)

Languaging is More-than-Linguistic

In this third section I outline some core perspectives of distributed language (DL), before proceeding to present my notion of the anthropocentric mistake and discuss some implications. The distributed language perspective, which I consider to be largely aligned with my approach based on the Umwelt theory, is dealt with using five key terms, namely *languaging*, *first-order languaging*, *movement*, *interactivity* and *enkinaesthesia*.

Languaging is a term originally coined by Humberto Maturana[35] to refer to complex behaviors oriented to the creation and sustaining of "consensual domains".[36] He held that all living systems *language.*[37] By contemporary proponents of the DL view the term is rather used to emphasise that language is an activity rather than some set of formal abstracta. In Cowley's words, "[l]anguage is *activity in which wordings play a part*".[38] "Rather than view language as an *object*",[39] DL enthusiasts tend to say, we should focus on first-order activity or human languaging.

A crucial distinction is Nigel Love's *first-order languaging* and *second-order language* (said to have originated in Love's work in 2004,[40] where there is talk of first-order "activity" and second-order "cognition"[41]). As Martin Neumann and

at least in our context, located where core experience meets mediated experience (and since these aspects often intermingle, the dividing line is not in plain sight).

[35] Maturana 1970.

[36] Thibault 2011, p. 215.

[37] Cowley 2014.

[38] Cowley 2011a, p. 4.

[39] *Ibid.*, p. 2.

[40] Love 2004.

[41] According to Paul Thibault (personal correspondence), the origin is really Love 1990.

Stephen J. Cowley point out, "linguists typically confuse language with second-order constructs".[42] In Paul Thibault's words, first-order languaging refers to "the organization of process on different scales that takes place when persons engage in talk together".[43] "First-order languaging crucially involves synchronized interindividual bodily dynamics on very short, rapid timescales of the order of fractions of seconds to milliseconds. [...] Persons in talk enact, exploit, respond to, and attune to such events in order to engage with others and to coconstruct their worlds with them".[44] Thibault further explains that "[f]irst-order languaging is a whole-body sense-making activity that enables persons to engage with each other in forms of coaction".[45] As we see, this notion encompasses *movement*. "Since human movements both enact and elicit interpretations", writes Sune Vork Steffensen, "we orient to norms (and judge people by how they do so)".[46] "While language can be mapped onto grammatical, semantic, discursive functions, human activity *is* whole bodied movement. As we orient to circumstances, and each other, we give a particular sense to the vagueness of (verbal) language. [...] While rooted in bodily movement, language is symbiotic: at times, dynamics dominate, at times, the verbal aspect of language".[47]

As Cowley notes, "[v]erbal patterns constrain bodily movements and the feeling of thinking as people co-ordinate the flow of activity. [...] Co-ordination becomes a means of embodying thoughts".[48] The motive of interindividual bodily dynamics overlaps with that of *interactivity*. Thibault observes that research in infant semiosis "shows very clearly that from the very earliest stages of the child's meaning-making, that is, well before the onset of language, the processes involved are in fact fundamentally dialogic and intersubjective".[49] Steffensen defines interactivity as "sense-saturated coordination that contributes to human action".[50] If it wasn't for the qualifying term *human*, one would think that this definition should make the term applicable in animal studies as well, since many animals are no less coordinated than ourselves, and perform wonderful coaction.

Interactivity points us further, to the notion of enkinaesthesia, coined by philosopher Susan Stuart.

> "Enkinaesthesia" is a neologism I will use to refer to the reciprocally affective neuromuscular dynamical flows and muscle tensions that are felt and enfolded between co-participating agents in dialogical relation with one another. Enkinaesthesia, like intersubjectivity and intercorporeality relates to notions of affect, but in this case it is with

[42] Neumann and Cowley 2013.

[43] Thibault 2011, p. 214.

[44] *Ibid.*

[45] *Ibid.*, p. 215.

[46] Steffensen et al. 2010, p. 210.

[47] *Ibid.*

[48] Cowley 2011a, p. 2.

[49] Thibault 2000, p. 294.

[50] Steffensen 2013.

> the affect we have on the neuro-muscular dynamical flow and muscle tension of the other, including other animals, through our direct and our indirect touch.[51]

Enkinaesthesia, then, is our felt sensitivity to the sensitivity of others – and a crucial aspect of interactivity, coaction, and social life. Enkinaesthesia arguably makes us human – and, indeed, animal.[52] Lived experience, in Stuart's view, is, first and foremost, enkinaesthetic.

Language and Self-Deception: The Anthropocentric Mistake

Identity is an intriguing thing. It is so obvious to us, who we are – or so it appears. Human identity is largely a linguistic phenomenon.[53] But fundamental as language is in constituting human cognitive reality, we *are* not entirely linguistic creatures. Man is not a sign. Man is not language. Man is not simply what it thinks it is (Man *is* not identity). Rather, Man is a creature who organizes ecological reality in linguistic categories – both perceptually and behaviourally.

It is very commonplace, therefore, to commit *the anthropocentric mistake*, namely to reason (erroneously) that human reality is practically all there is. We tend to think in terms of language, and in terms of language, all is language. All is *human* language – therefore all is human. What we do not realize when committing this mistake is that it is not only Man who judges, who categorizes, who organizes, who is different, and so forth.

The anthropocentric – or indeed linguistic – mistake, then, consists in mistaking human reality for reality as such.[54] Misjudging the nature of reality, we misjudge *our* nature – *living* nature – *human* nature. To put it bluntly, current mainstream views on language which are aligned with the anthropocentric mistake result in a string of distorted realities. They distort our view on consciousness, on experience, on knowledge/knowing, on reality, and on value, by making us believe that these are human phenomena only (or predominantly). As a result, philosophy of consciousness, phenomenology, epistemology and philosophy of science, ontology, ethics and aesthetics all underachieve in comparison with their innate potential.

---

[51] Stuart 2010, pp. 308–309. Indirect touch, writes Stuart, "can be achieved [e.g.] through a look where one becomes the object of someone else's subjective attention and experience" (*ibid.*, p. 309).

[52] Given that enkinaesthesia is, in a way, *felt togetherness* and thus implicitly social and potentially emphatic, it can even be said to be part of the groundwork of morality. In this sense the phenomenon of enkinaesthesia does not lack a normative dimension.

[53] In Tønnessen 2010 language, which is claimed to have the appearance though not substance of a total system, is described as one of three grand systems – "Nature, Language, the Economy – all of which apparently in quest of hegemony over our lives, as natural beings – linguistic creatures – economic stakeholders" (p. 383).

[54] For similar presentations of the notion of the anthropocentric mistake, cf. *ibid.*, p. 377 and Tønnessen 2011, pp. 325–326.

The classical Thomas theorem in sociology can shed light on the psychology of the anthropocentric mistake: "If men define situations as real, they are real in their consequences".[55] If people intuitively define human language and everything that can be associated with it as real, and Abram is correct in stating that we have developed an inability to "hear as meaningful anything other than human speech",[56] as cited in the motto of this section, then from a psychological perspective it makes perfect sense to disregard non-linguistic reality almost completely.

The anthropocentric mistake can be further clarified with reference to Cowley's notion of taking a language stance: "[H]earing 'words' is like seeing 'things' in pictures. This is described as taking a language stance. To defend the position, it is argued that, first, we learn to hear wordings and, later, to use 'what we hear' as ways of constraining our actions".[57] As described by Cowley, this implies "that humans depend on taking 'a language stance' or hearing utterances as if they really were little units (a view further encouraged by literacy)".[58]

This latter sentence resonates well with Abram's observation: "Only when a more thoroughly *phonetic* system of writing spreads throughout a culture do its members come to doubt the expressive agency of other animals and of the animate earth. Only in the wake of the *alphabet* does language come to be experienced as an exclusively human power".[59]

In Search of the Dark Matter of Our Enlightened Worlds

According to J. von Uexküll[60] everything that falls under the spell of the Umwelt is retuned and transformed until it has become a useful carrier of meaning, or it is totally neglected. As we have seen, language is a powerful framer of behavior and of perception. In the context of human beings, the Umwelt is quite fluid (i.e. amenable to change) both individually and temporally for society as such. As languaging and human practices develop, so do our respective Umwelten. What is gained in this process, and what is lost? What is certain is that nowadays language, language-derived practices and various media playing into our mediated Umwelten are becoming ever more dominant. What then of our actual encounters with other living beings? If reality as we perceive it is consistently linguistic, then what role do we have to assign to non-human nature?

---

[55] Thomas and Thomas 1928, pp. 571–572.

[56] Abram 1997, p. 27.

[57] Cowley 2011b.

[58] Cowley 2012a.

[59] Abram 2010, p. 17 (this observation was further developed in Abram 1997, where the philosopher analyses the connection between the emergence of written languages and the emergence of philosophy).

[60] Uexküll 1934–1940 [1956, p. 109].

Despite these tendencies toward alienation from nature, it remains the case, as we have seen earlier, that all languaging is underpinned by interbodily dynamics and sensual, carnal experience. And of course, any human doing is furthermore underpinned by an array of intercellular and ecological activities. We are just not always aware that this is the case – it belongs to the untold, the unseen which nevertheless sustains our conversations and our thoughts, our doings and our deeds.

In *Language: The Cultural Tool*, Daniel L. Everett[61] introduces the notion of "'dark cognitive and cultural matter' that appears in what is *not* said in discourse".[62] Though critical of aspects of Everett's book, Cowley nevertheless concurs that this is an interesting concept. "In Everett's idiom", he writes, "dark cultural matter imbues language with values". Cowley adds that "language shapes lives as individuals sensitize to dark cultural matter".[63]

Everett's point, or claim, is that any culture envelops much that is simply taken for granted. Therefore a full transcription of an everyday conversation will not spell out all there is to say about what two or more people have just talked about. What two persons both take for granted may be treated as given, when they speak. And it does indeed appear to be the case that volatile conversations are often characterized by uncertainty about what the other person is taking for granted.

How can we escape having a *tunnel vision of language* (seeing only what is in plain sight)? How do we contribute to shaping our own Umwelten in a healthy, sustainable, ecologically grounded manner? How can we co-create Umwelten that we are not all too ashamed to pass on to our children? We may have to reeducate ourselves. Learn how to see again. How can we study the "dark matter" of our enlightened worlds? Given all the dark matter underpinning and surrounding verbal practices, a foray into the hinterland of language – the land which sustains us – is definitively called for. Best of luck on that journey!

The Genesis and Modalities of Language

> *We get into the habit of living before acquiring the habit of thinking.*
>
> (Camus 1942 [1983, p. 8])

Origin and Evolution of Language

An *Umwelt trajectory* can be characterized as the course through evolutionary (or cultural) time taken by the Umwelt of a creature, as defined by its changing relations with the Umwelten of other creatures.[64] One way to portray the Umwelt trajectory

61 Everett 2012.

62 Cowley 2012b, p. 285, with reference to Everett 2012, p. 198.

63 Cowley 2012b, p. 285.

64 Tønnessen 2014.

of humankind in the most general terms possible would be to depict the human Umwelt in its aspect of emerging layers or aspects (cf. the core Umwelt, the mediated Umwelt, and the conceptual Umwelt). In the history of life in general, the core Umwelt is without doubt the initial Umwelt. It is equally clear that the next layer to emerge must have been the mediated Umwelt, followed by the conceptual Umwelt as the latest and most advanced aspect of Umwelt. But humankind must have had all three aspects of Umwelt from the outset, and several animal species likewise. So if we were to portray the Umwelt trajectory of humankind in these terms, we would have to go very far back in our pre-human evolutionary history. A macro-evolutionary event that is more characteristic of human existence is the emergence of languaging practices (followed, later on, by literacy).

As Sverker Johansson remarks, however, "there is no consensus on when the transition from non-language to language took place, nor any consensus on the species of the first language users".[65] Our subspecies, *Homo sapiens sapiens*, might not have been the first one to *language*, since other human subspecies (now extinct) might perhaps have developed the practice of languaging before us. Johansson examines whether Neanderthals had language, and asserts that "the preponderance of the evidence supports the presence of at least a spoken proto-language with lexical semantics in Neanderthals".[66] This conclusion, he writes, would be strengthened if genetic data suggesting that interbreeding between Neanderthals and modern humans took place were confirmed.[67] Just as there is no consensus on when language emerged, neither is there any consensus on "the nature of this transition – was it a sharp single-step leap [...] or a gradual evolution in many small steps"[68]? Noam Chomsky is among those who argue that the transition must have been sharp.[69]

"If language is not a purely mental phenomenon", writes Abram,[70] "but a sensuous, bodily activity born of carnal reciprocity and participation, then our discourse has surely been influenced by many gestures, sounds, and rhythms besides those of our single species" – including birds.[71] What is remarkable with regard to the evolution of language is that of the genes that have been identified as relevant for language abilities, "virtually all [...] are present also in animals. All known genes of language, in other words, are genes of the primary modelling system that we have inherited from our animal ancestors".[72] This is consistent with the view, shared by Chomsky and Sebeok, that language evolved as an exaptation, i.e. that the function of language has changed from one (e.g., cognitive modelling) to another (e.g., communication).[73]

---

[65] Johansson 2013, p. 35.

[66] *Ibid.*

[67] *Ibid.*, p. 57.

[68] *Ibid.*, p. 39.

[69] Chomsky 2010.

[70] Abram 1997, p. 82.

[71] Abram 2010, pp. 197–198.

[72] Barbieri 2012b, p. 458.

[73] But Chomsky, of course, takes language to be a language *faculty*, and his view is therefore, in this respect, fundamentally different from that which follows from an Uexküllian Umwelt perspective,

Hoffmeyer too shares this view, and builds on both Thomas Sebeok and Gregory Bateson[74]: "Implicit [in G. Bateson's theory] is the idea that [the verbal aspect of] language has not – at least to begin with – served any communicational purpose (similar in style to that of body language) whatsoever, but that it has more likely been associated with the development of a quite new type of inner, mental concept – let us call it a cognitive model".[75] As we have seen, the tripartite Umwelt model implies that the conceptual Umwelt must have emerged long before language. The modelling capabilities involved in predicative reasoning were arguably there for hundreds of millions (but not billions) of years before language evolved. This suggests that human language is a later, more commanding derivative of such capabilities. Just like predicative reasoning does for any animal endowed with it, language affords the human organism with the capacity to organise its Umwelt objects and factors more meticiously. Despite all the matchless characteristics of language, this suggests that the difference between language and other forms of predicative reasoning is in the end a matter of degree, or perhaps more fittingly of magnitude.

Acquisition of Language in Childhood

Besides Umwelt trajectories, the evolution and development of language can also be depicted in terms of an *Umwelt transition*,[76] i.e. a lasting, systematic change within the life cycle of a being from one typical appearance of its Umwelt to another. A human child arguably goes through several Umwelt transitions, or a very multifaceted one, as it learns to *language*. As Albert Camus says in his *Myth of Sisyphus*, "[w]e get into the habit of living before acquiring the habit of thinking" – and similarly, we arguably get into the habit of languaging before acquiring language.

With reference to Adolf Portmann's work, Barbieri neatly describes how being born prematurely (due to our short gestation period relative to lifespan compared with other mammals) affects our brain development and implicitly our capacity for language learning: "In all other mammals, the wiring of the brain takes place almost completely in the dark and protected environment of the uterus, whereas in our species, it takes place predominantly outside the uterus, where the body is exposed to the lights, the sounds and the smells of a constantly changing environment".[77] In effect, he suggests, the constraint of the birth canal "has split the foetal development of our brain into two distinct processes, one within and one without the uterus".[78]

---

or from the DL perspective.

[74] Bateson 1972.

[75] Hoffmeyer 1993 [1996, p. 101]. Hoffmeyer further asserts that "[t]hrough speech, human beings broke out of their own subjectivity because it enabled them to share one large, common *umwelt*. While pre-lingual creatures had recourse only to their own finite *umwelts*, speech had the benefit that it could turn the world into a mystically produced common dwelling place" (*ibid.*, p. 112).

[76] Tønnessen 2009.

[77] Barbieri 2012b, p. 457.

[78] *Ibid.*, p. 460.

This is crucial biological background for our species' character of being a generalist species, and for our understanding of language learning.

The Various Linguistic Modalities of the Human Umwelt

In point 4 of the platform for a *semiotics of being*,[79] I refer to *speechless Umwelten, spoken Umwelten* and *alphabetic Umwelten* as distinct categories of human Umwelten. Practically every human being, we may assert, experiences within his/her lifetime a transition from a speechless Umwelt to a spoken one – most persons further to a more or less alphabetic or pictographic one. Additionally, there are situations – states of mind – where we so to speak loose (or deliberately pause) our ability to speak, or to perceive in terms of language. These are border cases of the speechless and the spoken, some of them bordering on insanity.

Beyond the Anthropocentric Mistake: Languaging as if Nature Mattered

> *Today's intrepid researchers have yet to notice that the human body, in itself, is no more autonomous – and no more conscious – than an isolated brain. Sentience is not an attribute of a body in isolation; it emerges from the ongoing encounter between our flesh and the forest of rhythms in which it finds itself, born of the interplay and tension between the world's wild hunger and our own.*
>
> (Abram 2010, p. 110)

The recently deceased Estonian geologist and palaeontologist Ivar Puura (1961–2012) coined the notion of semiocide, which he defined as "a situation in which signs and stories that are significant for someone are destroyed because of someone else's malevolence or carelessness, thereby stealing a part of the former's identity".[80] "By wholesale replacement of primeval nature with artificial environments", writes Puura, "[a]t the hands of humans, millions of stories with billions of relations and variations perish".[81] As Timo Maran notes,

> Puura most correctly stresses that nowadays the phenomenon of semiocide is very widespread both in human culture and society as well as in relations between culture and nature. Unfortunately, semiotics appears to have overlooked this dark side of semiotic relations, as is evident from the lack of a conceptual framework and studies dedicated to this topic. [...] This is a question of the ethical responsibility of semiotics.[82]

---

[79] Tønnessen 2010.

[80] Puura 2013, p. 152; cf. Puura 2002.

[81] Puura 2013, p. 152.

[82] Maran 2013, p. 148.

Language is relevant here for two reasons. First, because when languages are going extinct, semiocide occurs and, second, because language can make us blind to the ongoing non-linguistic semiocide. The way we *language* around for example animals is telling of our relationship towards them. As Arran Stibbe notes, "the discourses we use to construct our conceptions of animals and nature have important consequences for the well-being of the animals and the ecosystems that support life".[83] If cognition is situated, embodied, extended and distributed, then we can engage in "thinking with animals"[84] in a literal sense. This chapter ends with three theses on the ethos of human-animal relations, which have implications for ethics, ontology and epistemology:

1. Language and languaging largely originated in human-animal co-action. Language did not emerge in a merely human setting.
2. In the modern era many people are inexperienced with regard to traditional human-animal encounters (and thus alienated with regard to nature).
3. In the future, it would be beneficial for people and animals alike if languaging practices around animals would entail less *anticipated muteness* and rely more on enkinaesthesia, "the entwined, blended and situated co-affective feeling of the presence of the other".[85]

Acknowledgements My thanks go to Nelly Mäekivi for pointing me to the quotes reiterated in the mottos of this article (Chesterton) and of section "I *Language*, Therefore I Model" (Coetzee). Thanks also to Paul Thibault and Stephen Cowley, for very stimulating discussions on the nature of language, cognition, human interactivity, agency, modern pitfalls etc. This work has been carried out thanks to the support of the research project Animals in Changing Environments: Cultural Mediation and Semiotic Analysis (EEA Norway Grants/Norway Financial Mechanism).

References

Abram, D. (1997). *The spell of the sensuous. Perception and language in a more-than-human world*. New York: Vintage Books.

Abram, D. (2010). *Becoming animal: An earthly cosmology*. New York: Pantheon.

Augustyn, P. (2013). What connects biolinguistics and biosemiotics? *Biolinguistics, 7*, 96–111.

Augustyn, P. (2015). Biology, linguistics, and the semiotic perspective on language. In E. Velmezova, K. Kull, & S. J. Cowley (Eds.), *Biosemiotic perspectives on language and linguistics* (pp. 169–189). Dordrecht: Springer.

Barbieri, M. (2012a). Organic codes and the natural history of mind. In L. Swan (Ed.), *Origins of mind* (Biosemiotics, Vol. 8, pp. 21–52). Dordrecht: Springer.

Barbieri, M. (2012b). Organic codes and the origin of language. In L. Swan (Ed.), *Origin(s) of design in nature* (Cellular origin, life in extreme habitats and astrobiology, Vol. 23, pp. 445–473). Dordrecht: Springer.

Bateson, G. (1972). *Steps to an ecology of mind*. New York: Ballantine.

[83] Stibbe 2012, p. 16.

[84] Daston and Mitman (eds.), 2005.

[85] Stuart 2010.

Camus, A. (1942 [1983]). *The myth of Sisyphus and other essays.* New York: Vintage.
Chang, H.-L. (2009). Semioticians make strange bedfellows! or, once again: "Is language a primary modelling system?". *Biosemiotics, 2009*(2), 169–179.
Chesterton, G. K. (1904). *G.F. Watts*. London: Duckworth.
Chomsky, N. (2010). Some simple Evo Devo theses: How true might they be for language? In V. Déprez, H. Yamakido, & R. K. Larson (Eds.), *The evolution of human language* (pp. 45–62). Cambridge: Cambridge University Press.
Coetzee, J. M. (1999). The lives of animals. In J. M. Coetzee, M. Garber, P. Singer, W. Doniger, & B. Smuts (Eds.), *The lives of animals* (pp. 15–72). Princeton: Princeton University Press.
Cowley, S. J. (2006). Language and biosemiosis: Towards unity? *Semiotica, 162*, 417–443.
Cowley, S. J. (2011a). Distributed language. In S. J. Cowley (Ed.), *Distributed language* (pp. 1–14). Amsterdam: John Benjamins.
Cowley, S. J. (2011b). Taking a language stance. *Ecological Psychology, 23*, 1–25.
Cowley, S. J. (2012a). *Distributed language: Cognition beyond the brain*. Proceedings of the humanities international forum. Moscow: Russian State University for the Humanities.
Cowley, S. J. (2012b). Linguistic fire and human cognitive powers. *Pragmatics & Cognition, 20*(2), 275–294.
Cowley, S. J. (2013). *Interactivity: Implications for language and cognition*. Paper presented at the workshop agency in health care: Phenomenology and experience, University of Stavanger, 15th April 2013.
Cowley, S. J. (2014). Bio-ecology and language: A necessary unity. *Language Sciences, 41*(A), 60–70.
Daston, L. J., & Mitman, G. (Eds.). (2005). *Thinking with animals: New perspectives on anthropomorphism*. New York: Columbia University Press.
Everett, D. L. (2012). *Language: The cultural tool*. New York/London: Pantheon Books (Random House USA)/Profile (UK).
Heidegger, M. (1977). Letter on humanism. In M. Heidegger, *Basic writings: From being and time (1927) to the task of thinking (1964)* (pp. 213–266). New York: Harper Collins.
Hodges, B. H. (2007). Good prospects: Ecological and social perspectives on conforming, creating, and caring in conversation. *Language Sciences, 29*, 584–604.
Hoffmeyer, J. (1993 [1996]). *Signs of meaning in the universe*. Bloomington: Indiana University Press.
Husserl, E. (1936–1939 [1970]). The origin of geometry. In E. Husserl, *The crisis of European sciences and transcendental phenomenology: An introduction to phenomenological philosophy* (pp. 353–378). Evanston: Northwestern University Press.
Johansson, S. (2013). The talking Neanderthals: What do fossils, genetics, and archeology say? *Biolinguistics, 7*, 35–74.
Lotman, J. M. (1991). *Universe of the mind: A semiotic theory of culture*. Bloomington: Indiana University Press.
Love, N. (1990). The locus of languages in a redefined linguistics. In H. G. Davis & T. J. Taylor (Eds.), *Redefining linguistics* (pp. 53–117). London/New York: Routledge.
Love, N. (2004). Cognition and the language myth. *Language Sciences, 26*, 525–544.
Maran, T. (2013). Enchantment of the past and semiocide. Remembering Ivar Puura. *Sign Systems Studies, 41*(1), 146–149.
Marcel, G. (1962). *Homo viator: Introduction to a metaphysic of hope*. New York: Harper & Brothers.
Markoš, A. (2002). *Readers of the book of life: Contextualizing developmental evolutionary biology*. New York: Oxford University Press.
Maturana, H. (1970). *Biology of cognition* (Biological Computer Laboratory [BCL] research report 9.0). Urbana: University of Illinois.
Neumann, M., & Cowley, S. J. (2013). Human agency and the resources of reason. In S. J. Cowley & F. Vallée-Tourangeau (Eds.), *Cognition beyond the brain: Computation, interactivity and human artifice* (pp. 13–30). Dordrecht: Springer.
Puura, I. (2002). Loodus meie mälus. *Eesti Loodus, 11*, 24–25.

Puura, I. (2013). Nature in our memory. *Sign Systems Studies, 41*(1), 150–153 [translation of Puura 2002 by E. Sütiste and T. Maran].
Sapir, E. (1949). The status of linguistics as a science. In D. G. Mandelbaum (Ed.), *Selected writings of Edward Sapir* (pp. 160–166). Berkeley: University of California Press.
Sebeok, T. A. (1987). Toward a natural history of language. *Semiotica, 65*(3–4), 343–358.
Sebeok, T. A. (1991). In what sense is language a "primary modelling system"? In M. Anderson & F. Merrell (Eds.), *On semiotic modeling* (pp. 327–340). Berlin: Mouton de Gruyter.
Sebeok, T. A., & Danesi, M. (2000). *The forms of meaning: Modeling systems theory and semiotic analysis*. Berlin: Walter de Gruyter.
Skinner, B. F. (1953). Operant behavior. In B. F. Skinner (Ed.), *Science and human behavior* (pp. 59–90). New York: The Free Press.
Steffensen, S. V. (2013). Human interactivity: Problem-solving, solution-probing, and verbal patterns in the wild. In S. J. Cowley & F. Vallée-Tourangeau (Eds.), *Cognition beyond the brain: Interactivity and human thinking* (pp. 195–221). Dordrecht: Springer.
Steffensen, S. V., Thibault, P., & Cowley, S. J. (2010). Living in the social meshwork: The case of health interaction. In S. J. Cowley, J. C. Major, S. V. Steffensen, & A. Dinis (Eds.), *Signifying bodies: Biosemiosis, interaction and health* (pp. 201–237). Braga: Portuguese Catholic University Press.
Stibbe, A. (2012). *Animals erased: Discourse, ecology, and reconnection with the natural world*. Middletown: Wesleyan University Press.
Stuart, S. (2010). Enkinaesthesia, biosemiotics, and the ethiosphere. In S. J. Cowley, J. C. Major, S. V. Steffensen, & A. Dinis (Eds.), *Signifying bodies: Biosemiosis, interaction and health* (pp. 305–330). Braga: Portuguese Catholic University Press.
Thibault, P. (2000). The dialogical integration of the brain in social semiosis: Edelman and the case for downward causation. *Mind, Culture, and Activity, 7*(4), 291–311.
Thibault, P. (2011). First-order languaging dynamics and second-order language: The distributed language view. *Ecological Psychology, 23*(3), 210–245.
Thomas, W. I., & Thomas, D. S. (1928). *The child in America: Behavior problems and programs*. New York: Knopf.
Tønnessen, M. (2003). Umwelt ethics. *Sign Systems Studies, 31*(1), 281–299.
Tønnessen, M. (2009). Umwelt transitions: Uexküll and environmental change. *Biosemiotics, 2*(1), 47–64.
Tønnessen, M. (2010). Steps to a semiotics of being. *Biosemiotics, 3*(3), 375–392.
Tønnessen, M. (2011). I, Wolf: The ecology of existence. In J. Servan & A. F. Aarø (Eds.), *Environment, embodiment and gender* (pp. 315–333). Bergen: Hermes Text.
Tønnessen, M. (2014). Umwelt trajectories. *Semiotica, 198*, 159–180.
Uexküll, J. von (1917). Darwin und die Englische Moral. *Deutsche Rundschau, 173*, 215–242.
Uexküll, J. von (1928). *Theoretische Biologie* (2nd ed.). Berlin: Verlag von Julius Springer.
Uexküll, J. von (1934–1940 [1956]). *Streifzüge durch die Umwelten von Tieren und Menschen: Ein Bilderbuch unsichtbarer Welten. Bedeutungslehre.* Hamburg: Rowohlt.
Uexküll, J. von (2010). *A foray into the worlds of animals and humans with a theory of meaning*. Minneapolis: University of Minneapolis Press [translation of Uexküll 1934–1940 (1956) by J. D. O'Neil].
Uexküll, J. von (2013). Darwin and the English morality. *Biosemiotics, 6*(3), 449–471 [translation of Uexküll 1917 by M. Tønnessen, English language editing by J. Beever].
Uexküll, T. von (1981 [1987]). The sign theory of Jakob von Uexküll. In M. Krampen, K. Oehler, R. Posner, T. A. Sebeok & T. von Uexküll (Eds.), *Classics of semiotics* (pp. 147–179). New York: Plenum.
Westling, L. (2014). The zoosemiotics of sheep herding with dogs. In K. Tüür & M. Tønnessen (Eds.), *The semiotics of animal representations* (pp. 33–52). Amsterdam/New York: Rodopi.
Zaliznjak, A. A., Ivanov, V. V., & Toporov, V. N. (1977). Structural-typological study of semiotic modeling systems. In D. P. Lucid (Ed.), *Soviet semiotics: An anthology* (pp. 47–58). Baltimore: Johns Hopkins University Press.

Deep Congruence Between Linguistic and Biotic Growth: Evidence for Semiotic Foundations

Jamin Pelkey

Abstract Language varieties undergo constant evolution, as do varieties of life. Both language and life unfold by semiosis – pervasive processes of growth in which relationships shared between the inherited past, the unstable present and the virtual future are organically intertwined. Although many recent attempts have been made to reunite biotic and linguistic evolution, contemporary treatments are mired in unexamined presuppositions inherited from twentieth century biological theory. Chief among these is the denial of implicit end-directed processes, that which biosemiotics finds to be the necessary condition of living systems – thereby providing semiotic foundations for human inquiry. After reviewing the history and problems of dialogue between linguistics and biology, I make two primary arguments in this essay, one a critique using historical evidence, the other a suggestion using empirical evidence. My critical argument is that crucial features of semiosis are missing from contemporary linguistic-biotic proposals. Entangled with these missing accounts is an analogous form of neglect, or normative blindness, apparent in both disciplines: the role of ontogeny in biological evolution and the role of diagrammatization in linguistic evolution. This linguistic-biotic analogy points to a deeper congruence with the third (and most fundamental) mode of evolution in Peirce's scientific ontology: "habit taking" or "Agapasm". My positive argument builds on this linguistic-biotic analogy to diagram its corollary membership in light of Peirce's "three modes of evolution": Chance (Tychasm), Law (Anancasm) and Habit Taking (Agapasm). The paper ends with an application involving complex correspondence patterns in the Muji language varieties of China followed by an appeal for a radically evolutionary approach to the nature of language(s) in general, an approach that not only encompasses both linguistic and biotic growth but is also process-explicit.

I am grateful to the editors of this volume for their invaluable criticism and suggestions that led to a more accurate and intelligible presentation of the paper argument. Remaining infelicities or errors are due to my own limitations.

J. Pelkey (✉)
Ryerson University, Toronto, ON, Canada
e-mail: jpelkey@ryerson.ca

E. Velmezova et al. (eds.), *Biosemiotic Perspectives on Language and Linguistics*, Biosemiotics 13, DOI 10.1007/978-3-319-20663-9_6

Keywords Biosemiotics • Linguistics • Ch.S. Peirce • Evolution • Habit-taking • Diagrammatization • Ontogeny • Language variation • Biolinguistics • Language change • Analogy • Ecology

Introduction

Attempts to understand linguistic and biological change have proceeded for more than two millennia via an interchange of ideas.[1] Cross-fertilization between philology and biology intensified during the eighteenth and nineteenth centuries; but, for reasons discussed below (cf. "Biotic and Linguistic Growth"), twentieth century influences discouraged this age-old dialogue. Recent studies[2] demonstrate that this hiatus was temporary. Just how the re-instantiation of dialogue between biology and linguistics should be framed in the twenty-first century, however, poses an open question, one that biosemiotics can help answer. Although the discussion below is focused on language in its root sense,[3] the scientific study of speech behaviour can benefit from a careful review of presuppositions at this intersection as well as Stephen Cowley[4] and others show.

Widespread disagreement on the nature, scope and applicability of biological models to linguistic (and cultural) change mark the current state of the dialogue (cf. section "Biological Analogies Gone Wild"). Some theorists promote widely divergent biological analogies for linguistic phenomena. Others suppress biological analogies in the hope of establishing more systematic domain-general approaches; but, over and beyond their differences, neither finds logical/ontological grounding. As I have argued elsewhere,[5] an architectonic system is needed that is capable of making the nature of domain-general evolution explicit. In other words, a model that embraces semiosis is required. Biosemiotics' grounding "on a strongly Peircean framework"[6] fills this gap.

With these issues in mind, I make two primary arguments in this essay, one a critique using historical evidence, the other a suggestion using empirical evidence. My critical argument (cf. especially section "Evolutionary Theory, Semiosis and Peircean Thirdness") is that crucial features of semiosis are missing from contemporary linguistic-biotic proposals, including basic accounts of the nature of process, the necessary role of future-oriented (pattern-solving) causality, and attention to modes of continuity or mediation.[7] Entangled with these missing accounts is

[1]Atkinson and Gray 2005, p. 524.

[2]E.g., Croft 2000 and 2008; Mufwene 2001 and 2005; Richerson and Boyd 2001 and 2005; Driem 2001 and 2008; Sterelny 2006; Mesoudi et al. 2006; Fitch 2008; Pelkey 2013.

[3]Cf. Sebeok 1986.

[4]E.g., Cowley 2007.

[5]Pelkey 2013.

[6]Kull et al. 2009, p. 168.

[7]I.e., "self-organizing" modes of process that mediate between inherited copying (e.g., "genotypic" analogues in language and culture) and ecological coupling (e.g., "phenotypic" analogues

an analogous form of neglect, or normative blindness, apparent in both disciplines – corresponding with the role of ontogeny in biological evolution and the role of diagrammatization in linguistic evolution. This linguistic-biotic analogy points to a deeper congruence with the third (and most fundamental) mode of evolution in Charles Sanders Peirce's scientific ontology: "habit taking" or "Agapasm".[8] My positive argument (sections "Evolutionary Theory, Semiosis and Peircean Thirdness" and "Deep Congruence") builds on this linguistic-biotic analogy to diagram its corollary membership in light of Peirce's "three modes of evolution"[9]: Chance (Tychasm), Law (Anancasm) and Habit Taking (Agapasm). Section "Further Evidence for Semiotic Foundations" supports these claims drawing on first-hand field work data gathered from the Ngwi languages of China (Burmic < Tibeto-Burman) to reveal modes of evolution that drive both biotic and linguistic growth. In contrast to mainstream accounts of evolution, this account develops an emerging Biosemiotic mandate by insisting that any theory of evolution should be grounded in explicit discussion of the nature of process.

Biotic and Linguistic Growth

In spite of Charles Darwin's own insistence in the *Descent of Man*[10] that evolution must also apply to human behaviour, social sciences shifted away from process thinking in the twentieth century and were little changed by advances in evolutionary theory.[11] The reasons for this neglect are numerous and complex (cf. Section "Challenges Facing Evolutionary Linguistics" summary below). Firstly, it will be helpful to consider the interdependent development of the biological and linguistic sciences in the late classical and romantic periods of Anglo-European thought.

Historical Cross-Fertilization Between Language Sciences and Life Sciences

Although Quentin D. Atkinson and Russell D. Gray describe "two millennia of coevolution between research in biology and historical linguistics",[12] the richest period of "mutual fencundation"[13] between the two disciplines spanned from the

in language and culture) – in short, processes that mediate between analogy and automation in linguistics (resp. ecology and phylogeny in biology).

[8] Peirce 1890–1892 [2010, p. 194].

[9] *Ibid.*, pp. 110, 194.

[10] Darwin 1882.

[11] Richerson and Boyd 2001.

[12] Atkinson and Gray 2005, p. 524.

[13] To borrow a phrase from John Deely (2007).

late eighteenth century to the late nineteenth century. During this time, the interplay between botany and philology was especially rich, but when it came to rigorous inquiry into the nature of patterned growth through space and time, philology clearly led the way from the 1780s to the 1860s. As a result, evolutionary concepts were being developed in the so-called "social" sciences long before Darwin.[14] When Max Müller retorted "I was a Darwinian long before Darwin",[15] he was already aware that the biological eclipse of linguistics was beginning to obscure the linguistic eclipse of biology that stood uncontested only decades earlier. The ascendancy of the biological model over the linguistic one by the end of the nineteenth century is undeniable, but biology's little recognized historical precedent must be emphasized in order to "change the relevance of past to present".[16]

In the immediate wake of the Darwinian revolution, the philologist August Schleicher had already filed something of an intellectual property complaint, arguing that Indo-European philologists were the true discoverers of evolution.[17] Even a century later, informed linguists[18] attribute such foresight to the eighteenth century philologist William Jones. In the words of Müller, "[l]ong before Darwin made the theory of evolution so widely popular, that idea had completely dominated the Science of Language. To speak of Darwin as the discoverer of evolution, has always seemed to me an insult to every student of philosophy".[19] Naturally, then, "Darwinian ideas of descent with modification were less revolutionary in linguistics than they were in biology. Phylogenetic understanding and methodology in linguistics had already developed rapidly before Darwin, and this continued throughout the nineteenth century".[20]

History affirms that philology was "an important source of inspiration for Darwin",[21] and Darwin himself, among others, supports a linguistic-biotic homology in a famous passage from *The Descent of Man*: "The formation of different languages and of distinct species, and the proofs that both have been developed through a gradual process, are curiously parallel".[22] These parallels are now a mere curiosity for most linguists, however; and, even when applied to linguistic phylogeny, still tend to be treated as groundless speculation or questionable analogies on loan from biology.[23] Even some seasoned historical linguists now falsely assume the comparative method to have originated in Biology[24]; and although variationists such as Salikoko

[14] Cf. discussion in Greenberg 1957; McMahon 1994; Alter 1999; Wyhe 2005; Atkinson and Gray 2005.

[15] Müller 1887, p. xi.

[16] Deely 2009, p. 142.

[17] Schleicher 1869 [1983, pp. 32–35].

[18] Greenberg 1957, quoted in Brosnahan 1961, p. 227.

[19] Müller 1887, p. xi.

[20] Atkinson and Gray 2005, p. 517.

[21] Fitch 2008, p. 373.

[22] Darwin 1882, p. 90.

[23] Cf. the corresponding discussion in Wyhe 2005.

[24] Cf., e.g., Rauch 1999, pp. 36, 45.

S. Mufwene may grant that "genetic linguistics can contribute to theories of evolution", most assume that nineteenth century historical linguists such as Schleicher were actually taking a "biological approach".[25] Recovering an evolutionary type of linguistics in the twenty-first century remains an uphill challenge.

Challenges Facing Evolutionary Linguistics

Twentieth century approaches to language and linguistics enforced (and, indeed, *invented*) various ahistorical synchronic analyses of linguistic data.[26] A general climate of positivism and dualism, along with entrenched worldviews of essentialism, mechanism and nominalism all mixed with anti-teleological thought to determine that discussions of linguistic or cultural change as modes of evolution should be viewed with suspicion. In fact, social scientists themselves "have often been downright hostile toward even considering cultural evolution in Darwinian terms".[27] This is at least partly due to the influence of nineteenth century philologists.

Although phylogenetic methods and concepts of heredity originated in eighteenth and nineteenth century historical linguistics,[28] philologists appear to have been distracted by the progressive Hegelianism of the day into thinking of languages as existing along a continuum from decay to improvement, to perfection.[29] This problematic baggage (think: "social Darwinism") contributed to unsavory associations between evolutionary theory and linguistic/cultural change. Meanwhile, "Darwin was pigeonholed as a biologist, and sociology, economics, and history all eventually wrote biology out of their disciplines. Anthropology relegated his theory to a subdiscipline, biological anthropology, behind the superorganic firewall".[30] As a result, the concept *evolution* is to this day conceptually (or dogmatically) fixed with biotic development.

Since we now assume biology to be the proper arena for evolution and language to be the special charge of the human and social sciences, insofar as the two may intersect, language is generally treated as proceeding from neo-Darwinian modes of genetic evolution. In other words, the phrase "language evolution" now seems to be concerned not with the ontological status of language but with its origins.[31] As a result, instead of focusing on patterns and processes of language growth, those interested in language evolution tend to focus on various prehistorical conditions: e.g., the development and function of the vocal tract, neural architecture, upright posture, tool use and (for many Biolinguists at least) genetics. These are the focuses

[25] Mufwene 2005, pp. 30, 32.

[26] Cf. Croft 2008, p. 220.

[27] O'Brien 2006, p. 359.

[28] McMahon 1994, p. 318.

[29] *Ibid.*, pp. 319–320; though for Schleicher at least any peak of perfection is followed by another stage of decay.

[30] Richerson and Boyd 2005, p. 17.

[31] This can be noted throughout the numerous contributions found in Tallerman and Gibson 2012.

of the discipline known as "Biolinguistics". Although such assumptions may sometimes be useful for understanding the nature of language, they also tend to distract from inquiry into language as a process (i.e., actual language ontology). In the biolinguistic school, language evolution is generally understood to mean the evolution of a (presumed) language faculty, not language as a mode of evolution. Of those who pursue the quest to understand language as a process, most default to various presuppositional traps, as I demonstrate further below. In short, cultural assumptions still largely ensure that evolutionary analogies must be drawn from the established categories and methods of contemporary Biology.

Biological Analogies Gone Wild

Numerous parallels have been drawn between linguistic and biological phenomena. Schleicher is perhaps the first to propose a multi-level analogy between the two, claiming that "[t]he rules now, which Darwin lays down with regard to the species of animals and plants, are equally applicable to the organisms of languages".[32] In addition to remarking on shared principles of gradualness, genealogy, and selective adaptation, Schleicher claims that species correspond to languages, races to dialects, breeds to subdialects and individuals to idiolects.[33] Indeed, whatever they may mean, such similarities are striking and continue to be elaborated, as can be noted in Table 1.

Table 1 Conceptual parallels between biological and linguistic evolution (Atkinson and Gray 2005, p. 514)

Biological evolution	Linguistic evolution
Discrete characters	Lexicon, syntax, and phonology
Homologies	Cognates
Mutation	Innovation
Drift	Drift
Natural selection	Social selection
Cladogenesis	Lineage splits
Horizontal gene transfer	Borrowing
Vegetative hybrids	Language creoles
Correlated genotypes/phenotypes	Correlated cultural terms
Geographic clines	Dialects/dialect chains
Fossils	Ancient texts
Extinction	Language death

[32] Schleicher 1869 [1983, p. 30].

[33] *Ibid.*, p. 32. In his own words, "[t]he species of a genus are what we call the languages of a family, the races of a species are with us the dialects of a language; the subdialects or patois correspond with the varieties of the species, and that which is characteristic of a person's mode of speaking corresponds with the individual" (*ibid.*).

Theorists who explore such analogies usually feel obliged to pin language to some specific sub-domain of biotic growth. William James selects "zoölogical evolution",[34] William Croft prefers botanical evolution,[35] but most theorists zoom in to more microbiotic levels – some claiming that linguistic phenomena are analogous to the cellular level[36] and others arguing for the genetic level[37] of biotic growth. For others, language is a parasite[38]; for others still, language is a virus.[39]

The disagreements over grounding analogies do not stop here. If language is a parasite or a "mutualist symbiont",[40] it is also supposedly an organism, or so the inevitable logic progresses – and not merely an organism but (and here the logic suffers) an organism dwelling in the human brain (cf. "the beast in the brain" discussion[41]). Mufwene,[42] however, insists: languages are not parasites qua organisms but viruses qua species. From Mufwene's perspective, idiolects (individual speakers), not languages, are analogous to biological organisms. Though this particular disagreement might be partially resolved by appealing to the embattled distinction between "Language" and "languages",[43] Noam Chomsky, the twentieth century champion of the former has determined that language is not an organism but an organ.[44] In short, biological analogies in linguistics have run amok. What is more, some language theorists mix and match biological analogies at so many different levels that we are left with no clear theory of how these analogies function together.[45] In spite of his own preferred analogy between language change and botanical growth, Croft warns against taking biologically grounded analogies as the basis for language theories:

> Although analogies or metaphors between biological evolution and language change can be fruitful, one does not know which parallels between the two domains are legitimate to draw and which are not, or even more important, which parallel structures must be present for the analogy/metaphor to make sense. In particular, it is common to assume that the mechanisms that cause variation and selection in biological evolution must be the same in other domains such as language change, yet the mechanisms are domain specific. What is required is a generalized theory of evolutionary change that subsumes biological evolution, language change, and other phenomena of evolutionary change such as cultural evolution.[46]

[34] James 1880, p. 441.

[35] Croft 2000.

[36] Sereno 1991.

[37] Dawkins 1976.

[38] Kortlandt 2003.

[39] Mufwene 2008.

[40] According to Driem 2008.

[41] *Ibid.*, p. 105 *sq.*

[42] E.g., Mufwene 2008.

[43] Also known as "competence" vs "performance", respectively, or the presumed language faculty vs its manifestations in different societies or circumstances.

[44] Chomsky 1980, p. 185.

[45] Croft 2010, p. 307.

[46] Croft 2008, p. 220.

Other contemporary thinkers concur (from different perspectives) arguing that "deviations from the biological case [...] do not necessarily invalidate an evolutionary approach to culture; they merely require novel treatments of cultural phenomena within a general evolutionary framework".[47]

But what is this "general evolutionary framework"? or rather, *which* general framework? Croft appeals to David L. Hull[48] who takes his cues from Richard Dawkins'[49] memetic theory of language (and cultural) evolution, a theory of cultural replicators that served as a catalyst for reviving biological analogies for language change in the late twentieth century.[50] Not only are such approaches founded on implicit Biological analogies themselves, but (more importantly) both are also heir to a number of questionable presuppositions that continue to guide contemporary Neo-Darwinian thought. It is at this crux that biosemiotic perspectives may well be indispensable for progress.

The Biosemiotic Synthesis

One key aim of the emerging biosemiotic synthesis[51] is "to understand the dynamics of organic mechanisms for the emergence of semiotic functions, in a way that is compatible with the findings of contemporary biology and yet also reflects the developmental and evolutionary history of sign functions".[52] As a result, biosemioticians find contemporary Neo-Darwinian biology to be "dependent on unanalyzed semiotic assumptions".[53]

Prominent among these presuppositions is the pervasive presence of "function" or "self-maintenance conditions"[54] in biological descriptions of living phenomena. These conditions are non-trivial and require that organisms be substantially defined by their needs – i.e., relationships with that which is extrinsic to them or absent in them. In other words, "[e]volution presupposes function, and not vice versa".[55] If specialized functions are in some way intrinsic to the very nature of evolution, it would be invalid to assume that specialized functions are merely the products of evolution. And yet the latter position is the current mainstream consensus.

Not bound to prevailing dogma, biosemiotics seeks to "provide a theoretical grounding" for teleological (end-driven) concepts. Since the pervasive role of function requires a teleological level of causation to be re-admitted to the natural (and social)

[47] Mesoudi et al. 2006, p. 345.

[48] Hull 1988.

[49] Dawkins 1976.

[50] Cf. also Mesoudi et al. 2006.

[51] Cf. Sebeok 2001; Hoffmeyer 2008.

[52] Kull et al. 2009, p. 170.

[53] *Ibid.*, p. 169.

[54] *Ibid.*

[55] *Ibid.*, p. 170.

sciences, it may be found inconvenient, or even threatening, to mainstream biological and linguistic theorists. On the other hand, this contribution and its implications provide grounding for a domain-general theory of evolution in which end-driven causation is "a natural property of the world at large".[56]

Evolutionary Theory, Semiosis and Peircean Thirdness

Biosemiotic's own grounding "on a strongly Peircean framework"[57] enables accounts of future-oriented causation to be neither naïve nor reactionary but natural – and informed by millennia of careful thought.[58] Built on his discoveries of three ontological categories[59] in nature and experience, Peirce's evolutionary cosmology[60] forms the "central nervous system" of his system building philosophy, without which some argue that his semiotic logic cannot be properly understood.[61] Relationships between this framework and the problems at hand can now be brought into sharper focus.

Semiosis and Domain-General Evolution

Many nineteenth century thinkers, including Charles S. Peirce, Herbert Spencer and Charles Darwin himself, assume evolution to be "true not of one class of phenomena but of all classes of phenomena".[62] Even two decades after the Darwinian revolution, M. Müller retorts: "How a student of the Science of Language can be anything but an evolutionist, is to me utterly unintelligible".[63]

Multiple contemporary theories seek to establish a "unified science of cultural evolution".[64] Kim Sterelny[65] overviews meme-based models, dual inheritance models, Boyd-Richerson (population genetic) models and others, including her own proposal; but no criteria emerge to ensure that biological and linguistic/cultural

---

[56] Hoffmeyer 2008, p. 51.

[57] Kull et al. 2009, p. 168.

[58] As reconstructed in Deely 2001.

[59] Discovered in mathematics, logic, chemical valence, phenomenology, and demonstrated to be at work in numerous other domains, these categories he discusses as Firstness (quality), Secondness (reaction) and Thirdness (mediation).

[60] Peirce 1890–1892 [2010].

[61] Thellefsen 2001.

[62] Spencer 1862, p. v (cf. pp. 144, 490). As the remainder of Spencer's book makes clear, this quotation refers prominently (though not exclusively) to evolution.

[63] Müller 1887, p. xi.

[64] Mesoudi et al. 2006.

[65] Sterelny 2006.

processes are both covered without defaulting to the dogmas of either science when making "unified" claims. Three features of Darwinian evolution, for instance, are widely discussed as domain-general: "variation, selection and inheritance".[66] Can these at least be identified as unifying features? Unless these aspects of evolution are examined at a presuppositional level, they are unlikely to be freed from the Neo-Darwinian assumptions of those who apply them.

From its inception "[n]atural selection, as conceived by Darwin, has been a mode of evolution in which the only positive agent of change in the whole passage from moner to man is fortuitous variation".[67] Although contemporary theories of evolution incorporate a replicating component to account for the maintenance of inherited features (and generation of further variation), at least three problems remain for mainstream (asemiotic) theories of evolution at this general level, each of which will be examined in the remainder of this paper: (1) the actual nature of process is left unexamined; (2) no mode of process is proposed that provides continuity or mediation between the extremes of random variation and mechanical replication; (3) no proposals are offered to account for future-oriented (pattern-solving) modes of causality required by the functional realities of life (and language).

Semiosis and Process-Explicit Evolution

Processual phenomena may seem to be part-and-parcel with evolution; but, as Terrence Deacon[68] describes in detail, inadequate theories of process are the Achilles heel of contemporary evolutionary theory. This is true in linguistics as much as in biology. Consider Croft's attempts to situate language change within a domain-general theory of evolution.

Following Dawkins,[69] Hull[70] and others, Croft argues that linguistic structures evolve through language *use*, explaining that in his model "linguistic replicators are [...] tokens of linguistic structures in utterances".[71] When replicating these tokens, speakers generate "variation in the production and comprehension of utterances".[72] Croft, however, deliberately avoids specifying "the mechanisms by which variation is generated".[73] Rather, "like all evolutionary biologists and most historical linguists, [he] rejects teleological mechanisms"[74] – instead, he cites such phenomena as

[66] Cf. Wyhe 2005, p. 97.

[67] Peirce 1890–1892 [2010, p. 190]. *Moner* is an archaic term meaning 'single celled organism'.

[68] Deacon 2012.

[69] Dawkins 1976.

[70] Hull 1988.

[71] Croft 2008, p. 222.

[72] *Ibid.*

[73] *Ibid.*

[74] *Ibid.*

"expressiveness" and "avoidance of misunderstanding" as stand-ins.[75] But what is understanding? And how is one to avoid the missing of it? And what is the nature of "expressiveness"? This problem is not unique to Croft.

Deacon[76] aims critical floodlights on precisely this issue: how can a legitimate theory of process legitimately provide no account of process? How can evolutionary theories of replicators (those which get copied) avoid open, direct and thorough discussion of the copying processes themselves? He remarks that "there is a curious irony in treating the only two totally passive contributors to natural selection – the genome and the selection environment – as though they were active principles of change".[77] This leaves us with a "self-referential loop"[78] in which "inanimate artifacts"[79] are somehow accepted as "patterns that contribute to getting themselves copied".[80] Such "highly non-trivial kinds of processes" aren't to be ignored.[81] Instead, theories of evolution require "a 'positive' (order-inducing) factor and not merely a multiplicative factor".[82] For Deacon, and other biosemioticians, this order-inducing factor is teleological not in the caricatured sense of spooky intervention or reified purpose acting backward from some distant future; rather, the order-inducing factor involves the emergence of integrative sign relations through constraints on information, the search (however vague or unwitting) for something *missing*.

From the changing morphology of finch beaks in the Galápagos, to the changing morphology of Germanic word structure in English, evolution is driven by a needs-based pursuit of better fit between population and environment. In the case of Darwin's finches, the need is for new sources of food locked in untapped resources, such as cactus seeds. In the case of English typology, the need is for more predictable regularity in grammatical paradigms, due to intensified language contact with non-native speakers. In each case we find an end-oriented pattern-solving activity involving "the *virtual influence* of the future upon the present changing the relevance of the past".[83]

Semiosis and Thirdness in Linguistics and Biology

Peirce[84] identifies three modes of evolution: (1) *Tychasm* ("evolution by fortuitous variation"), (2) *Anancasm* ("evolution by mechanical necessity") and (3) *Agapasm* ("evolution by creative love"). Agapasm he identifies as conspicuously missing

[75] Viz., "homunculi" cf. Deacon 2012, pp. 46–79.

[76] Deacon 2012.

[77] *Ibid.*, p. 132.

[78] *Ibid.*, p. 131.

[79] *Ibid.*, p. 132.

[80] *Ibid.*, p. 131.

[81] *Ibid.*, p. 437.

[82] *Ibid.*, p. 422.

[83] Deely 2008, p. 481.

[84] Peirce 1890–1892 [2010, p. 194].

from mainstream theories of evolution (with the possible exception, in his day, of Lamarckian ideas). Little has changed in this regard in the intervening century. This necessary mode of evolution can be thought of as the action of habit-taking and habit-breaking toward some general idea or developing function – a working need (whether witting or unwitting) to solve some puzzle, pattern or problem not yet in equilibrium between (e.g.,) organism/population and environment. In all cases, as discussed above, this is motivated by something absent in the organism, the individual and/or the population.[85] This mode would both mediate and motivate – not only natural selection[86] but also mechanical replication. Agapasm closely corresponds to "Synechism", or processual continuity[87]; and since "[c]ontinuity represents Thirdness almost to perfection",[88] Agapasm would be Thirdness in-process or the semiosis of mediation, that process which both brings together and moves between the selection of fortuitous variation on one hand and the mechanical replication of inherited features on the other in the pursuit of something missing.

Just as mediatory process is neglected in evolutionary theory, those processes in which mediation is most prominently at work suffer neglect in biological and linguistic research. In biology this corresponds with ontogenetic/developmental processes, or "ontogeny" – the growth of an organism from seed to maturity. In linguistics this corresponds to "diagrammatization" – the growing systematization of linguistic patterns based on perceptions of resemblance (iconicity) and the potential for these perceptions to fit with, or reorganize previously recognized part-whole relationships[89] – or so I will argue below. First let us consider ontogeny.

The relative neglect of ontogeny in biological accounts is not a new problem,[90] but complaints continue to be registered.[91] Yet, in the words of zoologist Charles Otis Whitman, "[a]ll that we call phylogeny is today, and ever has been, ontogeny itself. Ontogeny is, then, the primary, the secondary, the universal fact. It is ontogeny from which we depart and ontogeny to which we return".[92] Stephen Jay Gould agrees: "Evolutionary changes must be expressed in ontogeny, and phyletic information must therefore reside in the development of individuals".[93]

Although ontogeny is now widely (if reluctantly) accepted as a key source of phylogeny (according to some the sole source),[94] and although evolutionary developmental (a.k.a. "evo-devo") biologists seek to devote attention to organism/population

---

[85] Cf. also Deacon 2012, pp. 1–17.

[86] Cf. also *ibid.*, p. 136: "Natural selection could not have produced the conditions that made natural selection possible".

[87] That which Peirce once claimed as his "one contribution of value" (Peirce 1866–1913 [1931–1958], CP [= *Collected Papers*] 5.415, 1905 [= a manuscript of 1905]).

[88] *Ibid.*, CP 1.337, 1886.

[89] Cf. Jakobson 1965 [1987]; Shapiro 2002; Nöth 2008.

[90] Cf. complaints in Whitman 1910 [1919]; Gould 1977.

[91] Cf., e.g., Adams and Pedersen (eds.), 2000; Wimsatt 2006, p. 364.

[92] Whitman 1910 [1919, p. 178].

[93] Gould 1977, p. 2.

[94] Hall 1999, p. 13.

growth, widespread acknowledgement that ontogeny involves end-directed processes is still forthcoming. And yet, as I will outline further below, ontogeny closely corresponds with Peircean evolution (i.e., Thirdness in-process or Agapastic Synechism: cf. above), which recognizes from the outset continuity between all things. As Whitman notes: "Ontogeny teaches us, then, that there are no disconnected jumps in its processes [...] subtle internal processes that bind all the external form-changes into one unbroken sequence. The invisible work going on beneath the surface follows steadily in a definite direction, culminating at the appropriate times and places in all of the outer and inner form and structure characters peculiar to the species".[95] Contrary to Whitman's early claim,[96] phylogeny cannot be reduced to ontogeny; rather, the two are interdependent.[97] The same can be said of the relationship between ontogeny and environmental-coupling factors.[98] In short, "[a]ll three modes of evolution are composed of the same general elements".[99] But, ontogeny/agapasm "exhibits them the most clearly".[100]

Just as ontogeny has tended toward neglect in biology, diagrammatization has been neglected in linguistics.[101] At the most general level, the two are not distinct. Just as Agapasm is manifest in the "disposition [...] to catch the general idea [...] and thus to subserve the general purpose",[102] diagrammatization is manifest in processes that lead to discovering a "fuller realization of the values specific to one's type",[103] – either of which might be a way of discussing the general nature of ontogeny in biology.

While Peirce's type-token distinction has been absorbed into contemporary linguistic theories,[104] the place and purpose of the distinction within the broader doctrine of signs has been all but ignored. Diagram tokens (whether linguistic or non-linguistic) are "iconic sinsigns"; diagram types are "iconic legisigns". A diagram is an "icon of intelligible relations"[105] that "facilitates reasoning possibilities".[106] Diagrams are by no means restricted to visual signs and are manifest at every level of speech activity and language organization. Diagram tokens are always variable

---

[95] Whitman 1910 [1919, p. 176].

[96] *Ibid.*, p. 178.

[97] Cf. Rieppel 1990.

[98] Hoffmeyer 2008, pp. 102–108.

[99] Peirce 1890–1892 [2010, p. 194].

[100] *Ibid.*

[101] Cf. the discussion in Jakobson 1965 [1987]; Shapiro 2002; Nöth 2008; Pelkey 2013. As Frederik Stjernfelt (2007) and Winfried Nöth (2008) note, the term *diagram* in this sense encompasses relations within and between embodied cognitive types at numerous levels, including schemas, prototypes, constructions, blends, gestalts, concepts and general cognitive models.

[102] Peirce 1890–1892 [2010, p. 194].

[103] Shapiro 2002, p. 118.

[104] Nöth 2002, p. 5.

[105] Peirce 1866–1913 [1931–1958], CP 4.531, 1903.

[106] Stjernfelt 2007, p. ix.

and available for selection, whether in comprehension or production, to serve some general end. These include idiolectal phones, novel utterances, nonce formations, live speech events and the like.

Diagram types involve real relations – perceived and remembered resemblances that are cognitively organized into integrated part-whole schemas – but are not themselves existing things.[107] Iconic legisigns grow out of iconic sinsigns, following the end-directed repetition of sufficiently similar tokens, resulting in taking up a new habit or "self-organization".[108] Thus, in linguistics, diagram types should be understood to encompass distinctions at all levels, including phonemes, lexemes, concepts, gestalts, schemas, constructions and the like.[109] As with ontogeny, such processes may be largely subliminal but are governed by vague (i.e., "rhematic") alertness to the potential for enhanced equilibrium – the ongoing search for a better fit between perceived resemblances (in speech or memory) and the efficient relations that hold between part-whole schemas, in order to serve some end related to enhanced communication. As we find illustrated in the growth of English morphology mentioned above, the perception of regular morphemes such as /-s/ to mark plurality continues to lead to the extension of this resemblance so that it comes to include more and more previously irregular plurals, such as *oxen* > *oxes*, thus enhancing the equilibrium of the overall system.

Deep Congruence

The prospect of a deep congruence between biotic and linguistic growth is now primed for exploration: (1) environmental-coupling processes introduce chance variation, revealing surprising gaps and suggesting new habits; (2) iconic replication processes entrench inherited and acquired habits into law like patterns; (3) these processes are mediated through space and time as new habits are taken up and old ones dispensed with for the sake of realizing a better fit according to some general cognitive type. In this section I consider the relevance these rough descriptions hold for biology and linguistics, arguing that, in biology, they apply to Ecology, Phylogeny and Ontogeny, respectively; while in linguistics, they apply to Analogy, Automation and Diagrammatization, correspondingly. Peirce summarizes them as "Chance", "Law", and "the tendency to take habits"[110]: modes of evolution that are actually domain-general. With these distinctions and relations in mind (summarized in Table 2), an evolutionary account of language and linguistics is possible – one that avoids the worst unanalyzed presuppositions (i.e., asemiotic pitfalls) of mainstream biology.

A growing number of contemporary studies in biology implicitly illustrate the intertwining relationships between these three modes of evolution. This can be

[107] Peirce 1866–1913 [1931–1958], CP 4.447, 1903.

[108] Cf. Hoffmeyer 2008, p. 62.

[109] Nöth 2008; Stjernfelt 2007.

[110] Peirce 1890–1892 [2010, p. 110].

Table 2 Deep congruence between linguistic and biotic growth

Peircean...	Mode of evolution	Evolution by...	Biotic...	Linguistic...
Tychasm	Chance	"Fortuitous variation"	Ecology	Analogy
Anancasm	Law	"Mechanical necessity"	Phylogeny	Automation
Agapasm	Habit taking	"Creative love"	Ontogeny	Diagrammatization

noted, to cite a handful of specific cases, in research ranging from the evolution of bats,[111] to primate lactation,[112] to botanical fructification,[113] to vertebrate neurology,[114] to mongoose middle ear development.[115] In their collection of perspectives focused on the re-assessment of bat evolution, for instance, Rich Adams and Scott Pedersen[116] draw attention to overlooked niche habitats (ecology) that can be described at various early stages of bat development (ontogeny). These result in remarkably distinctive juvenile characteristics that often differ as dramatically as those commonly held to distinguish various bat species (phylogeny). Thus, the exclusive comparison of adult members of various bat species is shown to be short-sighted for purposes of understanding the actual nature of bat evolution. They insist, rather, that such complex patterns of growth require an "integrative biology", one which "utilizes multi-disciplinary approaches to establish a more complete and, therefore, insightful interpretation of an organism's biology".[117] Such complex integration is rare in linguistic treatments; but we have much evidence that the same three modes of evolution are working together in linguistic processes.[118] For the sake of focus and further validation, it will be helpful to consider each mode of linguistic evolution in relative isolation before turning to an integrated illustration.

Evolutionary Chance: Linguistic Analogy

Linguistic chance involves a vast array of contextual factors, frequently contact-based (external) and/or gestural/articulatory (internal), in driving language variation through time. Mufwene, who champions an "ecological" account of language evolution, states that "the communicative activities that produce language evolution are largely determined by the socio-economic ecologies in which speakers evolve,

[111] Adams and Pedersen 2000.

[112] Milligan 2007.

[113] Leins and Erbar 2010.

[114] Creutzfeldt 1995.

[115] Gishlick 2008.

[116] Adams and Pedersen 2000.

[117] *Ibid.*, p. 1.

[118] For explicit treatments cf. Pelkey 2011 and 2013; for implicit treatments cf., e.g., Shapiro 1991, and 2002; Bybee 2010.

which is similar to saying that *the ecology rolls the dice in evolution*".[119] Indeed, as in biology, so in linguistics: with ecological-coupling comes chance variation. Language variation is pervasive at every level of linguistic structure, within the production and comprehension of every speaker, and between all groups of speakers, in ways that are fractal scalable, potentially approaching infinity.[120] Darwin, concurs: "We see variability in every tongue, and new words are continually cropping up".[121] Variation alone is not evolution, but the analogical selection of chance variation is.

Analogy pervades linguistic communication.[122] "Speakers/signers understand each other not because they use identical systems", notes Mufwene, "but because similar minds deriving similar patterns from similar data can 'read' each other".[123] Similarity, however, is semiotic: a semiotic relation of resemblance between an object and its sign vehicle according to the experience, expectations and needs of an interpretant. Language is pervasively iconic,[124] and analogy is iconicity in process. Most variation goes unnoticed and may seem inconsequential, but more remarkable variation (experienced first as "rhematic indexical sinsigns") evokes surprise. If selected and replicated, a token variation comes to be a shared symbolic type useful for communication between lects of a speech variety.

Evolutionary Law: Linguistic Automation

Linguistic law involves the automation of a given habit or diagram type through mimetic replication[125] to the degree that the original innovation becomes a linguistic fact, an identifiable part of the whole at the level of symbolic pattern, in relation to and in contrast with other linguistic facts. These facts are mimetically replicated and inherited, and often seem to lose association with their original motivation (as in the lexicalization of *goodbye* from the utterance *God be with you*).

Paradoxically, such replication also involves variation, both in production and in comprehension[126] that may lead to further habituation[127] and/or provide grounds for new analogies. Naturally, "with every adaptation, there are innumerable other arbitrary properties potentially brought into play".[128] Nevertheless, the regularity of sound change and the clear presence of historical splits persist in spite of this

---

[119] Mufwene 2005, p. 30; italics mine. – *J.P.*

[120] Kretzschmar 2010.

[121] Darwin 1882, pp. 90–91.

[122] Cf. Anttila 2003.

[123] Mufwene 2014, p. 15.

[124] Not only in terms of its production and comprehension, but also (and especially) in terms of its organization (cf. Nöth 1999 and 2008 for further verification and clarification of this claim).

[125] I.e., the action of dicent indexical sinsigns.

[126] Croft 2008.

[127] Cf. the discussion in Bybee 2010, pp. 50–53, 75.

[128] Deacon 2012, p. 424.

variation. Just as most language varieties are maintained until ecological pressures become unbalanced,[129] most automated forms persist until they no longer fit the patterned needs of the individual and/or community.

Darwin makes the argument that "The survival or preservation of certain favoured words in the struggle for existence is natural selection".[130] Here the tell-tale metaphor of competition and struggle, may easily distract from our need to account for implied modes of process. The natural selection of new words begins with analogy, according to some general pattern (of diagrammatization). The preservation of such words, within the macro-diagram depends both on the ongoing goodness of fit and various frequency-induced fossilization effects as "the human brain adjusts to repeated access by creating shortcuts".[131]

Evolutionary Habit Taking: Linguistic Diagrammatization

As discussed above, linguistic habit-taking (necessarily including habit breaking[132]) involves the "self-organization" or future-oriented pattern solving of language relationships, working "to render inefficient relations efficient [...] to establish a habit or general rule".[133] This mode is typified in the lifetime growth of polylectal individuals and populations through space and time. As Deacon reminds us, "[v]ariations do not exist in the abstract; they are always variations of some organism structure or process or their outcome".[134] As variations are selected, for purposes of implicit pattern solving (or diagrammatization) they enter into phases of habituation, gradually becoming automated, institutionalized and fossilized.

Ordinary examples of such fundamental process include negotiation of meaning in live conversation, evolution of speech pragmatics, development of new vocabulary, linguistic uses of communication media, empathic speech comprehension, the slow development of grammatical paradigms (e.g., pronoun systems), and embodied modeling of spatial relations typified in grammar. Consider the development of new vocabulary as a succinct example of diagrammatization. When a new student of linguistics begins to learn the technical jargon of articulatory phonetics, he simultaneously takes on new habits and breaks old ones. His new habits of embodied perception correspond to new lexical titles, such as "interdental" and "palatal". His old habits of less-focussed attention are also marked by phrase-level circumlocutions such as "with my tongue in-between my front teeth" and "on the roof of my mouth", respectively. The embodied semantics of such new concepts (or prototypes) are

---

[129] Bailey 1982.

[130] Darwin 1882, p. 91.

[131] Bybee 2010, p. 50.

[132] To take up a new habit is in some sense to break an old habit. Thirdness involves "a habit of taking and laying aside habits" (Peirce 1866–1913 [1931–1958] CP 6.101, 1902).

[133] Peirce 1866–1913 [1931–1958], CP 8.332, 1904.

[134] Deacon 2012, p. 422.

slowly organized in relation to each other (e.g., further forward vs further back, or part vs whole), as an integrated paradigm set. This is accomplished both through trial and error processes of analogy and through mimetic repetition of relationships that slowly come to be more and more efficient and automated. Nevertheless, the intervening, or mediating, stages of learning, and any future modifications of the lexical paradigm that result (however slight) require the presence of a modelling activity that can be reduced to neither the guesswork of analogy nor the automation that proceeds from mimetic repetition. This modelling activity is diagrammatization.

Many other linguistic phenomena such as recursion, borrowing, semantic shift and language acquisition also mediate between fluctuation and regularization. Advances in grammaticalization and lexicalization theory,[135] show that processes of semantic bleaching that enable greater grammatical functionality and processes of semantic enrichment that allow for fresh lexical versatility are operative between analogy and automation.[136] Without a third element in language theory, we can neither understand linguistic systems nor explain them. Exploration of mediatory process has become a priority and a necessity.

Further Evidence for Semiotic Foundations

The deep congruence mapped out above may indeed point to semiosis as the common ground of biotic and linguistic growth. To better support these distinctions and their interrelationships, and in order to clarify them through a condensed illustration, I offer evidence drawn from my research on the Muji languages[137] of Southwest China.[138]

The Muji languages of southeastern Yunnan Province are marked by a redistribution of lexical tones in syllable classes that were historically coda-final,[139] according to a mirror-image pattern in which High>Low and Low>High, conditioned by inverse manner of articulation in the proto-syllable onset. In the Phuma language, 'sweep', for instance, is pronounced /ɕi^{33}/, and 'shake' is pronounced /ɬɯ^{33}/. Since both descend from the high-checked tone class,[140] this shared pitch is the expected reflex of each. Low-falling tones featured on words like /ɕi^{21}/ 'kill' and /vjɛ^{21}/ 'pig' are also expected since these forms descend from the low-checked tone class. In fact, this particular binary distinction is inherited from the proto language stage – preserved via replication through space and time for more than 1,000 years. Long-term preservation of linguistic features is one aspect of evolutionary Law: the process of linguistic automation at work in ways that go beyond conscious awareness or deliberate control.

135 E.g., Brinton and Traugott 2005.

136 Pelkey 2013.

137 Tibeto-Burman > Burmic > Ngwi > Southeastern.

138 For a further, more detailed empirical study in which these relations are made explicit, cf. Pelkey 2013.

139 Pelkey 2007 and 2011, pp. 293–300.

140 Cf. Bradley 1979.

Other lexical reflexes break this pattern, however: /tɕ^{h}i^{21}/ 'pinch' and /kɑ^{21}/ 'stir' feature low tones even though they descend from the high-checked tone class; while /ʔv̩^{33}/ 'hang' and /na^{33}/ 'demon' feature high tones even though they descend from the low-checked tone class. This distinctive mirror-image reversal cries out for explanation.[141] Closer inspection reveals that the inverse redistribution is conditioned by [+obstruent] (i.e., non-continuant) syllable onsets. In other words, syllables that began with stops exchanged pitch value, and syllables involving fricatives or glides in the onset preserved the original tone. Strong evidence from the Laghuu language (a peripheral Muji variety currently only spoken in Vietnam) indicates that the low-checked tone values shifted first in this environment, followed in other Muji languages by high-checked tone values.

In order for this to occur, Chance variation of L>H on token utterances must have been provisionally selected, only to spread by analogy to other lexical utterances in this class, whereupon the analogy was taken up (selected) as a newly developing Habit for some implicit end such as an identity-based dialect distinction. This, naturally, would have had unintended consequences relative to the overall system. Underdifferentiation in the checked-tone classes, for instance, would have prompted a correlative shift or "flip-flop" of H>L to maintain equilibrium (e.g., homophony avoidance), thus illustrating the close cooperation between the work of linguistic Analogy and linguistic Diagrammatization.

This pattern is unattested elsewhere in Ngwi (or in the rest of Tibeto-Burman for that matter), a fact that points to evolutionary Chance, the selection of random variation according to some implicit absence – in this case, a likely blend of diverging ethnic identity reflected in emerging dialect differentiation, later followed by the need for pattern equilibrium as pitch values began to merge ambiguously. The correlative pattern is now highly regular among the Muji languages, suggesting a reinitiation of evolutionary Law in the wake of the innovation – the subsequent automation or fixing of efficient relations.

The pattern also features a significant exception, not discussed here,[142] which has its own internal regularity (Chance within Law). That there is a pattern at all (much less patterns within patterns) is the result of evolutionary Habit Taking or "self-organization" relative to the overall system. This aspect is irreducible to mechanical necessity (i.e., Law: the blind copying of replicators) since one outlying Muji language does not share the innovation, and Laghuu speakers, who emigrated early to Vietnam, show evidence of only the first half of the split (i.e., L>H). Thus, three distinct but intertwining modes of evolution are necessary not only to account for the full complexity of the innovation but also to make progress in understanding it.

[141] A dicent indexical legisign.

[142] Cf. Pelkey 2007 and 2011, pp. 293–300.

Coda: On the Need for a Semiotic Theory of Evolution for Linguistics and Biology

A radically evolutionary linguistics is overdue[143] – one consonant with biosemiotic developments but not tied to mainstream biological ideology. In other words, it is time to ground the study of linguistic (and biotic) growth in a domain-general theory of evolution that is process-explicit. Whether we wish to work technically or wish to understand the nature of language in general, it will be necessary to move beyond the endless generation of factual trivia multiplied by bald descriptions and statistical manipulations of linguistic data as ends in themselves. The interrelationships between Chance (Analogy), Law (Automation) and Habit Taking (diagrammatization) in language change – and the congruence these share with biological ecology, phylogeny and ontogeny – should open a way. Jesper Hoffmeyer sees in the biosemiotic approach to the life sciences the potential for a revitalization of human inquiry in general.[144] Kalevi Kull agrees, noting that biosemiotics should be understood as an enhancement of biology rather than as a mere commentary upon it.[145] Applied to linguistics, this, more than anything else, should motivate further development and application of the relationships mapped out above: an open opportunity to enhance our understanding of language, a possibility for revitalizing the science of language. Linguistics would no longer be bound to isolated descriptions of speech that oscillate between minimalist regularity and bewildering variation; instead a growing understanding of language processes and linguistic relations may well emerge.

This incomplete contribution is offered to that end.

References

Adams, R. A., & Pedersen, S. C. (Eds.). (2000). *Ontogeny, functional ecology, and evolution of bats*. London: Cambridge University Press.

Alter, S. G. (1999). *Darwinism and the linguistic image*. Baltimore: Johns Hopkins University Press.

Anttila, R. (2003). Analogy: The warp and woof of cognition. In D. J. Brian & R. D. Janda (Eds.), *The handbook of historical linguistics* (pp. 425–440). Oxford: Blackwell.

Atkinson, Q. D., & Gray, R. D. (2005). Curious parallels and curious connections: Phylogenetic thinking in biology and historical linguistics. *Systematic Biology, 54*(4), 513–526.

Bailey, C.-J. N. (1982). *On the Yin and Yang nature of language*. Ann Arbor: Karoma.

Bradley, D. (1979). *Proto-Loloish* (Scandinavian Institute of Asian studies monograph, Vol. 39). London: Curzon Press.

Brinton, L. J., & Traugott, E. C. (2005). *Lexicalization and language change*. Cambridge: Cambridge University Press.

[143] Cf. also Rauch 1999, p. 48.

[144] Hoffmeyer 2011, p. 203.

[145] Kull 2011, p. 226.

Brosnahan, L. F. (1961). Language and evolution. *Lingua, 9*, 225–236.

Bybee, J. (2010). *Language, usage and cognition*. Cambridge: Cambridge University Press.

Chomsky, N. (1980). *Rules and representations*. New York: Columbia University Press.

Cowley, S. J. (2007). Distributed language: Biomechanics, functions, and the origins of talk. In C. Lyon, C. L. Nehaniv, & A. Cangelosi (Eds.), *Emergence of communication and language* (pp. 105–127). London: Springer.

Creutzfeldt, O. D. (1995). *Phylogenetic, ontogenetic, and functional development of the cerebral cortex*. Oxford: Oxford University Press.

Croft, W. (2000). *Explaining language change: An evolutionary approach*. Essex: Longman.

Croft, W. (2008). Evolutionary linguistics. *Annual Review of Anthropology, 37*, 219–234.

Croft, W. (2010). Review of Mufwene, Salikoko S. (2008), Language evolution: Contact, competition and change. *World Englishes, 29*(2), 306–311.

Darwin, C. R. (1882). *The descent of man and selection in relation to sex* (2nd ed.). London: John Murray.

Dawkins, R. (1976). *The selfish gene*. Oxford: Oxford University Press.

Deacon, T. W. (2012). *Incomplete nature: How mind emerged from matter*. New York: W.W. Norton & Company.

Deely, J. (2001). *Four ages of understanding: The first postmodern survey of philosophy from ancient times to the turn of the twenty-first century*. Toronto: University of Toronto Press.

Deely, J. (2007). *Intentionality and semiotics: A story of mutual fecundation*. Scranton: University of Scranton Press.

Deely, J. (2008). From semiosis to semioethics: The full vista of the action of signs. *Sign Systems Studies, 38*(2), 437–491.

Deely, J. (2009). *Purely objective reality* (Semiotics, communication and cognition, Vol. 4). Berlin: Mouton de Gruyter.

Driem, G. van (2001). *Languages of the Himalayas: An ethnolinguistic handbook of the greater Himalayan region, containing an introduction to the symbiotic theory of language* (Vol. 2). Leiden: Brill.

Driem, G. van (2008). The language organism: Parasite or mutualist? In R. Derksen, J. Schaeken, A. Lubotsky, J. Wiedenhof, & S. Siebinga (Eds.), *Evidence and counter-evidence: Essays in honour of Frederik Kortlandt* (Studies in Slavic and general linguistics, Vol. 33, pp. 101–112). Amsterdam: Rodopi.

Fitch, W. T. (2008). Glossogeny and phylogeny: Cultural evolution meets genetic evolution. *Trends in Genetics, 24*(8), 373–374.

Gishlick, A. K. (2008). *The ontogeny, phylogeny, and ecology of the herpestid auditory bulla*. (PhD dissertation). Yale University, New Haven.

Gould, S. J. (1977). *Ontogeny and phylogeny*. Cambridge, MA: Belknap.

Greenberg, J. H. (1957). Language and evolutionary theory. In J. H. Greenberg, *Essays in linguistics* (pp. 56–65). Chicago: University of Chicago Press.

Hall, B. K. (1999). *Evolutionary developmental biology* (2nd ed.). Berlin: Kluwer.

Hoffmeyer, J. (2008). *Biosemiotics: An examination into the signs of life and the life of signs*. Scranton: University of Scranton Press.

Hoffmeyer, J. (2011). Astonishing life. In P. Cobley, J. Deely, K. Kull, & S. Petrilli (Eds.), *Semiotics continues to astonish: Thomas A. Sebeok and the doctrine of signs* (pp. 191–206). Berlin: Mouton de Gruyter.

Hull, D. L. (1988). *Science as a process: An evolutionary account of the social and conceptual development of science*. Chicago: University of Chicago Press.

Jakobson, R. (1965 [1987]). Quest for the essence of language. In K. Pomorska & S. Rudy (Eds.), *Language in literature* (pp. 413–427). Cambridge, MA: Belknap Harvard.

James, W. (1880). Great men, great thoughts, and the environment. *Atlantic Monthly, 46*, 441–459.

Kortlandt, F. (2003). The origin and nature of the linguistic parasite. In B. Bauer & G.-J. Pinault (Eds.), *Language in time and space* (pp. 241–244). Berlin: Mouton de Gruyter.

Kretzschmar, W. A., Jr. (2010). Language variation and complex systems. *American Speech, 85*(3), 263–286.
Kull, K. (2011). The architect of biosemiotics: Thomas A. Sebeok and biology. In P. Cobley, J. Deely, K. Kull, & S. Petrilli (Eds.), *Semiotics continues to astonish: Thomas A. Sebeok and the doctrine of signs* (pp. 223–250). Berlin: Mouton de Gruyter.
Kull, K., Deacon, T. W., Emmeche, C., Hoffmeyer, J., & Stjernfelt, F. (2009). Theses on biosemiotics: Prolegomena to a theoretical biology. *Biological Theory, 4*(2), 167–173.
Leins, P., & Erbar, C. (2010). *Flower and fruit: Morphology, ontogeny, phylogeny, function and ecology*. Stuttgart: Schweizerbart.
McMahon, A. M. S. (1994). *Understanding language change*. Cambridge: Cambridge University Press.
Mesoudi, A., Whiten, A., & Laland, K. (2006). Towards a unified science of cultural evolution. *Behavioral and Brain Sciences, 29*, 329–383.
Milligan, L. A. (2007). *Nonhuman primate milk composition: Relationship to phylogeny, ontogeny, and ecology*. (PhD dissertation). University of Arizona, Tuscon.
Mufwene, S. S. (2001). *The ecology of language evolution*. Cambridge: Cambridge University Press.
Mufwene, S. S. (2005). Language evolution: The population genetics way. In G. Hauska (Ed.), *Gene, Sprachen und ihre evolution* (pp. 30–52). Regensburg: Universitätsverlag Regensburg.
Mufwene, S. S. (2008). *Language evolution: Contact, competition and change*. London: Continuum.
Mufwene, S. S. (2014). Language ecology, language evolution, and the actuation question. In T. A. Åfarli & B. Mæhlum (Eds.), *The sociolinguistics of grammar* (pp. 13–36). Amsterdam/ Philadelphia: John Benjamins.
Müller, M. (1887). *The science of thought*. London: Longmans, Green and Co.
Nöth, W. (1999). Peircean semiotics in the study of iconicity in language. *Transactions of the Charles S. Peirce Society, 35*(3), 613–619.
Nöth, W. (2002). Charles Sanders Peirce, pathfinder in linguistics. *Interdisciplinary Journal for Germanic Linguistics and Semiotic Analysis, 7*(1), 1–14.
Nöth, W. (2008). Semiotic foundations of natural linguistics and diagrammatic iconicity. In K. Willems & L. de Cuypere (Eds.), *Naturalness and iconicity in language* (Iconicity in language and literature, Vol. 7, pp. 73–100). Amsterdam: John Benjamins.
O'Brien, M. J. (2006). Archaeology and cultural macroevolution. *Behavioral and Brain Sciences, 29*, 359–360.
Peirce, C. S. (1866–1913 [1931–1958]). *The collected papers of Charles Sanders Peirce*. (Ch. Hartshorne, P. Weiss & A. W. Burks [Eds.], 8 Vols). Cambridge, MA: Harvard University Press.
Peirce, C. S. (1890–1892 [2010]). The *Monist* metaphysical project. In Peirce Edition Project (Eds.), *Writings of Charles S. Peirce: A chronological edition* (Vol. 8, pp. 83–205). Bloomington: Indiana University Press.
Pelkey, J. (2007). Muji's mirrored merger: Correlative redistribution of checked tone classes in a new newly defined Burmic cluster. *Mon-Khmer Studies Journal, 37*(1), 179–196.
Pelkey, J. (2011). *Dialectology as dialectic: Interpreting Phula variation* (Trends in linguistics: Studies and monographs, Vol. 229). Berlin: Mouton de Gruyter.
Pelkey, J. (2013). Analogy, automation and diagrammatization: Ngwi-Branch polygrammaticalization of Tibeto-Burman *lak. *Studies in Language, 37*(1), 144–195.
Rauch, I. (1999). *Semiotic insights: The data do the talking*. Toronto: University of Toronto Press.
Richerson, P. J., & Boyd, R. (2001). Built for speed, not for comfort: Darwinian theory and human culture. *History and Philosophy of the Life Sciences, 2001*(23), 423–463.
Richerson, P. J., & Boyd, R. (2005). *Not by genes alone: How culture transformed human evolution*. Chicago: University of Chicago Press.
Rieppel, O. (1990). Ontogeny – A way forward for systematics, a way backward for phylogeny. *Biological Journal of the Linnean Society, 39*(2), 177–191.
Schleicher, A. (1869 [1983]). Darwinism tested by the science of language. In E. F. K. Koerner (Ed.), *Linguistics and evolutionary theory: Three essays* (Amsterdam studies in the theory and history of linguistic science, Vol. 6, pp. 3–72). Amsterdam: John Benjamins.

Sebeok, T. A. (1986). Communication, language and speech: Evolutionary considerations. In T. A. Sebeok, *I think I am a verb. More contributions to the doctrine of signs* (pp. 10–16). New York: Plenum.

Sebeok, T. A. (2001). Biosemiotics: Its roots, proliferation, and prospects. *Semiotica, 134*, 61–78.

Sereno, M. I. (1991). Four analogies between biological and cultural/linguistic evolution. *Journal of Theoretical Biology, 151*(4), 467–507.

Shapiro, M. (1991). *The sense of change: Language as history*. Bloomington: Indiana University Press.

Shapiro, M. (2002). Aspects of a Neo-Peircean linguistics: Language history as linguistic theory. In M. Shapiro (Ed.), *The Peirce seminar papers: Essays in semiotic analysis* (Vol. 5, pp. 108–125). New York: Berghahn Books.

Spencer, H. (1862). *First principles*. London: Williams and Norgate.

Sterelny, K. (2006). The evolution and evolvability of culture. *Mind & Language, 21*(2), 137–165.

Stjernfelt, F. (2007). *Diagrammatology: An investigation on the borderlines of phenomenology, ontology, and semiotics* (Synthese library, Vol. 336). Dordrecht: Springer.

Tallerman, M., & Gibson, K. R. (2012). *The Oxford handbook of language evolution*. Oxford: Oxford University Press.

Thellefsen, T. (2001). C.S. Peirce's evolutionary sign: An analysis of depth and complexity within Peircean sign types and Peircean evolution theory. *SEED: Semiotics, Evolution, Energy, and Development, 1*(2), 1–45.

Whitman, C. O. (1910 [1919]). *Orthogenetic evolution in pigeons* (O. Riddle [Ed.], *Posthumous works of Charles Otis Whitman*, Vol. 1). Washington: The Carnegie Institute of Washington.

Wimsatt, W. C. (2006). Generative entrenchment and an evolutionary developmental biology for culture. *Behavioral and Brain Sciences, 29*, 329–383.

Wyhe, J. van (2005). The descent of words: Evolutionary thinking 1780–1880. *Endeavour, 29*(3), 94–100.

Part II
Observations and Case Studies

Verbal Patterns: Taming Cognitive Biology

Stephen J. Cowley

Abstract Linguists classically focus on phenomenologically salient units or verbal patterns. In biolinguistics, these are "explained" by positing a brain that grows a system that identifies/generates linguistic forms (a "language faculty"). The paper offers an alternative: individuals become skilled in linguistic action by using cultural resources to extend their embodiment. Language and languages are heterogeneous and distributed. Although the verbal is salient, its basis lies in coordinated biosemiotic activity. In illustrating this perspective, the paper builds on two case studies of real-time events. These show that people link fine inter-bodily coordination with skills in orienting to utterances as types – they use cultural patterns to constrain biosemiosis. As people become strategic actors, they rely on embodiment (and, of course, brains) to develop skills based on *taking a language stance*. By imaginatively separating language from activity, they both tame biosemiotic powers and transform the brain's functional organisation. There is no need for language genes, neural spandrels or undiscovered physical principles. Wittgenstein's view that language connects living human bodies within *forms of life* can thus be extended by means of empirical and observational work.

Keywords Biosemiotics • Distributed cognition • Enactivism • Systemic cognition • Interpersonal communication • Coordination • Social interaction • Distributed language • Languaging • Embodied cognition

Where Is Language?

The hypnotic effects of verbal patterns induce us to picture human languages in terms of verbal patterns and, of course, we are likely to think of verbal patterns as specifying meanings. If not wary, we may even ascribe the meanings to the hypnotic effects of languages that are constituted by verbal patterns. That, of course, is

S.J. Cowley (✉)
Centre for Human Interactivity and the COMAC Cluster, University of Southern Denmark, Slagelse, Denmark
e-mail: cowley@sdu.dk

E. Velmezova et al. (eds.), *Biosemiotic Perspectives on Language and Linguistics*, Biosemiotics 13, DOI 10.1007/978-3-319-20663-9_7

circular. Yet, in spite of circularity, just such an approach dominates mainstream linguistics. The scientific study of language is all too often restricted to the study of forms that arise in acts of saying and, conversely, how the results allegedly reflect not only what is said but also what is meant. Of course, this persists because, in human forms of life, people are bound to regard acts of speech and construals as public events that arise as we communicate. From this perspective, language is separate from cognitive biology.

One well-known school challenges this view. Building on Noam Chomsky's generative grammar,[1] some linguists deny that language co-evolved with human communication.[2] Dismissing the external or E-language, language is ascribed a hypothetical inner *language faculty*. On a *biolinguistic* view, an internal or I-language grows in the brain. Using comments of Chomsky's, Prisca Augustyn[3] connects biosemiotics to this neurocentric view. Human genetics, she believes, allow cognitive processing to draw on semiotics. This paper uses another view of cognitive biology in its approach to language. Far from ascribing language to a mental organ, it is seen as extending primate biosemiotic abilities. While neutrally enabled, language spreads across bodies, societies and space: it is multi-scalar or *distributed*. As people *language*, neuro-dynamics connect phenomenological experience, life-span events, history and, crucially, semiotic processes. Language is based in, not words and genes, but the evolutionary history of semiosis. While computation and textual use of language have a formal basis, they serve to extend embodiment. Even today talk and meaning depend on meshing bodily dynamics with wordings: human understanding binds symbolic aspects of language to iconic and indexical modes of neurophysiological activity.

Why a Distributed View?

Emphasis on the distributed nature of language[4] arose from challenging the "code" view of mainstream linguistics.[5] Figures as diverse as Ferdinand de Saussure, Leonard Bloomfield, Zellig Harris, Noam Chomsky, Burrhus Fredric Skinner, Michael Halliday, George Lakoff and Michael Tomasello all identify language with the words and rules that they were taught at school. By contrast, on the distributed view, language is traced to a history of coordination that transforms human biofunctionality. The neurobehavioural results enable humans to integrate bodily

[1] In Chomsky 1957, this was presented as a descriptive model; however, by the time of publishing *Aspects* (Chomsky 1965), it was said to some kind of inner reality. While the theory has changed greatly over the years, Chomsky retains the view that scientific linguistics has discovered a neural language organ.

[2] Cf. Jenkins 2000.

[3] Augustyn 2015.

[4] Cf. Cowley 2007 and (ed.), 2011.

[5] Cf. Love 2004; Kravchenko 2007.

dynamics with speech gestures[6] and use utterance acts in treating (partly shared) situations as meaningful. The distributed perspective thus denies that language is reducible to semantic, phonological, syntactic and morphological "forms". While necessary for developing writing systems and in language teaching, "forms" are to be recognised as theoretical constructs. Rather than invoke behaviourism or cognitivism, language is a dialogical activity that prompts people to develop linguistic skills. It is therefore a category mistake to posit an inner process to "explain" language. Far from using a neural or mental "faculty" (or I-language), people need strategic ways of interlacing language, action, perception and thought. Using a principle of ecological assembly,[7] people make what they can of social rules and all the other resources of the life-world. Since language meshes action-perception with thinking, it shapes context as, together, people construe circumstances. Remarkably, this applies *even if nothing is said*. As I look out of the window and see (as it happens to be the case) a train, language shapes perception. Though the train is no more than a salient part of the surrounding, the consequences of looking are verbally constrained: they prompt me to pick out something that is likely to be familiar to a reader. Wordings call forth a familiar world of objects and events based on how we act and perceive. However, neither wordings nor verbal patterns determine anything. Most certainly, they cannot influence how populations act-perceive and how individuals think. Rather, they serve bonding functions, acting to ensure that lives cohere within communities. Embodied and embedded acts of utterance unite speaking, hearing and action. "Language exists", Mikhail Bakhtin suggests, "only in the dialogic imagination of those who make use of it".[8] As imaginative activity, language is irreducible to forms; rather, it shapes living human beings. Wordings merely constrain what William James calls *the thinking that goes on.*[9] In short, picturing languages as verbal patterns that specify meanings lead us astray. This hypnotic effect occurs because it is so tempting to accept the commonplace that language depends on verbal patterns. Patently, however, this view is circular. Unlike computers, human infants have no need to ground their "words" in an objective world: they are learned as part of activity and, of course, their function is inseparable from action. From the start, infant activity is construed in terms of wants and beliefs that a caregiver uses to sustain consistent modes of action (and thought). Later, these simple forms of understanding become intermeshed with language: people come to say what they want and believe. As a result, language both allows individuals to act under collective control and enables them to develop as persons. This gives rise to what Ludwig Wittgenstein appositely calls *forms of life*: complex social practices during which people feel, think, speak and act by linking language with cognitive biology.[10]

---

[6] Rączaszek-Leonardi and Keslo 2008.

[7] Clark 2008.

[8] Bakhtin 1963 [1984, p. 183].

[9] James 1890, p. 225.

[10] Wittgenstein 1953 [1958, p. 226].

Human activity can be described as semiosis. This occurs, among other things, as language spreads in space and time both during talk and when they engage with text, computers and their technological extensions. Human sense-making connects circumstances as people orient to social (and verbal) routines as well as one-off events. In pursuing the distributed view, studies described elsewhere bring home how dynamics shape real time construal.[11] Species-specific *languaging* is influenced by wordings that take on a particular sense for each person concerned. *First-order languaging* is defined as, face-to-face (non-ritualised) activity in which wordings play a part. Its basis lies in sense saturated coordination or *interactivity*[12] that allows people to coact as they echo voices, doings, sayings and ways of using common expectations. For Charles Sanders Peirce, therefore, even man is a sign.[13] This striking claim is vindicated in how software designers encourage "users" to present themselves as semiotic beings. They are encouraged to develop habits in projecting likely events, acting, monitoring the results and re-engaging with the system. Language and interactivity or cognitive biology thus ground semiosis. Pursuing this, the paper focuses on the unfolding of two complex social events. It highlights how the persons concerned use a dialogical imagination to integrate their embodiment with how verbal patterns serve to appraise and manage the changing circumstances of their actions.

First-Order Languaging

As the life-world comes to be seen as irreducible to information processing, new importance falls on *languaging*. Qualifying Humberto Maturana's use of the term,[14] I limit its application to action where wordings play a part – or to the human world. It is striking that Maturana's biological perspective[15] preceded the neuroscientific challenge to the view that neural activity resembles machine code. Rather than posit a language "faculty", languaging was seen as structural coupling between environments and living beings. It is thus a form of communication bound up with attention, perception, action and learning. Languaging gives rise to selections that make up an individual's lived world. When traced to organism-environment relations, language can emerge independently of discontinuities in natural selection, a spandrel or mysterious physical principles. Its basis is embodiment, iconic and indexical activity that, in our species, is also phenomenological. People create a "consensual domain" as, in our terms, they use biosemiotic skills to create and construe the

[11] The analytical details of the simulation are described in detail in Steffensen et al. 2010; the South African interaction is the main focus of Cowley 2001. Finally, Stephen Cowley links the two incidents in a paper on intercultural communication (Cowley 2012a). While the analysis remains much the same, in this context, the interpretation is substantially extended.

[12] Kirsh 1997; Steffensen 2011; Cowley and Vallée-Tourangeau 2013b.

[13] Peirce 1931, §34.

[14] Maturana and Varela 1992.

[15] Maturana 1988.

wordings and practices of a familiar world. In *Homo sapiens*, feeling and thinking fall under verbal constraints. Given skills in orienting to these second-order patterns, people come to accept the agreements in judgement that underpin all kinds of social order. In an English speaking environment, therefore, they see *trees* as trees. Yet, while having a general aspect, as Maturana realised,[16] linguistically grounded experience is connotational. An individual not only construes what, on a given occasion, is meant by *tree* but also learns to discriminate what counts as trees in different situations. Human languaging allows self-creating, self-maintaining autopoietic systems to thrive in a constrained environment. In this sense, like other living systems, they depend on language. In the terms of the semiotician Thomas A. Sebeok,[17] they rely on a primary modelling system.

Most linguists (including biolinguists) ignore this plausible link between language and what Maturana calls the structural coupling of living systems. However, many trace language to coordinated use of the body. In independent work, Alton Becker demonstrates that acts of utterance always mean something to someone.[18] In short, their particularity undermines any code view. In making real-time events/construals part of what he too calls *languaging*, Becker concurs that understanding has a public aspect. However, within this dwelling-place, as Martin Heidegger puts it,[19] language speaks through us. General meanings disambiguate a situation (for a person) while the sense of an utterance – and situation – is particular. As the case studies show, even barely "linguistic" acts (e.g., "**ye:::s**") make utter sense. In linking culture to embodied coordination, Roy Harris suggests that "biomechanical" constraints are necessary to language.[20] Building on this observation, Nigel Love contrasted what linguists usually describe (verbal patterns) with embodied "first-order" activity.[21] On this view, mainstream tradition – Saussure, Bloomfield, Skinner, Chomsky and Lakoff etc. – are trapped by conceptual confusion. First-order bodily activity (languaging) is conflated with second-order products (imagined counterparts to wordings). In Per Linell's terms,[22] written language bias prompts the erroneous view that form-based patterns arise from a language system or faculty. Using a covert analogy with texts, mainstream linguists overlook *dialogicality* or the creation and construal of linguistic signs. While languaging occurs in space-time, it evokes the not-here and not-now. For Linell[23] mainstream models go wrong in conflating the situated aspects of language with its non-local resonances. In social life, people create/construe physical events that evoke expectations, norms and traditions. In dialogical imagination, wordings grant utterances a particularity derived from the astounding precision with which we concert activity. Human lan-

[16] Maturana 1978.

[17] Sebeok 1991.

[18] Becker 1988.

[19] Heidegger 1959 [1971].

[20] Harris 1998.

[21] Love 2004.

[22] Linell 2005.

[23] Linell 2009.

guage is polyphonic – a voice (or text) echoes others: it is iconic, indexical and, yet, symbolic. In terms of the distributed perspective, speaking while moving unites the general with the particular. In the next two sections, therefore, this symbiotic view of language is illustrated with respect to instances of first-order languaging.

The Case Studies

The first case study reports on a discussion which took place in the 1990s within a South African non-governmental organization (NGO) whose mission was to retrain Black teachers. A (male) senior teacher, Musa (M), comes to discuss a transfer to a place called Jozini with a White female administrator (Daphne, D). She had already heard about his upcoming request and, correctly, anticipates what he wants. She thus picks up *his* point of view (he had bad luck in the draw). The teacher accepts this view of the situation

1 D: hi musa you wanna see me is it about going to Jozini
2 M: ye:::::::::::::::s
3 M: ye::::s eeh I had a bad luck eeh
4 D: d'you have a bad luck with the draw okay musa
5 M: ye:::s

Instead of directly *saying* that she will help, she uses an indirect strategy. In ways that may strike a reader as odd, she tells him how to act – to do like two other ladies by writing his name on a list in the blue block (or group f). As becomes apparent in 10, he does so.

6 D: will you plea(se) put your name here
I've I've jus I've just said to two ladies who has just come to see me as well
7 M: yes
8 D: hmm # I can't make any promises alright musa which group are you in
9 M: umm group ef

Thus, in the next utterance, she offers implicit reassurance.

10 D: group ef okay okay that's fine okay your name is there in the blue block so I can't miss that name I've gotto look at it alright

However, Musa does *not* thank her. Ignoring reassurance that she will look at it (and, by conventional implication, act on the basis of his request), he pleads in a respectful Nguni way (saying little, speaking slowly and using a deep voice) – saying only "**please eeeh**".

11 M: please eeeh

Daphne is unmoved; she repeats her indirect reassurance (in 12) and, in so doing, fails to adapt to his pace or the Nguni custom of showing respect by saying little. It is thus perhaps not surprising that she fails to elicit his thanks but, as shown below, triggers further strained interaction:

12 D: but I can not make any promises alright
13 M: mmm
14 D: and I'm I'm not gonna be able to l look at this

15 M: when uh will you you be final
16 D: Monday morning
17 M: Monday morning
18 D: yes
19 M: ah

Having not been thanked, in 14, she indirectly suggests that there is nothing to be done. Musa displays understanding by asking when she will finalise the decision. Instead of showing gratitude, he repeats her utterance and, once again, meets her reply with a respectful *ah*. At this point a colleague, Lynette, is moved to contribute to the conversation. She says (of Daphne):

20 L: she always does her best with everyone

She makes explicit to Musa that Daphne is doing her best – and, by implication, being fair (favouring neither him nor the two women who came earlier). But Musa knew this – did he not? After all, in accepting that she would not finalise the decision until Monday, he shows his grasp of the process. So what is going on? To give another view, I turn to an examination answer written by a black South African student months after having heard – and discussed – the recording in an academic seminar. To these ears, the conversation exemplifies:

> [w]hat we come across and see in our daily lives. [...] There was a big problem with the interaction because of language boundaries i.e. Daphne was an English speaker and Musa a Zulu speaker. The problem of language led into Musa being offered answers and options, not given a chance to choose for himself and just accepting and acknowledging everything. I do not blame Daphne for this, she could not switch in register or even code-switch and Musa too could not express himself in English. The interaction did not become productive, as Musa was not well informed about what he wanted and Daphne on the other side wanted to get rid of him as soon as she could.[24]

Beyond the *said*, much more is happening. However, the nature of the "big problem" is not clear (even if familiar from "our daily lives"). Perhaps the reader will gain from reconsidering the narrative. As noted, Lynette acted in a way that (from her point of view) is fair: using a "block system", she agrees to consider Musa for transfer to Jozini. He understands and, Nguni style, shows due respect. However, Lynette feels moved to say that Daphne always does her best. In the examinee's terms, Musa proceeds by "just accepting and acknowledging everything".

21 M: ye:::s
22 M: ye::s and I I also yo::u see:::: you see:: why

Offering repeated respectful, low-voice drawled versions of "ye:::s" he concurs and, then, using another indirect strategy, starts to repeat his plea. However, he does not get to the end of his account of why he wants to go to Jozini (in that, as noted, speaking slowly enacts respect). Lynette interrupts:

23 L: if she can't do it then you know that God can't do it/[25]

---

[24] Cowley 2001, p. 180. In fact Daphne is Afrikaans speaking and, of course, the analysis is limited. It is striking that the student pictures her Whiteness and fails to identify her accent; further, in my view Musa, a senior English teacher, expresses himself well in Black South African English. However, his style is typical of a man from an oppressed group.

[25] In the transcript, slashes (/) indicate overlap.

This too seems to be an attempt to get him to understand the situation and, perhaps, to show gratitude. Daphne laughs:

24 D: haha/
25 M: I'm so much willing to go there.

Far from laughing, Musa again repeats his desire for the transfer. Although all three are skilled in the local *lingua franca*, English, the legacy of living apart (apartheid) generates an ugly tone. The problem is neither miscommunication nor non-understanding: they re-enact behaviour that occurs between oppressed and oppressor groups.[26] The administrators speak to Musa as if he was a child – and he responds *like* a child. Verbally, this appears in comparing Daphne's good grace with that of a (most likely) shared God. However, it is most audible in how speaking "**she always does her best with everyone**" is rhythmically integrated with the beat of the previous four utterances (shown in bold).

16 D: Monday morning
17 M: Monday ***mor*** *ning*
18 D: ***yes***
19 M: ***ah***
20 L: ***she*** *always does her*
best *with*
ev *eryone*

She chimes in with a striking mode of articulation. Showing exquisite timing, she changes the tone by seeking to render explicit what has occurred. She tells Musa – almost directly – that Daphne is doing her best. Her action is especially striking because, a moment before, the talk had seemed to be coming to a close. In 17, Musa's rhythm had picked up on Daphne's "**Monday morning**", elicited confirmation and he had, still showing respect, signed off with "*ah*". Lynette's attunement is striking because of a metrical/intonational mismatch[27] that uses the syllable timing of her first language (Afrikaans). This displays that she is speaking *to* Musa (prominent initial "**she**" sounds marked in English) while standing up *for* Daphne (prominent, but softly spoken, *always*). Further, the syllable-based style allows loudness to parallel metrical patterning. (She speaks the first two feet loudly [to Musa] and the last softly [to Daphne].) The metrical organization can be shown by using updated classical notation.[28]

20 L: // ***she al****ways //* ***does*** her ***best*** *//with everyone//*
//— ~ ~ // ~ ~ — // ~ ~ ~ —//

Unexpected prominence on "**she**" (spoken on a low-falling tone) is striking: this allows her to emphasise Daphne's goodness, reassure (always) and offer solidarity (with everyone). Both parties hear "their" message. Musa sounds (slightly) reas-

[26] This alludes to Georg Wilhelm Friedrich Hegel's insight that collective forces provide a setting for master-slave relations (Hegel 1807 [1967, p. 65]).

[27] Cf. Pike 1945.

[28] The transcription shows prominent syllables in bold — and how the utterance ends with suppression of loudness. In the accompanying metrical gloss, longer syllables are shown as '—' and shorter ones as '~'. Thus while '**she**' is both prominent and long, the first syllable on '**always**' is prominent and short.

sured as he ups the loudness and smilingly repeats 'yee:::s, yeee::::s' before restating his wish. Daphne, by contrast, is silent. Perhaps, it is this – plus Musa's again repeated non-display of gratitude that prompts Lynette to *go on* by explicitly supporting her colleague to the extent that she is comparable with God.

First-order languaging uses modulations in voice dynamics to perform a *cognitive* role: the phonetics of one burst of speech influence what follows (and the thinking that goes on). Many implicatures depend less on verbal content than voice dynamics. In echoing a historical context, I show their subtlety – and, yet, I have only begun to show how finely people use acoustic and gestural resources.[29] Nonetheless this level of detail clarifies why no neural model of "forms" can "explain" the events. In spite of failure to meet each other's expectations, Musa, Daphne and Lynette are all highly competent in their own languaging and, thus, each experiences events in a particular sense. More theoretically, unless the events are deemed "non-linguistic", they show that human language does not reduce to the "use" of verbal patterns. Further, given the role of bodies, it cannot be "explained" by models of genes and brains. As for infrahuman species, languaging is embodied or biosemiotic activity.

Timing shapes communication because, to affect B, A's behaviour must impact on B's attending. Human language is therefore necessarily temporal. In its classic manifestations of speech, it arises as people control the airstream while using the muscles of the vocal folds to modulate the vibrations used in voicing of vowels and many consonants. These physical changes in time generate pulses of energy that are modulated by the vocal organs.[30] This results in speech whose timing is inseparable from bodily movement (and, especially, gesture). Further not only are prosodic and gestural-expressive aspects of language actual movements but these are also heard as verbal patterns. However, timing is not only a vocal skill: as illustrated in the second case study sense-making also draws on how visible movements are timed. The events occur in a high fidelity simulator, a safe setting where doctors learn about emergency situations (cf. Fig. 1). In this training scenario, while a senior doctor takes the role of a nurse (on the left), a junior doctor (on the right) is expected to take charge of the case (through diagnosis and administration of pharmaceuticals). The event occurs early in the simulation where, perhaps because of nerves, the doctor fails to carry out a physical examination. Rather, after greetings, he enquires about the patient. The nurse-facilitator picks up the medical chart and seeks out relevant facts. Thus, we might expect a verbal description of the patient's condition: in fact, the act of utterance functions as a Zeitgeber: it contributes to prompting the doctor to take the patient's pulse. This appears on the video still that is presented in Fig. 1.

What brings about the junior doctor's action? Crucially, pulse-taking both interrupts the doctor's course of action and, strikingly, parallels what is articulated. It arises under dual control. On the "surface" these routine events consist in a greeting,

---

[29] Cf. Cowley 1994 and 2010; Thibault 2011a; Steffensen 2013.

[30] While mainstream linguists claim that brains (or minds) identify and recode linguistic features (viz. as "forms"), the well-established ecological aternative is that, on the one hand, we make and track phonetic gestures (cf. Fowler 2010) and, on the other, use rich phonetic memory (cf. Port 2010).

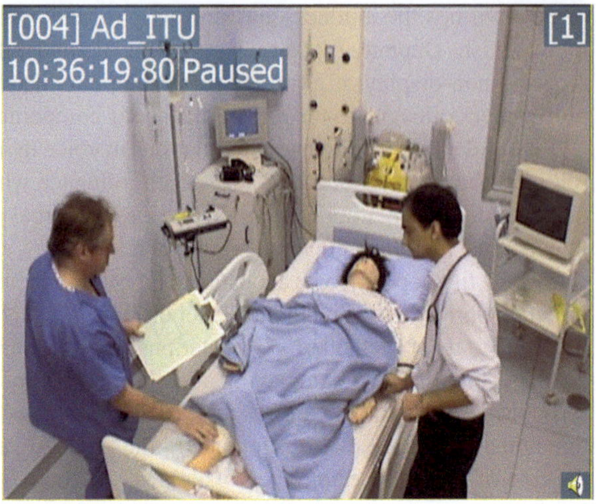

Fig. 1 Culmination: taking the pulse

introduction of the patient and, in 6, an enquiry about the patient's condition ("**What happened?**"). When scrutinised, they can be traced to *exactly* how the nurse's bodily response draws on movements accompanying the voicing of 8–9.

1. N: Hello
2. D: Hello
3. N: Hi. (.) I'm n[urse (.) Smith
4. D: [Doctor (xxx)
5. N: Hello hi. I work night here. (.) I look after (.) mrs. Kennedy here.
6. D: uhum, what happened?
7. N: (0.8)
8. N: Well, all I pretty know uh is that u- (.) um she- she had some
9. N: (0.7) orthopaedic surgery on this [leg here=

The main cue has little, not nothing, to do with what can be seen in the transcript. The cue lies in the pacing of events that include a 800 millisecond pause that is followed (in 8–9) by filled and unfilled ones ('uh', 'u- (.)) and a 700 ms silence. In replying to the question, the experienced facilitator slows the action or, colloquially, gives the doctor "time to think". Turning to a pico-scale, key moments are shown below. In picking up on context – on *how* the nurse moves – the doctor is prompted by *the act of hearing* that the patient has had orthopaedic surgery. As this is said, the nurse acts as if he were taking the pulse; he touches the bandaged leg. Embodied interactivity thus overrules the said (cf. Figs. 2, 3, 4, and 5 below) by prompting the doctor to act in a mimetic fashion.

In presenting the patient, the nurse speaks slowly: "**Well, all I pretty know uh is that u- (.) um she- she had some (0.7) orthopaedic surgery on this [leg here]**". Wordings serve, above all, in coordinating attention. As the nurse says 'well' (Fig. 2), his action culminates in touching the chart (a gestural "stroke"). The doctor follows the nurse's gaze onto the object (Fig. 3). Serving as a Zeitgeber, the speech-gesture movements enable the doctor to size up the situation. He finds a common perspective as the nurse says "**uh is that u:::**". Like a caesura, the second "**u::**" is a non-gestural beat which, as it turns out, leads into "**she had some …**" During

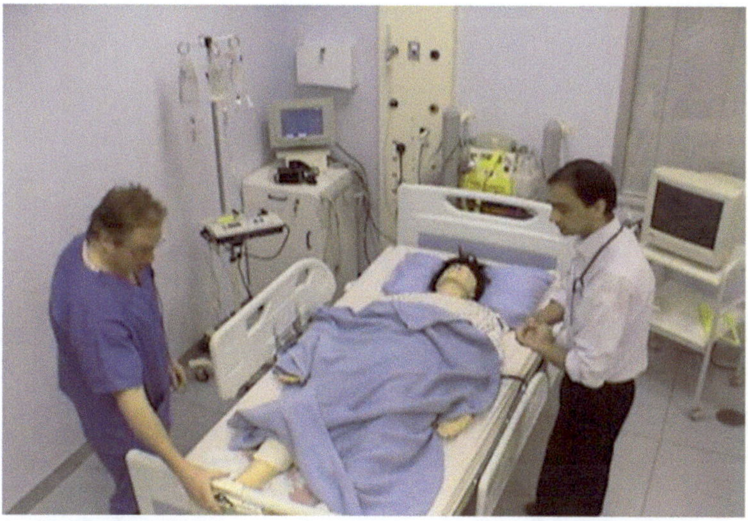

Fig. 2 **"Well all I pretty know"**

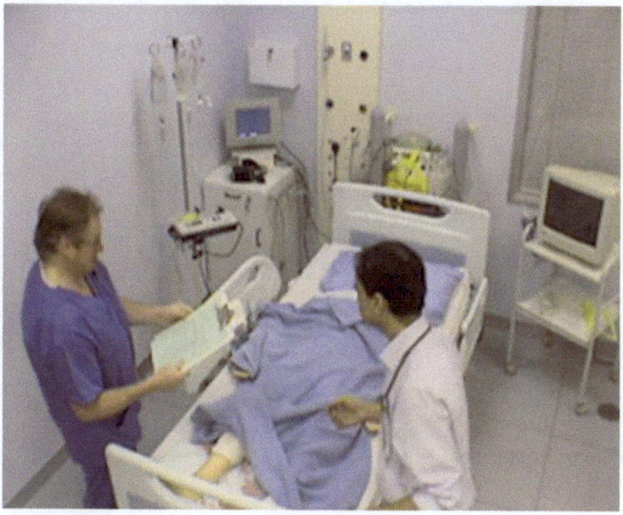

Fig. 3 **"uh is that u- she had s'm"**

another long pause, the gesture's pre-stroke phase (up to first syllable of "**orthopaedic**"), the doctor's body begins to sway (Fig. 4). He mimics the nurse's gestural stroke as the nurse utters the prominent syllable of "**orthopaedic**" – and touches the patient's leg (Fig. 5). The doctor moves to the patient and carries out a minimal physical examination. Mimicking the nurse, he takes the patient's pulse and, as he does so, reorients gaze. By the next beat ("**on this leg here**"), as shown initially, the

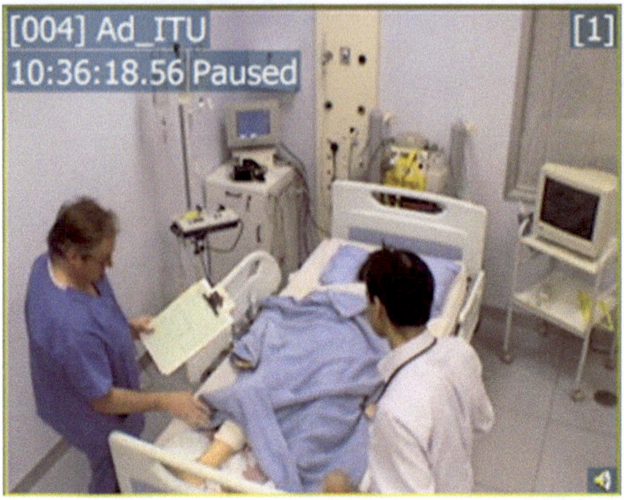

Fig. 4 **"(.)orthopedic surgery"**

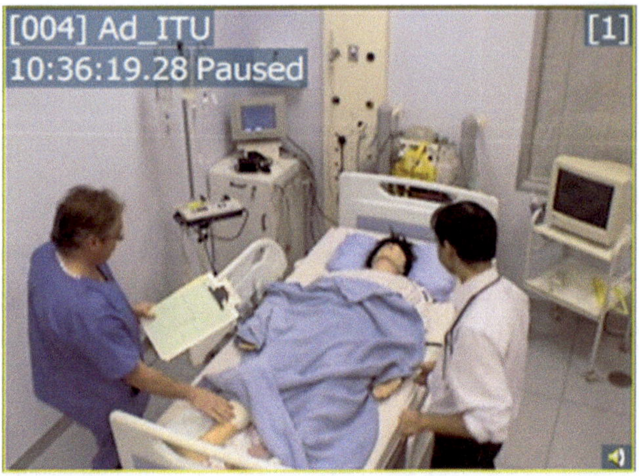

Fig. 5 **"(.) ortho pedic surgery"**

doctor is again attending to the nurse (Fig. 1). The words actually spoken – together with bodily synchrony – give coaction a temporal rhythm. The result arises in moving together as perception prompts creative mimesis. This aspect of human communication uses interactivity or an individual's changing sense of *how* to gesture and articulate syllabic patterns. Physical words enact pico-scale events that prompt the doctor to orient to patient care.[31]

[31] The pico-scale captutes how syllables are articulated, faces moved and gestures made – typically using dynamics of 40–200 milliseconds. By comparison, a stressed syllable lasts about 200–300

Given their rapidity, events like those described elude any brain-based model. Far from depending on inner processes, people use body's communicative and cognitive power in adapting to what happens. At this instant, biosemiotic attunement stands in for inference. In turning to first-order languaging, investigation turns to how people engage by connecting wordings with actions. The thinking that goes on connects circumstances with the said as people draw on each other's manifest expectations. People demonstrably generate synergies – ways of acting that, otherwise, would not have arisen. Indeed, it is when persons are influenced by each other's movements that they come up with the thoughts and feelings that drive events. In what follows, therefore, I ask how we might use biosemiosis to rethink aspects of first-order languaging that occur beyond the reach of verbal patterns.

Beyond Symbols: Part of the Game

Like visible movement, verbal patterns become salient to speakers of a language. No doubt this is why they dominate both writing systems and Western theories of language. However, in first-order languaging, the verbal often serves merely to orient affect and attention to a common focus. At such moments, wordings direct (often) subtle actional and perceptual moves. If we are not to be distracted by abstract models, attention must be given to how thinking is enacted. As Timo Järvilehto shows,[32] focusing on *results* offers a radically different perspective on mind and behaviour. Indeed, in the case studies, people use pico-scale events to attune to whole body activity. However, this is *not* always so: often, verbal patterns are more prominent. Even in talking to oneself, people listen and learn from their voices.[33] Indeed, because generalities (or future causes) influence human action, verbal patterns attain influence: in Biblical exegesis, reciting the Qu'ran or legal and scientific practice, inscriptions are treated as (relatively) fixed. When said to depend on "language", appeal is made to written language bias. Far from relying on verbal patterns "in themselves", people take a special attitude to "what is said". Where wordings dominate, Cowley argues, they *take a language stance.*[34] They draw on skills in construing utterances as instances of peculiar types (e.g., of words that are [un]true). That is beyond debate. In this context, however, the point is that, during much first-order languaging, people rely exclusively on neither verbal patterns nor trust in abstract types. Synergies between bodies enable people to attune to each other's ways of attuning. Strikingly, this is intrinsic to expertise and, just as

ms and the time-span is often treated as the window of consciousness; for example, pauses of 200+ milliseconds can be heard. Pico-scale contrasts with the micro-scale used by above all, most who work in the field of *Conversation Analysis* and gesture/nonverbal behaviour.

[32] Järvilehto 2009.

[33] Cf. Cowley 2014.

[34] Cowley 2011.

crucially, consistent with the claim that the evolution of human intelligence has a mimetic basis.[35]

Not only does "body language" (or nonverbal behaviour) communicate but, on a distributed view, the dynamics are *partly* constitutive of thinking. Human coordination enacts physical changes as people draw on parameters to control action. As in many biological systems, physical changes trigger *possible* goal-states. While autopoietic, organisms also draw on physical and social constraints. For example, a bacterial population use collectively engendered constraints as they move. Simple forms of structural coupling predate brains. Of course, more sophisticated coordination appears in vertebrates. Horses, for example, *learn* from a felt, two way, anticipatory relation. As they get to know their riders, they share understanding through what Susan Stuart calls *enkinaesthesia.*[36] Much is gained from concerting bodily dynamics. In social mammals, play nurtures anticipatory modes of action. In dynamical systems theory, neural "frustrations"[37] are said to be released as organisms appraise circumstances. Of course, frustrations also arise in the world beyond the skin – much of what is said and done seeks to avoid their effects. Although sometimes goal-directed, as David Kirsh and Paul Maglio show,[38] action is often epistemic. In computer games, for example, simple moves depend on orienting to norms: expertise plays out as sense-saturated coordination or interactivity.[39] This is why software packages encourage habit-taking; they prepare people for future benefits. Further, human life is embedded in social institutions that favour the use of available external resources. Just as in human-computer interaction, sense-saturated coordination is shaped by skilled action and expertise. This interactivity is necessary to language because it links felt anticipation (or enkinaesthetic events) to statistical phenomena. Given a sensorimotor basis, the said evokes connotations that prime for what is likely. In careful study of reading, the process is shown to be anticipatory[40]: sense-making enables a reader to project what may follow. These ideas underpin how pico-scale events contribute to cognition and communication[41] and, specifically, undergird cognitive event analysis.[42]

While first-order dynamics can be measured, persons also use non-local parameters. Using experience, they draw on recurrent patterns or, simply, what is familiar. In the case studies, whereas contingencies prompt the doctor and nurse to attune, at the NGO, Musa and the ladies fail to do so. On such occasions, dynamics come to the fore. In everyday life, however, such cases may be rare. Much of what we do and say is routine activity based in phenomenological or micro-scale events. Action uses meshed neural control hierarchies: temporal (and other) phonetic aspects of "**she**

[35] Cf. Donald 1991 and 2007; Cowley 2012b.

[36] Stuart 2010.

[37] Wallot and Orden 2010.

[38] Kirsh and Maglio 1994.

[39] Kirsh 1997.

[40] Cf. Järvilehto et al. 2009.

[41] E.g., Cowley 1994; Steffensen and Cowley 2010; Thibault 2011a, b.

[42] Steffensen 2013.

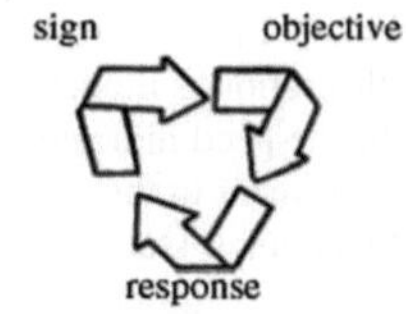

Fig. 6 Alexander's relabelled Peircean triad (reproduced with permission)

always does her best with everyone" pick up and trigger (partly) shared effects. Equally, the "nurse's" touching the mannequin (while saying "**orthopaedic**") sets off Zeitgebers that serve mimesis. Events afford opportunities to realise values.[43] Thus, whereas Lynette reassures Musa while also showing solidarity with Daphne, the doctor acts professionally by checking if the patient is alive. Not only do social affordances fit nature's normative order but, crucially, they call forth human values. Shifting sensitivity in perceiving affordances affects an organism's (changing) "objectives". In human life, at least, it matters that much behaviour is lived as purposeful. The idea appears not only in biosemiotics but also, for example, in Daniel Hutto and Erik Myin's "teleosemiotics".[44] In broad terms, it is captured by Victoria Alexander's[45] relabelling of Peirce's triad (cf. Fig. 6) as showing sign-objective-response relations.

The simple model serves not only to highlight what systems achieve but, in allowing comparison across cases, shows broad application. Thus, applied to protein synthesis in the paradigm case of "organic coding",[46] a second-messenger becomes a sign. Its effects set off folding that contributes to an organic "objective" as transcribed DNA synthesises a protein (response). Indeed, the model's strength is also its weakness: while unable to clarify *how* the process is accomplished, it captures a general pattern. Precisely because mechanism is ignored, the model can easily be generalised to, for example, how a cockpit manages a plane's speed. As Edwin Hutchins showed in his classic work,[47] the "objective" depends on a distributed pilot-cockpit system whose human part attunes to *precisely* when an airspeed indicator indexes a "salmon bug" (cf. Fig. 7[48]).

When the salmon bug is reached (a pink marker at 240 mach), the pilot enacts the objective by acting to extend the flaps and slats. This "response" reduces the plane's speed. In Alexander's terms, an objective can be managed by either an RNA complex or a coupled pilot-cockpit system. At other times, objective-based responses may exploit conditions in a beehive, weather, or silent thought. However, since biosemiosis enacts a web of criss-crossing processes,[49] the objectives that shape systemic actions (Alexander's "responses") rarely depend on individual intentions.

---

43 Cf. Hodges 2007; Hodges et al. 2012.

44 Hutto and Myin 2012.

45 Alexander 2013.

46 Barbieri 2003.

47 Hutchins 1995.

48 From *ibid.*, p. 273.

49 Lotman 1984 [2005].

Fig. 7 Speed bugs (this illustration is modelled on the airspeed indicator instrument in the McDonnell Douglas MD-80, as described by Tenney 1988. Reproduced with permission from Hutchins 1995)

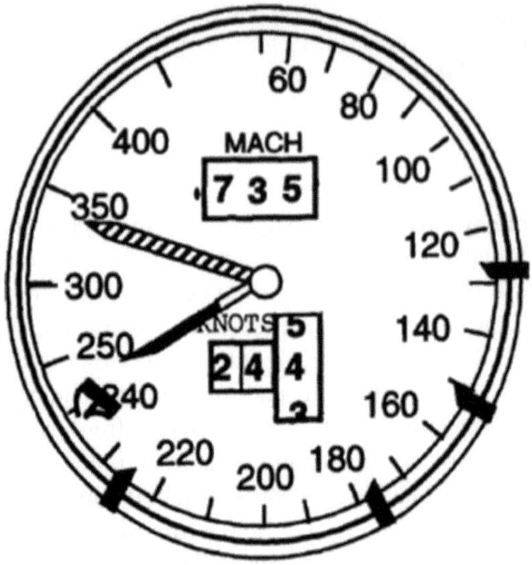

Much depends on, not life history, but a lineage and/or social strategies that exploit body-world relations. Importantly, this shows that Maturana's "structural coupling" functions as an abstract description of iconic-indexical processes. In at least some cases, structural coupling is biosemiotic. Indeed, this helps clarify how, often without intending to do so, *people reach* objectives that can trigger normative – or culturally tuned values. Often it is sufficient to show sensitivity to qualities and relations, colour, and affect to act in ways that link interactivity with practices and verbal patterns. Even though philosophers often seek explanations in terms of natural kinds – objects, events and causal relations – this is likely to be mistaken. This is because, far from being part of the natural world, these pertain to how nature appears from a language stance. With history, human lives draw heavily on structures beyond symbols: our ecology is extended by bodies, social institutions and technology. Since artifacts and institutions contribute to our lives, Alexei Sharov[50] views our world as a pragmasphere. Human modes of engagement are increasingly dominated by how we perceive the species-specific counterpart to fundamental physical reality. For, as Sharov brings home, the creative force of nature is logic.

The Pragmasphere: The Role of First-Order Languaging

Whereas nineteenth century disciplines like psychology, linguistics and chemistry posited an "object" of enquiry, semiotics focuses on relations. In biology, the focus falls on the dynamics of living systems and, above all, their complexity. While there

[50] Sharov 2010.

are many views of evolution, Sharov[51] stresses how a lineage of agents exploits the functional interplay of utility and logic. Physics thus provides constraints within which life evolves and, as evolutionary products, humans come to regard its constraints as logical. An agent is a system that (from an observer's perspective) is capable of goal directed behaviour by virtue of how it is connected "horizontally, hierarchically and genealogically".[52] Accordingly, agency evolves in the context of achievable effects and sustained values. Although blind to underlying function, living systems rely on replicable sequences of actions based, presumably, on how bodies master functional information. Actions and metabolic processes result from selection based on functional value. Agents use functional signs/interpretants: simple agents depend on mechanisms often likened to lock-and-key devices. In protein synthesis, organic coding uses objectives that include the adaptor-molecules (RNA complexes) that give rise to metabolism. However, even that process depends on a whole cell in an environment that functions within a multicellular system. Semiosis occurs both as agents encode/regulate or as they control events at a boundary with the environment and also as they contribute to a goal that given for a hierarchy of agents.

Living systems make much use of hierarchical organization: this is especially marked in organisms that perceive and, using a CNS, learn. In this case, Sharov argues, basic signs that prompt perception and learning serve to *aggregate* functional information.[53] While able to pick up affordances, agents attend to cues that appear meaningful. Thus, something remarkable occurs: an organism can ready itself for later actions by relating current circumstances to earlier events. Using statistical learning, brains favour the anticipatory action that characterises first-order languaging. Thus, while verbal patterns disambiguate, much depends on habit and the specifics of a situation. Humans adapt to the familiar by construing circumstances and, thus, develop powers of discrimination. The semiosis-saturated nature of embodiment sharpens perception: a way of touching a leg or vocal chiming resonate between people. Far from being word-based interpretation (or decoding) this is based on using iconic-indexical behaviour. However, the depth of the interpretation is not explicable by determinate cues: simple events can evoke complex expertise.

A Biosemiotic View of the Case Studies

In the simulated emergency, the nurse's iconic-indexical behaviour triggers action that might well have occurred in a hospital ward. The doctor *knows* that he should check the patient's pulse. In the simulation, under stress, he fails to activate the routine. Relying on higher-level knowledge (as in medical school), he seeks out facts – he relies on the language stance. Thus he asks "**what happened?**" In the

[51] *Ibid.*

[52] *Ibid.*, p. 1052.

[53] Sharov 2010.

situation (as opposed to the simulation), the "nurse" facilitates appropriate behaviour. He speaks in an inarticulate way, bringing home that what is on the chart is not, at this moment, the most relevant affordance. In touching the patient's leg, he both shows that the patient has had orthopaedic surgery and also gets the doctor to attend to the mannequin. The "nurse's" hand-to-leg movement *is* functional information (in Sharov's sense). It regulates the junior doctor's actions as a contingency displays the situation's *logic*. To find out what is wrong, he moves *down* the hierarchy of control. Using mimesis, non-conscious sensitivities come to the fore. Thus, instead of focusing on what the nurse says – wordings – situated dynamics trigger the doctor to move into a professional role. He realises a goal – in Alexander's terms,[54] semiosis is mediated by an objective: overcoming frustration, he uses medical skills to establish that the patient is alive (he feels the mannequin's pulse). A biosemiotic view thus allows first-order languaging – and thinking – to arise as interactivity meshes with wordings. Knowledge arises in concert as, together, parties use circumstances to collaborate. Without any need to be explicit, the trainer brings about an affirmation of the trainee's skills.

In the South African NGO, a different logic applies. The big problem comes to the fore when Lynette utters "**she always does her best with everyone**" (and compares Daphne with God). By being explicit, she sounds patronising, an effect enhanced by unexpected prominence on the initial "**she**" and use of a slow, soft syllable-based rhythm. She sounds as if she is talking to a (big) child. In this way, events depend on more than the words actually spoken.[55] In the NGO, tension rises, in part, because of failure to meet/acknowledge embodied perception expectations. Having failed to elicit gratitude, Daphne makes no attempt to grasp Musa's perspective. Rather, she moves *up* the control hierarchy by invoking (likely) shared values. Speaking as a Christian, she draws on the commonplace that Black Africans are often devout. Space permits no further analysis of the "big problem" that connects languaging with second-order logic. Accordingly, let us be generous and suppose that Lynette intends to help Musa. Indeed, in spite of cultural disharmony, her tone does have a noticeable effect. Having offered more Nguni style agreement with a low-voiced and respectful "**ye::::s**", he follows up with a recycling and an attempt to explain. He does not hear her speech as offensive (for the examinee, such events are "part of our daily lives"). Nonetheless, rather than let him speak, she moves to the cultural level (one of shared beliefs); if it is in God's power for him to go to Jozini, it is in Daphne's too. While more could be said, the point is that most of what is said depends on – not inner processes – but cognitive dynamics.

Enkinaesthesia and participatory sense-making contribute to the talk. The bodies concert, the affective is partly anticipatory and, given an admixture of tension and politeness/respect, the parties exert hierarchical control. Equally, wordings and the social order influence the outcomes. Since wordings contribute to face-to-face activity, a biosemiotic view traces understanding to indexical-iconic dynamics. In dis-

---

[54] Alexander 2013.

[55] The same idea is central to Wittgenstein's view that language has much in common with musical themes and that a sentence (or word) can be heard in a particular sense (Wittgenstein 1953 [1958]).

tributed terms, first-order languaging is whole-body movement. The words that are actually spoken are *one* aspect of activity that relies heavily on shifts in attention. Situational particulars change together with the multiple objects that contribute to signs/responses. In both settings, it seems that stress induces people to shift attention up or down a control hierarchy: in the medical case, they experience empathy and, for a moment, rely on motor mimicry.[56] In the NGO, the opposite occurs: the ugly encounter features lack of interpersonal connection and, in its place, an attempt to be explicit by establishing (dubious) religious links.

The Boundaries of Language

Where the boundaries of the verbal are taken to identify the boundaries of the world, there is a danger that first-order languaging will vanish. Indeed, when linguists focus on form/meaning, they lose sight of how people create and track understanding. However, the case studies show the importance of boundary events: human meaning-making uses a continuing flow of pico-scale coordination. Through *saying things*, language alters attention, perception and action: thinking meshes with non-verbal experience. Hearing how people speak is, in Wittgenstein's sense,[57] crucial to *how we go on*. Language-use is far too impoverished a concept to suffice to clarify why people act, feel and think as they do. If linguists are not merely to describe form-based patterns and functions, they can turn to events at the boundaries of language. Interaction and understanding depend on connecting linguistic (and other forms of) knowledge with affect and our experience of embodiment. If we are concerned with how language contributes to humanity, biosemiotic phenomena matter. No alternative to mainstream linguistics can be built on transcriptions, analysis of linguistic forms or statistical relations between invariant verbal patterns and (perceived) wordings. From a distributed perspective, attention to languaging must be central to the language sciences.

The case studies show that biosemiotic processes link verbal patterns to how people deal with hierarchies of (putative) objects. People depend on anticipatory coupling between bodies, pico-scale activity and thinking – events that affect the fringe of conscious experience. So what does it imply for everyday views of words, languages and, indeed, the conception of human language? Although the distributed view retains the folk "intuition" that language is (partly) verbal it reverses standard priorities. Second-order or verbal patterns are perceived against a foreground of first-order activity: embodiment relies on (constraining) forms and functions. To the extent that populations (or linguists) agree, models can indeed be developed to *describe* languages, language-varieties and even ways of speaking. In diachronic

[56] Such phenomena appear in, for example, watching sport or during a first date: the simulation centre may be a valuable learning environment because people act under stress and thus relate closely to each other (in "facilitation-based learning").

[57] Wittgenstein 1953 [1958, Sections 179–181].

linguistics, the same logic captures aspects of language-families and even linguistic super systems. Each has a population-level complexity – it describes *abstracta* or social constraints. These draw on what people *believe* language to be and, importantly, how these beliefs impact on the practices of the human life-world. However, they do not build on psychology: descriptions of language-systems show little about mind (let alone the brain!). Rather, they show the transformational power of taking a language stance. Indeed, the case studies matter because they show how peripheral "form" can be when people depend on orienting to each other as they go about their lives while relying on how language shapes action and perception. In this way, classic linguistic models marginalise people, thinking and understanding. Rather than scrutinise linguistic action, language is treated like a machine code.[58] However, quasi-mechanistic views leave aside human embodiment or the role of activity in languages, persons and cultural achievements. They cannot show how, given its grounding in biosemiotics, language resembles dance, music, law and religion. It is metabolic activity that draws on non-local or culturally derived pattern. On the distributed view, language is a species-specific mode of cognition. For, in spite of tradition, a person's intelligence is *not* brain centred. As argued by James Hollan, Edwin Hutchins and David Kirsh,[59] human cognition has three main characteristics:

1. Its main processes are social;
2. Its main processes arise as we (together and alone) connect internal and external resources;
3. By linking these resources, we exploit the products of past events (including verbal patterns) in ways that impact on later events.

Though rooted in the first-order dynamics (speech and visible expression), these are heard as wordings: these evoke other voices, meanings and what is absent. Lest this be seen as a truism, the reader is invited to draw something that he or she can see (say, a lampshade or a tree). Yes, dear reader, please stop reading and draw! For, when a person articulates *what* they perceive, non-local patterns come to the fore. In drawing, while skilled actors use movements to conjure up form, those with less experience draw familiar shapes. They focus on, not the seen, but imaginary features: people mask any local details and their own traces. A linguistic counterpart is uttering, say, "**my name is Stephen**" or "**propositions are picture like**". Like a thing drawn, language is all too readily identified with *what* is perceived – what is *not* here (e.g., a statement or a set of words). Just as the picture masks physical details, no inscription captures precisely how wordings are (or could be) articulated. Indeed, even in looking at marks, first-order activity connects the here and the not-here. Dynamics – human interactivity – evoke voices/objects that inform experience. If an utterance is clear, or the sketch is good, it may later seem to be "the same": the maker masks circumstances, materials and what prompted the act. Polyphony and shared experience offer foreground, a second-order domain that

[58] Love 2004; Kravchenko 2007.

[59] Hollan et al. 2000.

reinforces certainties – including the folk intuitions that grant familiar views of language and the world.

Recalling the work by Hollan, Hutchins and Kirsh,[60] to ensure that the products of past events (including verbal patterns) impact on later events an observer must treat them as "unchanged" (on at least two occasions). However, language lacks any observer-product invariance. Only written language bias tempts one to see "**my name is Stephen**" (or "**vgyulasy**")[61] as other than an inscription. In consistently failing to grasp this basic fact, the written or sketched is erroneously seen as *prior* to creation of pixelated patterns. Careful consideration of how products of past events impinge on later ones leads to an unexpected finding. Dynamic activity falls under various available constraints. Although we depend on making/perceiving physical changes (articulation), we tend to think about this in terms of producing static forms (what can be said/seen). While language statics – its verbal aspect – can be extended by institutions and e-technologies, their role is not to be exaggerated. Their grounding is always in first-order or biosemiotic timescales. Patterns merely anchor functional information that has potential value for collaborating with others. As semiotic creatures, we gain skills in using this functional information. Given human forms of life, we use wordings to reach agreements in judgement and make sense of experience. Indeed, the symbiotic nature of language is necessary to making sense in a human life-world. In spite of appearances, it is non-trivial to recognise that language is distributed.

Languaging: Cognitive Biology

First-order languaging is cognitive biology in action. It arises as (non-ritualised) face-to-face activity in which wordings tame much older biosemiotic processes. Using the case studies I have shown that events depend on neither words nor intuitions about meaning but, rather, a frustration-based dynamic that sets off indexical-iconic activity. This applies even now. My meaning-making arises as I use biosemiotic skills to make inscriptional marks; you use biosemiotic activity to deal with reliably copied (and corrected) versions of these inscriptions (and their successors). To make sense for each other, to the extent that we can, we trust each other to see/mean the inscriptions as signs: however, we also rely on skills in using the visible marks to stand in for working modelling systems. These skills depend on a peculiar attitude or, alternatively, *taking a language stance.*

Both languaging and making use of a language stance are temporal processes whose salient results are verbal patterns. In construing this differently, both the general public and biolinguists blind themselves to the symbiotic nature of language. However, while the general public regard it as non-biological, biolinguists

[60] *Ibid.*

[61] Like "**my name is Stephen**", "**vgyulasy**" is an inscription; if the former invites more confident acts of construal, this depends on a reader's biosemiotic skills.

prefer to invoke a mysterious I-language.[62] By failing to see that first order language depends on available functional information, they make the simple logical error of attributing language to genes and the brain. However, the case studies show that embodied living beings, not brains, mix languaging with using a language stance. Cognitive biology serves to construe thoughts, visible marks or first-order events as wordings. People learn to attend to aspects of the world – vague thoughts, patterns of pixels or phonetic/visible gestures: we use the said to develop individual understanding. Unlike much practical action, languaging draws on a dialogical imagination or, in Maturana's sense, is connotational. Overlooking this, the public struggle to "explain" utterances like "**she always does her best with everyone**" or, indeed, "**ye::::s**". Biolinguists ignore its sub-verbal nature. However, by once its importance is acknowledged languaging is found to enact perception-action: a doctor is moved to pulse-taking as *this* gesture co-occurs with *that* syllable of *orthopaedic*. Conceptual analysis cannot explain judgements that enact forms of life. As we language, we modulate use of the language stance as we rely more and less on biosemiotic skills. Human language is thus quite unlike the languaging of other species. While based on local features and skills in real-time coordination, much depends on its non-local aspects (forms, voices and second-order constructs). For Sebeok,[63] this contrast is to be described in terms of primary and other modelling systems.

Stance-taking depends on treating utterances as utterances of something. Over time, skills in "repeating what is said" give rise to human rationality.[64] By regarding this as a developmental achievement, we come back to the challenge to mainstream views. Neurocentrism builds on the lay person's views and thus ignores history. The hypnotic effects of verbal patterns induce even linguists to picture human languages as verbal patterns and, when they turn to semantics, to trace meanings to languages that constitute arrangements of verbal patterns. They mistakenly separate language from living human beings. As a result, linguists get trapped by their models. They oppose a descriptive approach to models which purport to "explain" utterance perception and production. On the one view, language is non-biological and, on the latter, living beings use a mysterious "system". The problem has been known for 60 years – Fred Householder opposed hocus pocus linguistics to a God's truth view.[65] Wittgenstein cleared the conceptual ground for an alternative.[66] He traced language to the agreements of judgements or certainties that shape human forms of life, events that shape an individual's "natural history". In terms of this paper, language emerges in ontogeny as biosemiotic activity moves infants to action and, then, once they learn to take a language stance, they are able to develop the strategies of social

[62] Ignoring neurophysiology, Chomsky posits that a language "organ" can be detected by means of the formal analysis of verbal patterns. A cheeky response was that, if this were true, he would deserve a Nobel prize in Medicine.

[63] Sebeok 1991.

[64] Neumann and Cowley 2013.

[65] Householder 1952.

[66] Wittgenstein 1980.

actors. As discussed elsewhere,[67] the skills needed for the language stance emerge late in the first year and depend on how children learn from concerting with caregivers. Gradually, they come to hear utterances as utterances of something and thus discover the power of linguistic reflexivity. They pretend, name objects, ask questions, make paraphrases and challenge others. Eventually, they may learn to think hypothetically, focus on the said and gain an individual grasp of aspects of the world.

Given the richness of iconic-indexical understanding, the language stance is crucial in becoming a *person* who performs various roles. In modern societies, self-development is channelled, to a large extent, by exposure to many kinds of literacy. The resulting written language bias has led many to reduce language and languages verbal patterns. If this is combined with appeal to an organism, it seems natural to suppose that these are manipulated in a brain. By challenging neurocentrism the *distributed view* opens up horizons. In the first place, traditions – and forms of life – become the basis for stance taking and, thus, the establishment of meaning and truth. Further, first-order languaging is inherently biosemiotic. By hypothesis, embodiment suffices to allow people to individuate, develop relationships and enact both individual and collective lives. A population's ways of acting exert control over our individuals and, as Heidegger suggests,[68] we gain experience of *what questions grant*. Humans develop individual-collective agency. Mimicking the social nature of ants, termites, mole-rats and meerkats, humans live a unique kind of eusociality. As communities and individuals, people accord much weight to what is, can and cannot be said and done. This depends on the language stance. It has a down-side too. For one thing, it detaches us from the lived environment and our biosemiotic nature. Further, it tempts us to revere (or fetishise) what language makes salient – verbal patterns and textual or technological extensions. Though needed to bring forth the new, the language stance also favours conformity, collective blindness and ostentatious display. It biases us towards languaging that shapes tools, institutions and technologies. Often, these diminish biosemiotic modes of engagement with the world; by treating life as mediated we risk coming to experience it as less than lived.

References

Alexander, V. N. (2013). Creativity: Self-referential mistaking, not negating. *Biosemiotics, 6*(2), 253–272.

Augustyn, P. (2015). *Biology, linguistics, and the semiotic perspective on language*. Dordrecht: Springer.

Bakhtin, M. M. (1963 [1984]). *Problems of Dostoyevsky's poetics*. Minneapolis: University of Minneapolis Press.

Barbieri, M. (2003). *The organic codes*. Cambridge: Cambridge University Press.

[67] Neumann and Cowley 2013b.

[68] Heidegger 1959 [1971].

Becker, A. L. (1988). Language in particular: A lecture. In D. Tannen (Ed.), *Linguistics in context* (pp. 405–426). Norwood: Ablex.
Chomsky, N. (1957). *Syntactic structures*. The Hague: Mouton.
Chomsky, N. (1965). *Aspects of the theory of syntax*. Cambridge, MA: MIT Press.
Clark, A. (2008). *Supersizing the mind: Embodiment, action and cognitive extension*. Oxford: Oxford University Press.
Cowley, S. J. (1994). *The place of prosody in Italian conversations*. (Ph.D. dissertation). University of Cambridge.
Cowley, S. J. (2001). Prosody and pedagogy in a democratic South Africa. *Southern African Linguistics and Applied Language Studies, 19*(3–4), 179–196.
Cowley, S. J. (2007). The cognitive dynamics of distributed language. *Language Sciences, 29*(5), 575–583.
Cowley, S. J. (2010). Beyond symbols: Interaction and the enslavement principle. In J. Streeck (Ed.), *New adventures in language and interaction* (pp. 47–70). Amsterdam: John Benjamins.
Cowley, S. J. (2011). Taking a language stance. *Ecological Psychology, 23*(3), 185–209.
Cowley, S. J. (Ed.). (2011). *Distributed language*. Amsterdam: John Benjamins.
Cowley, S. J. (2012a). Cognitive distribution and human communication. In S. A. Pesina, I. R. Pulexa, J. L. Vtorušina, & L. V. Pavlova (Eds.), *Aktual'nye problemy lingvistiki i mežkul'turnoj kommunikacii: Materialy Vserossijskoj naučno-praktičeskoj konferencii s meždunarodnym učastiem, posvjaščennoj 65-letiju fakul'teta lingvistiki i perevoda* (pp. 6–16). Magnitogorsk: Magnitogorskij Gosudarstvennyj Universitet, issue V.
Cowley, S. J. (2012b). Mimesis and language: A distributed view. *Interaction Studies, 13*(1), 17–40.
Cowley, S. J. (2014). Human language and sensorimotor contingency. In J. M. Bishop & A. O. Martin (Eds.), *Contemporary sensorimotor theory* (pp. 235–251). Dordrecht: Springer.
Cowley, S. J., & Vallée-Tourangeau, F. (2013a). Systemic cognition: Human artifice in life and language. In Cowley, Vallée-Tourangeau (Eds.), 2013, pp. 255–273.
Cowley, S. J., & Vallée-Tourangeau, F. (Eds.). (2013b). *Cognition beyond the brain: Computation, interactivity and human artifice*. Dordrecht: Springer.
Cowley, S. J., Major, J. C., Steffensen, S. V., & Dinis, A. (Eds.). (2010). *Signifying bodies: Biosemiosis, interaction and health*. Braga: Portuguese Catholic University Press.
Donald, M. (1991). *The origins of the modern mind*. Cambridge, MA: Harvard University Press.
Donald, M. (2007). The slow process: A hypothetical cognitive adaptation for distributed cognitive networks. *Journal of Physiology, 101*(4–6), 214–222.
Fowler, C. A. (2010). Embodied, embedded language use. *Ecological Psychology, 22*(4), 286–303.
Harris, R. (1998). *Introduction to integrational linguistics*. Oxford: Pergamon.
Hegel, G. W. F. (1807 [1967]). *The phenomenology of mind*. New York: Harper Torch Book.
Heidegger, M. (1959 [1971]). *On the way to language*. New York: Harper & Row.
Hodges, B. H. (2007). Good prospects: Ecological and social perspectives on conforming. *Language Sciences, 19*(1), 584–604.
Hodges, B. H., Steffensen, S. V., & Martin, J. E. (2012). Caring, conversing, and realizing values: New directions in language studies. *Language Sciences, 34*(5), 499–506.
Hollan, J., Hutchins, E., & Kirsh, D. (2000). Distributed cognition: Toward a new foundation for human-computer interaction research. *ACM Transactions on Computer-Human Interaction, 2*, 174–196.
Householder, F. W. (1952). Review of methods in structural linguistics by Zellig S. Harris. *International Journal of American Linguistics, 18*, 260–268.
Hutchins, E. (1995). How a cockpit remembers its speeds. *Cognitive Science, 19*(3), 265–288.
Hutto, D. D., & Myin, E. (2012). Radicalizing enactivism: *Basic minds without content*. Cambridge, MA: MIT Press.
James, W. (1890). *The principles of psychology* (Vol. I). New York: Dover Publications.

Järvilehto, T. (2009). The theory of the organism-environment system as a basis for experimental work in psychology. *Ecological Psychology, 21*(2), 112–120.

Järvilehto, T., Nurkkala, V.-M., & Koskela, K. (2009). The role of anticipation in reading. *Pragmatics and Cognition, 17*(3), 509–526.

Jenkins, L. (2000). *Biolinguistics. Exploring the biology of language*. Cambridge: Cambridge University Press.

Kirsh, D. (1997). Interactivity and multimedia interfaces. *Instructional Science, 25*(2), 79–96.

Kirsh, D., & Maglio, P. (1994). On distinguishing epistemic from pragmatic actions. *Cognitive Science, 18*, 513–549.

Kravchenko, A. V. (2007). Essential properties of language: Why language is not a digital code. *Language Sciences, 29*(5), 650–621.

Linell, P. (2005). *The written language bias in linguistics: Its nature, origins and transformations*. London: Routledge.

Linell, P. (2009). *Rethinking language, mind and world dialogically: Interactional and contextual theories of sense making*. Charlotte: Information Age Publishing.

Lotman, J. M. (1984 [2005]). On the semiosphere. *Sign System Studies, 33*(1), 205–229.

Love, N. (2004). Cognition and the language myth. *Language Sciences, 26*(6), 525–544.

Maturana, H. R. (1978). Biology of language: The epistemology of reality. In G. Miller & E. Lenneberg (Eds.), *Psychology and biology of language and thought: Essays in honor of Eric Lenneberg* (pp. 28–62). NewYork: Academic.

Maturana, H. R. (1988). *Ontology of observing: The biological foundations of self consciousness and the physical domain of existence*. In Conference workbook: Texts in cybernetics. American Society for Cybernetics Conference, Felton, 18–23 October 1988.

Maturana, H. R., & Varela, F. J. (1992). *The tree of knowledge; the biological roots of human understanding*. Boston: Shambhala.

Neumann, M., & Cowley, S. J. (2013). Human agency and the resources of reason. In Cowley, Vallée-Tourangeau (Eds.), 2013, pp. 13–30.

Peirce, C. S. (1931). *The collected papers of Charles Sanders Peirce* (Ch. Hartshorne & P. Weiss [Eds.], vol. 5). Cambridge, MA: Harvard University Press.

Pike, K. L. (1945). *The intonation of American English*. Ann Arbor: University of Michigan Press.

Port, R. (2010). Rich memory and distributed phonology. *Language Sciences, 32*(1), 43–55.

Rączaszek-Leonardi, J., & Keslo, S. (2008). Reconciling symbolic and dynamic aspects of language: Toward a dynamic psycholinguistics. *New Ideas in Psychology, 26*(2), 193–207.

Sebeok, T. A. (1991). *A sign is just a sign*. Bloomington: Indiana University Press.

Sharov, A. A. (2010). Functional information: Towards synthesis of biosemiotics and cybernetics. *Entropy, 12*(5), 1050–1070.

Steffensen, S. V. (2011). Beyond mind: An extended ecology of languaging. In Cowley (Ed.), 2011, pp. 185–210.

Steffensen, S. V. (2013). Human interactivity: Problem solving, solution probing and verbal patterns in the wild. In Cowley, Vallée-Tourangeau (Eds.), 2013, pp. 195–221.

Steffensen, S. V., & Cowley, S. J. (2010). Signifying bodies, health and non-locality: The aftermath. In Cowley, Major, Steffensen, Dinis (Eds.), 2010, pp. 331–355.

Steffensen, S. V, Thibault, P. J., & Cowley, S. J. (2010). Living in the social meshwork: The case of health interaction. In Cowley, Major, Steffensen, Dinis (Eds.), 2010, pp. 207–244.

Stuart, S. A. J. (2010). Enkinaesthesia, biosemiotics and the ethiosphere. In Cowley, Major, Steffensen, Dinis (Eds.), 2010, pp. 305–330.

Tenney, D. C. (1988). Bug speeds pinpointed by autothrottless meanless jockeying but more thinking. *Professional pilot,* (December 1998), 96–99.

Thibault, P. J. (2011a). First-order languaging dynamics and second-order language: The distributed language view. *Ecological Psychology, 23*(3), 210–245.

Thibault, P. J. (2011b). Languaging behaviour as a catalytic process: Steps towards a theory of living language (parts I and II). *Public Journal of Semiotics, 3*(2), 2–155.

Wallot, S., & Orden, G. van (2010). Grounding language performance in the anticipatory dynamics of the body. *Ecological Psychology, 23*(3), 157–184.
Wittgenstein, L. W. (1953 [1958]). *Philosophical investigations* (2nd ed.). Oxford: Blackwell.
Wittgenstein, L. W. (1980). *On certainty*. Oxford: Blackwell.

Part III
Biosemiotics and Biolinguistics

Biolinguistics and Biosemiotics

Winfried Nöth

Abstract The paper surveys the fields of biolinguistics and biosemiotics, outlines their domains of common interest, and discusses the differences between their research programs. It shows that the two interdisciplines have developed in parallel, carry a similar academic prestige, overlap in their scope of topics of inquiry, and have common roots in the history of evolutionary and genetic biology. Whereas biolinguists restrict themselves to the study of language, biosemioticians are interested in the study of organisms in general, wherefore the biosemiotic research program is closely associated with theoretical biology. The differences are not only differences between the general and the specific but also between theoretical foundations. Biolinguistics has its foundation in Chomsky's linguistics, in particular in his "Minimalist Program", and it has a high interdisciplinary interest in neurolinguistics, genetics and the behavioral and brain sciences. Biosemiotics, by contrast, is founded on a research program that extends semiotics to a theory of sign processes in culture and nature. The paper concludes with considerations about the influence of Peirce's semiotics on Chomsky's biology of language.

Keywords Biolinguistics • Biosemiotics • Biology • Language • Semiotics • N. Chomsky • Ch.S. Peirce

Biolinguistics and biosemiotics are two sister sciences of common lineage, which overlap in their domains of research. Nevertheless, it seems as if the two siblings have taken little notice of each other until very recently.[1] Were they separated at birth or have they become alienated since then? What do they have in common? The present paper can only suggest a few answers to such questions, which are worth a research project of its own.

[1] Augustyn 2009 and 2013; Barbieri 2010; Swan 2011.

W. Nöth (✉)
Catholic University of São Paulo, São Paulo, Brazil
e-mail: noeth@uni-kassel.de

E. Velmezova et al. (eds.), *Biosemiotic Perspectives on Language and Linguistics*, Biosemiotics 13, DOI 10.1007/978-3-319-20663-9_8

What Biosemiotics and Biolinguistics Have in Common

Biolinguistics and biosemiotics show remarkable parallels in their history and prehistory. They have common roots and many common research interests.

Parallels: Beginnings, Development, and the State of the Art

Biolinguistics and biosemiotics are about the same age, have partly the same origins and they have gone through parallel developments. Both interdisciplines enjoy international prestige and have succeeded in securing a firm place among the academic disciplines at the crossroads of life sciences and humanities within a few decades.

Neither biolinguistics nor biosemiotics were heard of in the current sense before the 1960s or 1970s, respectively, although the research topics of the two interdisciplines had been studied earlier under other designations. The editors of *Biolinguistics* give the following information about the genealogy of the designation of their interdiscipline in the first issue of their journal:

> The term "biolinguistics" first appears, to our knowledge, as part of a book title, the *Handbook of Biolinguistics*, published nearly 60 years ago (Meader and Muyskens 1950). The book advocates (as the authors put it) a modern science of biolinguistics, whose practitioners "look upon language study […] as a natural science, and hence regard language as an integrated group of biological processes […]. This group seeks an explanation of all language phenomena in the functional integration of tissue and environment" (Meader and Muyskens 1950, p. 9). The term "biolinguistics" resurfaces in 1974 as part of a report on an interdisciplinary meeting on language and biology (Piattelli-Palmarini 1974), attended by Salvador Luria and Noam Chomsky, and organized by Massimo Piattelli-Palmarini, under the sponsorship of the Royaumont center for a Science of Man.[2]

Without any apparent connection to these two terminological precursors, the term *biolinguistics* also appeared in East Germany, in its German variant *Biolinguistik*, in the title of a paper by Joachim-Hermann Scharf in 1975. However, before the turn of the century, the term was rarely used, if at all. In French its first occurrence seems to be in the title of a paper by Jacques Ninio, in 1990. The term *biosemiotics* first appears in sporadic usages employed by Friedrich S. Rothschild (in 1962 and 1968), Juri Stepanov (in 1971), Marcel Florkin (in 1974), Walter A. Koch (in 1974),[3] and Rudolf Jander (in 1981) before it became the name of a research field of its own from the late 1980s onwards.[4]

The topics of both interdisciplines were first studied under other names, usually expressions with *biology* as one of their constituents and *language*, *communication*, or *semiotics* as the other. The immediate precursors of modern biolinguistics can be found in studies carried out under the designation of *biology of language*. Widely

[2] Boeckx and Grohmann 2007, p. 2.

[3] Cf. Koch 1974, p. 318.

[4] Cf. Nöth 2000, p. 254; Kull 1999.

acknowledged as a classic of modern biolinguistics are two books with titles of this kind: Eric Heinz Lenneberg's book on *The Biological Foundations of Language* of 1967, and Philip Lieberman's *Biology and Evolution of Language* of 1984. There are good reasons to consider the date of Lenneberg's book's publication, 1967, as the birth date of modern biolinguistics.[5] In fact, the term *biology of language* remained a synonym of *biolinguistics* for many years. In the immediate succession of Lenneberg and Lieberman, early studies in biolinguistics continued to be published under titles such as *The Biology of Language*[6] or *Biological Foundations of Language*.[7]

Among the immediate precursors of the term *biosemiotics* is *biocommunication*, used in the title of Günter Tembrock's remarkable book of 1971 on animal communication.[8] This is the field of research for which Thomas A. Sebeok had earlier introduced the designation of *zoosemiotics*.[9] Zoosemiotics and biosemiotics are not always sharply distinguished from one another. Logically, the former can be conceived as a branch of the latter since the study of biological sign processes evidently includes the study of animal communication. However, there is also a tendency to define biosemiotics more narrowly in contrast to zoosemiotics as the study of microbiological sign processes.[10] Nevertheless, if we take the two volumes issued under the titles of *Biosemiotica I* and *II* as paradigmatic of its scope,[11] the research field of biosemiotics comprises a very broad spectrum of topics ranging from cellular sign processes and genetic codes to the evolution of human sign use up to the emergence of verbal language.

There are also remarkable parallels between the two disciplines as to their development since their beginnings and their current state of the art. In both fields of research, we now find programmatic surveys and in depth studies of the respective research fields. In biolinguistics, the current state of the art is covered comprehensively by Lyle Jenkins, Talmy Givón, W. Tecumseh Fitch, Anna Maria Di Sciullo and Cedrik Boeckx, C. Boeckx and Kleanthes K. Grohmann.[12]

The state of the art in biosemiotics is well documented in the *Introduction to Biosemiotics* edited by Marcello Barbieri,[13] the collective volumes *Biosemiotics*,[14] *Biosemiotica I* and *II*,[15] in the works by Joachim Schult,[16] Jesper Hoffmeyer,[17] in the

[5] Jenkins 2000, p. 3; Fitch 2009, p. 284.

[6] Walker 1978; Puppel 1995.

[7] Ballmer 1982; Suchsland 1992.

[8] Tembrock 1971; cf. Sebeok 1968b.

[9] Sebeok 1968a and 1972.

[10] Cf. Nöth 2000, p. 254.

[11] Sebeok 1999; Hoffmeyer and Emmeche 1999.

[12] Cf. Jenkins 2000; Givón 2002; Fitch 2009; Di Sciullo and Boeckx 2011; Di Sciullo 2012; Boeckx and Grohmann 2013, correspondingly.

[13] Barbieri 2007a.

[14] Sebeok and Umiker-Sebeok (eds.), 1992.

[15] Sebeok 1999; Hoffmeyer and Emmeche 1999.

[16] Schult 2004.

[17] Hoffmeyer 1993 [1996] and 2008.

survey articles by Kalevi Kull[18] and M. Barbieri, and in the *Essential Readings in Biosemiotics* edited by Don Favareau.[19]

Last but not least, both interdisciplines have managed to establish themselves in academia by means of periodicals. Since 2007, *Biolinguistics* is an open access internet journal which serves as a forum for "the exploration of issues related to theory formation within the biolinguistic program of generative grammar as well as results drawn from experimental studies in psycho- and neurolinguistics or cognition at large",[20] and since 2005 biosemioticians have had their own periodical, first, the *Journal of Biosemiotics* and since 2008, *Biosemiotics*.

What is the scope of biolinguistics and of biosemiotics, respectively? Let us first consider the recurrent topics of research and the interdisciplinary connections of the two research fields (in sections "Parallels: Beginnings, Development, and the State of the Art" and "The Scope of Biosemiotics"). Both research fields are evidently concerned with biological foundations, determinants, or roots of their respective domains, one domain being language, the other consisting of sign processes and sign systems in general. Since language is a sign system and semiotics is the study of signs and systems of signs, biolinguistics should be a branch of biosemiotics. In reality, however, there is only an overlap between the two research fields and most publications in biolinguistics are not based on biosemiotic premises.[21] The two research fields are not **a priori** rigidly defined in their extent, but the topics subsumed under each of them in the publications that carry the names of the respective disciplines permit the following outline of the two research fields.

The Scope of Biosemiotics

Recurrent topics of biosemiotics, as it presents itself in the papers of *Biosemiotica I* and *II*, *Biosemiotics*, the *Introduction to Biosemiotics* edited by Barbieri, Favareau's *Essential Readings*, and elsewhere, deal with

- microbiological and molecular sign processes,[22]
- cellular semiosis (sign processes within and between cells[23]),
- processes of immunological semiosis,[24]

[18] Kull 1999.

[19] Favareau 2010.

[20] As formulated online in the journal's "Editorial Policies" (http://tinyurl.com/k47h8gw; website consulted in September 2014).

[21] Sebeok 1999.

[22] Kawade 1996.

[23] Florkin 1974; Sercarz et al. 1988; Barbieri 2003 and 2007; Bruni 2007.

[24] Eco 1988; Prodi 1988a, b; Sercarz et al. 1988.

- endosemiosis vs exosemiosis, i.e., sign processes that take place within organisms and between organisms,[25]
- genetics, "the grammar of genes", in particular "how the genetic code resembles the linguistic code",[26]
- neurosemiotics,[27]
- phytosemiosis (sign processing by and in plants[28]) and semiosis in symbiosis, parasitism, and mimicry,[29]
- the semiotics of nature in general[30] and ecological aspects of biosemiosis in particular,[31]
- physical bases of biosemiotic processes[32] and the role of semiosis in the emergence of life from lifeless matter,[33]
- biological evolution,[34] communication,[35] and the origins of semiosis in general,[36]
- evolutionary roots of language,[37] biosemiotics and biolinguistics,[38] language and life,[39]
- evolutionary roots of culture, literature, and the arts[40] and the "poetics of nature",[41]
- artificial life,[42]
- transdisciplinary connections with cybernetics,[43] information theory,[44] and the theory of self-organizing systems,[45]

[25] This is a distinction first drawn by Th.A. Sebeok (1972, p. 163), Uexküll et al. (1993), J. Hoffmeyer (2008, pp. 213–264).

[26] As the subtitle of López-Garcia 2005 puts it; Pollack 1994; Barbieri 2003.

[27] Nöth 2000, p. 259; Roepstorff 2004; Kull et al. 2008, p. 50.

[28] Krampen 1981 and 1992; Witzany and Baluška 2012.

[29] Nöth 2012b.

[30] Nöth and Kull 2001; Hoffmeyer 2005 and 2010; Nöth 2008.

[31] Nöth 1998; Nielsen 2007.

[32] Pattee 1997 and 2001.

[33] Hoffmeyer 1993 [1996] and 2008; Weber 2009.

[34] Kull 1992; Andrade 1999; Hoffmeyer 1993 [1996] and 2008.

[35] Sonea 1992.

[36] Nöth 1994.

[37] Koch 1991; Deacon 1997; Hoffmeyer 1993 [1996, pp. 97–112] and 2008, pp. 265–314; Katz 2008; Barbieri 2007b.

[38] Augustyn 2009; Barbieri 2010.

[39] Emmeche and Hoffmeyer 1991.

[40] Koch 1983, 1986a, b, 1989 and 1993; Coletta 1999.

[41] Weber 2011.

[42] Emmeche 1992; Etxeberria and Ibañez 1999.

[43] Brier 1999.

[44] Nöth 2012a.

[45] Vijver 1999.

- basic concepts of semiotics, such as sign, semiosis, cognition, intelligence,[46] signal, symptom,[47] meaning,[48] signification,[49] self-reference,[50] information,[51] or intentionality,[52] in light of biosemiotics.

The Scope of Biolinguistics

An authoritative definition of biolinguistics has been proposed by Noam Chomsky. Biolinguistics studies internal languages ("I-languages") in the following way: "The biolinguistic perspective regards the language faculty as an 'organ of the body', along with other cognitive systems. Adopting it, we expect to find three factors that interact to determine I-languages attained: genetic endowment (the topic of Universal Grammar), experience, and principles that are language- or even organism-independent. Research has naturally focused on I-languages and UG, the problems of descriptive and explanatory adequacy."[53]

With its programmatic restriction to how knowledge is encoded by a language organ, the scope of biolinguistics is narrower than that of biosemiotics. Which aspects of language are in its focus, and what is the interdisciplinary scope of biolinguistics? Answers to these questions can be found in Jenkins's study entitled *Biolinguistics: Exploring the Biology of Language*: "Evidence has been drawn from studies of: universal and comparative grammar (syntax, semantics, morphology, lexicon, phonetics, phonology), acquisition in children, psycholinguistic tests, perceptual studies, articulatory and acoustic phonetics, brain injuries and diseases (aphasias, aprosodias, etc.), split brains, language-isolated children (Genie), developmental disorders (Laura), electrical activity (e.g., ERPs), imaging (PET, MRI, etc.), genetic disorders (sporadic and familiar), twin studies, language in the deaf (sign language), language in the blind, linguistic savants, pidgin and creole languages".[54]

Besides linguistics proper, neurophysiology and neurolinguistics,[55] on this account, genetics and the behavioral and brain sciences are close to biolinguistics. However, Jenkins's list of the interdisciplinary connections of biolinguistics is by no means complete. Among the disciplines whose research results other biolinguists have consulted are evolutionary and comparative historical linguistics,[56]

[46] Hoffmeyer 2008.

[47] Staiano-Ross 2012.

[48] Cowley 2008.

[49] Hoffmeyer 1993 [1996, pp. 1–10] and 2000.

[50] Hoffmeyer 1993 [1996, pp. 39–51]; Goudsmit 2009.

[51] Salthe 2007; Barbieri 2012.

[52] Schult 1992; Deely 2007.

[53] Chomsky 2005, p. 1.

[54] Jenkins 2000, pp. 228–229.

[55] Ahlsén 2006.

[56] Scharf 1975; Bichakjian 1995; Hauser 1996; Larson et al. 2010; Di Sciullo and Boeckx 2011.

paleoanthropology and comparative anatomy,[57] sign language studies,[58] ethology and animal communication studies,[59] especially in apes and various bird species, cultural anthropology, cognitive science, as well as cell and molecular biology.[60] The *Cambridge Handbook of Biolinguistics* outlines the interdisciplinary scope of biolinguistics by dividing the research field into three domains, (1) language development (psycholinguistics of language acquisition and bilingualism), (2) mind, brain, behavior (cognitive and brain sciences, neurosciences, aphasiology, genetics), and (3) language evolution (including biological and human evolution in general as well as evidence from primatology and bird song studies).[61]

Overlap, Differences, and Common Ground

Although the survey of the topics and affiliations of biolinguistics and biosemiotics presented above testifies to common interests and some overlap between the two fields, differences must not be ignored. Such differences are apparent in the relevant definitions of the two interdisciplines and the premises of the research programs by the founders and leading representatives of the two interdisciplines.

The Biolinguistic Research Program

There is little disagreement about the basic assumptions and premises of the biolinguistic research program. The core belief of biolinguists, according to Fitch, is "that the human capacity to acquire and use language is an aspect of human biology, and that it can thus be profitably studied from a biological perspective".[62] His résumé that "the central research topic in biolinguistics is a characterization and explanation of the human capacity to acquire and use language"[63] is in full accordance with the much earlier outline of the goals of the same research program, which Barbara von Eckardt formulated in the form of the following questions: "What is the genetic program underlying the uniformity in human language capacity, the course of language acquisition in children, and the apparent diversity of natural languages?"[64]

57 Lieberman 1984.

58 Armstrong et al. 1995.

59 Suchsland 1992, pp. 103–142; Győri 1995.

60 Fitch 2009.

61 Boeckx and Grohmann 2013.

62 Fitch 2009, pp. 283–284.

63 *Ibid.*, p. 287.

64 Eckardt Klein 1978, p. 3.

Manfred Bierwisch specifies the biolinguistic program by substantiating the claim for the biological nature of the human language faculty with three arguments: (1) The human language faculty is species-specific, it has genetic roots, and it develops in critical phases. (2) Evidence for the biological nature of language comes from language disturbances caused by brain lesions, which is proof that the human language faculty is due to certain cerebral mechanisms. (3) Language is acquired with a remarkably incomplete, heterogeneous, and sometimes even mistaken verbal input.[65]

The founding father of the biolinguistic research program is Chomsky, whose programmatic manifestos of the biolinguistics research program are his treatises *Cartesian Linguistics* (1966) and *Language and Mind* (1968). According to Jenkins, Chomsky's guidelines for biolinguistic research can be summarized in five programmatic questions: "(1) What constitutes knowledge of language? (2) How is this knowledge acquired? (3) How is this knowledge put to use? (4) What are the relevant brain mechanisms? (5) How does this knowledge evolve (in the species)?"[66]

The Biosemiotic Research Program

"Biosemiotics can be defined as the science of signs in living systems", states K. Kull[67] succinctly, while Claus Emmeche presents the following outline of a more comprehensive research field: "*Biosemiotics* proper deals with sign processes in nature in all dimensions, including (1) the emergence of semiosis in nature, which may coincide with or anticipate the emergence of living cells; (2) the natural history of signs; (3) the 'horizontal' aspects of semiosis in the ontogeny of organisms, in plant and animal communication, and in inner sign functions in the immune and nervous systems; and (4) the semiotics of cognition and language. [...] Biosemiotics can be seen as a contribution to a general theory of evolution".[68]

In contrast to N. Chomsky, who conceives of "the study of language as part of biology",[69] and C. Boeckx and Massimo Piatelli-Palmarini, who propose that biolinguistics and linguistics be seen as two "natural sciences",[70] biosemiotics is not a branch of biology for C. Emmeche, but "it is a branch of general semiotics".[71] J. Hoffmeyer, too, rejects the view of biosemiotics as a natural science. In his opinion, biosemiotics is more closely related to a "process philosophy, which considers substance (matter) not as life's fundamental entity but rather as an intermediate stage of an emergent *process*" and which is "principally anchored in the evolutionary philosophy of Charles S. Peirce".[72]

[65] Bierwisch 1992, pp. 8–11.

[66] Jenkins 2000, pp. 1, 228.

[67] Kull 1999, p. 386.

[68] Emmeche 1992, p. 78.

[69] Chomsky 2007, p. 14.

[70] Boeckx and Piatelli-Palmarini 2005.

[71] Emmeche 1992, p. 78.

[72] Hoffmeyer 2008, p. 4.

The undisputed founder of the biosemiotic research program is Th.A. Sebeok (1920–2001), although he himself reminds us that it was Charles Morris (1901–1979), who, in his book *Signs, Language and Behavior* of 1946, had already postulated that progress in semiotics "rests finally upon the development of a genuine science of signs, and that this development can be most profitably carried on by a biological orientation".[73]

Like Chomsky, Sebeok has his background in linguistics, and like Chomsky, Sebeok is in favor of a "biological approach" to the study of signs.[74] However, Sebeok cannot subscribe to the view that biosemiotics is a branch of biology because the spheres of life and signs, Juri Lotman's biosphere and semiosphere,[75] are coextensive: "The criterial mark of all life is semiosis; and [...] semiosis presupposes life. Accordingly, the bailiwick of biology may be viewed as equivalent to 'natural semiotics' [...] or biosemiotics".[76]

Sebeok's biosemiotics is not directed towards affirming the uniqueness of the human language faculty. In the debate between the essentialists and the evolutionists, in which we find biolinguists generally taking the essentialist side, biosemioticians are usually found on the evolutionist side. The former argue that language is essentially "different from other forms of communication and that language separates humans from other species",[77] whereas the latter postulate continuity in the growth of sign processes and systems.[78] Furthermore, whereas biolinguistic research begins with the origins of language, the biosemiotic research program begins with the origins of life.[79]

The current biosemiotic view about the relation between biology and semiotics, documented in the first of eight theses of a joint manifesto of the biosemioticians K. Kull, Terrence W. Deacon, C. Emmeche, J. Hoffmeyer, and Frederik Stjernfelt, can be read as a homage to Sebeok, when its very first thesis states that "the semiosic-nonsemiosic distinction is coextensive with the life-nonlife distinction, i.e., with the domain of general biology".[80] For Sebeok, the semiotic threshold between the non-semiotic and the semiotic world is the threshold between life and lifeless things.[81] For him, this is a threshold between information and semiosis. In evolution before the origins of life we only find information (the ongoing increase of entropy), whereas semiosis begins with the origin of life.[82]

---

[73] Sebeok 2001, p. 3.

[74] Sebeok 1994, pp. 5–9.

[75] Sebeok 2001, p. 158.

[76] *Ibid.*, p. 10.

[77] Messer 1995, p. 174.

[78] Cf. Bichakjian 1995.

[79] Nöth 1994.

[80] Kull et al. 2009, p. 168.

[81] Sebeok 1986, p. 15.

[82] *Ibid.*

The eighth programmatic thesis on biosemiotics, which states that "organisms create their umwelten",[83] shows the hand of another precursor of modern biosemiotics, Jakob von Uexküll (1864–1944), the author of an ecological *Theory of Meaning*.[84] Environment, according to Uexküll,[85] is not a world exterior to the organism, but rather a subjective *Umwelt*, consisting of an inner world, as given by the organism's perception and specific operational world of practical interaction, with the environment. *Umwelt*, in this sense, is the way in which the environment is represented to the organism's mind, and it comprises the scope of the organism's operational interaction with its environment. Because of the species-specific differences between organisms, their different needs, capacities, and perspectives of their environment, there are as many kinds of *umwelt* as there are species (or even organisms). Every species and every organism can only perceive whatever the biological structure of its receptors, its brain, and its specific perspective of its environment allows it to perceive.

Further sources of inspiration of Sebeok's biosemiotics are Peirce's as well as Lotman's semiotics, in particular Lotman's theory of the semiosphere (cf. above) and of modeling systems.[86] Adapting Lotman's theory of culture as a secondary modeling system to the broader scope of a semiotics that begins in the organic world and with reference to Peirce's premise that "not only thought is in the organic world, but it develops there",[87] Sebeok postulates that modeling begins with mental representations in all organisms so that it "permeates the entire organic world".[88] *Modeling* and semiosis are hence practically synonyms, but humans model at three levels, whereas animals model only at one. Sebeok defines modeling in anima semiosis and in human cognition as *primary. Secondary modeling*, by contrast, begins with human language and its unique syntactic potential (an acknowledgement of Chomsky's biolinguistic claim), whereas *tertiary modeling* is the characteristic of "true culture".[89]

The seventh of the programmatic theses on biosemiotics states that "semiosis is a central concept for biology"[90] and thus reveals its foundation in the semiotics of Peirce (1839–1914). Sebeok gives with the following definition of semiosis: "In Peirce's usage, semiosis, or 'action of a sign', is an irreducible triadic process, comprising a relation between (1) a sign, (2) its object, and (3) its actual or potential interpretant.[91] Peirce particularly focuses upon the way that the interpretant is produced, and thus what is involved is understanding, or teleonomic (i.e., goal-directed) interpretation of a sign".[92] This is why semiotics cannot be a branch of biology and

[83] Kull et al. 2009, p. 172.

[84] Cf. Uexküll 1928 [1973] and 1940; Kull 2001.

[85] Uexküll 1940, pp. 158, 334.

[86] Cf. Sebeok and Danesi 2000.

[87] Peirce 1866–1913 [1931–1958], CP (= *Collected Papers*) 5.551, 1905 (=a manuscript of 1905).

[88] Sebeok 1994, pp. 126–127.

[89] *Ibid.*

[90] Kull et al. 2009, p. 171.

[91] Peirce 1866–1913 [1931–1958], CP 5.473, 1907.

[92] Sebeok2001, p. 17.

neither can biology be a branch of semiotics. Life and semiosis are intimately intertwined, so that "a full understanding of the dynamics of semiosis may, in the last analysis, turn out to be no less than the definition of life".[93]

Rudimentary Semiosis in the Realm of Plants

What Peirce means by semiosis as the action of a sign is not always well understood. Although semiosis has indeed to do with interpretation, Peirce does not define it as the agency of an *interpreter* or *code-maker*, as Barbieri[94] and others see it, who have adopted Morris's view of semiosis as the agency of a sign maker. With Peirce, the notion of the "action of the sign" has to be taken literally. The sign, and not some interpreter, is the agent in semiosis.[95] The agency of semiosis is one of mediations between the object represented by the sign and interpretant, which is the semiotic effect of the sign. Furthermore, processes of semiosis involve teleology or purpose, a mode of causality which begins at the microbiological level.[96] Peirce goes so far as to say that such processes involve *mind*, when he states: "The microscopist looks to see whether the motions of a little creature show any purpose. If so, there is mind there".[97]

Let us illustrate Peirce's theory of semiosis in nature further with an example of phytosemiosis. As early as 1865, Peirce had begun to reflect on affinities between the biological dissemination of plants and processes of semiosis and representation. However, these first associations between biological reproduction and semiotic mediation were still rather hypothetical. The argument was only that a plant propagating itself is "somewhat like" a medium standing for something: "Everything may be comprehended or more strictly translated by something; that is, has something which is capable of such a determination as to stand for something through this thing; somewhat as the pollen-grain of a flower stands to the ovule which it penetrates for [the] plant from which it came since it transmits its peculiarities of the latter".[98] Before 1900, Peirce could not yet affirm that plants are semiotic agents because his definitions of sign, representation, and the representamen still postulated the criterion of an interpreting mind. In 1873, Peirce argues that phenomena of

[93] Sebeok 1985, p. 69.

[94] Barbieri 2008a, b and 2010, p. 205. Barbieri does not quote Morris, but his definition of semiosis as "the production of signs" (Barbieri 2008a, p. 577) or as the result of the agency of a "code-maker" who "is the agent of semiosis, whereas signs and meanings are its instruments" is certainly in line with Morris's definition of semiosis "as a process in which something is a sign to some organism" (Morris 1946, p. 366) as far as the question of the agency in the process of semiosis is concerned (the question as to who is the agent in a sign process; cf. Nöth 2009).

[95] Cf. Nöth 2014a.

[96] Santaella 1999.

[97] Peirce 1866–1913 [1938–1958], Peirce 1982, CP 1.269, 1902.

[98] P. 333, (a manuscript of 1865).

an inanimate nature are signs only if understood as such by an interpreting mind. A weather-cock, for example, "is a sign of the direction of the wind", but usage of the word **sign** applied in this case "is an indirect one", for: "unless there be some way or other which shall connect words with the things they signify, and shall ensure their correspondence with them, they have no value as signs of those things". A thing "is not actually a sign unless it is used as such; that is unless it is interpreted to thought and addresses itself to some mind".[99] In 1897, the interpreting mind is a real interpreter. Here, "a sign, or representamen, is something which stands to somebody for something in some respect or capacity. It addresses somebody, that is, creates in the mind of that person an equivalent sign".[100]

With his extension of the concept of representamen, a quasi-synonym of "sign", in 1902, to processes in the absence of human minds, Peirce could now affirm what he had merely hypothesized in 1873, namely that the faculty for biological self-reproduction makes a sunflower a representamen: "If a sunflower, in turning towards the sun, becomes by that very act fully capable, without further condition, of reproducing a sunflower which turns in precisely corresponding ways toward the sun, and of doing so with the same reproductive power, the sunflower would become a Representamen of the sun".[101] The process of semiosis described here characterizes the sunflower as a representamen; its object is the sun, and its interpretant is the flower's offspring. The sun is the object represented by the plant because it determines it to turn towards the sun. The flower's offspring is its interpretant because the daughter-flower stands in the same relation to the sun as its mother stood and because the daughter is determined by its mother to behave in the same way as she used to behave.

Peirce thus seems to be more specific as to the agency of plants in processes of semiosis: the sunflower exemplifies the agency of a representamen representing an object and translating its message to its offspring. Nevertheless, instead of saying that these plants *are* representamens which are not signs, Peirce restricts himself to saying that there are "possibly" representamens which are not signs, and instead of concluding that the sunflower *is* a representamen of the sun he only says, in the above quote, that it "would become a Representamen of the sun". This way of avoiding an early commitment to insights which have meanwhile been advanced in biosemiotics may be read as an exemplification of Peirce's principle of fallibilism: instead of raising the new insight immediately to the status of a certainty, he foresees the necessity of further research into the questions raised by his hypothesis.

Peirce comes to the conclusion that "possibly there may be Representamens that are not Signs", not without adding the additional reservation that "thought is the chief, if not the only, mode of representation".[102] Again, Peirce still uses the modal adverb "possibly" to express some fallibilistic uncertainty as to the possibility of semiosis in a nature without thoughts of minds. In 1906, he finally attributes even

[99] Peirce 1866–1913 [1931–1958], CP 7.356, 1873.

[100] *Ibid.*, CP 2.228, 1897.

[101] *Ibid.*, CP 2.274, circa 1902.

[102] *Ibid.*, CP 2.274, circa 1902.

thought to non-human nature,[103] when he writes that "thought is not necessarily connected with a brain. It appears in the work of bees, of crystals, and throughout the purely physical world".[104]

Chomsky, Peirce, and the Biology of Language

Prisca Augustyn argues that there are three bridges able to connect Chomsky's biolinguistic program with Sebeok's biosemiotics.[105] The first two are in Chomsky's references to two topics of equal interest to biosemiotics, ethology, and the Uexküllian notion of *umwelt*. How far these occasional references can justify a significant affinity between biolinguistics and biosemiotics must be left open here.

The third bridge is Peirce's logic of *abduction*, to which Chomsky makes several explicit references in a good number of his papers in the context of reflections on language learning.[106] Is Peirce's logic of abduction a cornerstone of Chomsky's biolinguistic program? Already Chomsky's early remarks on abduction were critical. In *Language and Mind*, Chomsky expresses his "opinion" that Peirce's arguments are "not very persuasive",[107] and his interest in the logic of abduction was apparently short. Trevor Pateman explains why and when Chomsky abandoned the model of abductive language learning.[108]

The most significant incompatibilities between Peirce's semiotics and Chomsky's biolinguistic program are probably two. First, while syntax is the most important module of the human language faculty in the narrower sense according to the biolinguists, pragmatics is in the center of the Peircean semiotic approach to language.[109] Second, while biolinguists focus on genes, the human brain, and the physiology of speech production, Peircean semiotics has its focus on the agency of the sign, to which it attributes a life of its own which is not the sign maker's life.[110] The complementarity of the scopes of the two research fields should be a challenge for more intense interdisciplinary collaboration between biolinguists and biosemioticians.

By means of a provocative thought experiment, Peirce presents the following reasons why the language competence of humans cannot only be accounted for by the way human brains have developed genetic forms that are missing in the brains of other animals: "A psychologist cuts out a lobe of my brain [...] and then, when I find I cannot express myself, he says, 'You see your faculty of language was local-

[103] Cf. Santaella 1994.

[104] Peirce 1866–1913 [1931–1958], CP 6.551.

[105] Augustyn 2009 and 2013.

[106] Wirth 1993.

[107] Chomsky 1968 [2006, p. 80].

[108] Pateman 2003. For Peirce's theory of language learning, cf. Nöth 2014b.

[109] Cf. Nöth 2011.

[110] Cf. Nöth 2009 and 2014a.

ized in that lobe.' No doubt it was; and so, if he had filched my inkstand, I should not have been able to continue my discussion until I had got another. Yea, the very thoughts would not come to me. So my faculty of discussion is equally localized in my inkstand. It is localization in a sense in which a thing may be in two places at once".[111] Peirce's argument is that the human language faculty is not embodied in brains and tongues alone. The *umwelt* – here exemplified by the writer's inkstand – and external signs play an equally important role. Ideas are not produced by brains, and thought is not only limited to inner thought. It lives on in external embodiments in which it continues to act in semiosis.

References

Ahlsén, E. (2006). *Introduction to neurolinguistics*. Amsterdam: John Benjamins.

Andrade, L. E. (1999). Natural selection and Maxwell's demons. *Semiotica, 127*(1–4), 133–149.

Armstrong, D. F., Stokoe, W. C., & Wilcox, S. E. (Eds.). (1995). *Gesture and the nature of language*. Cambridge: Cambridge University Press.

Augustyn, P. (2009). Uexküll, Peirce, and other affinities between biosemiotics and biolinguistics. *Biosemiotics, 2*(1), 1–17.

Augustyn, P. (2013). What connects biolinguistics and biosemiotics? *Biolinguistics, 7*, 96–111.

Ballmer, T. T. (1982). *Biological foundations of linguistics communication: Towards a biocybernetics of language*. Amsterdam: John Benjamins.

Barbieri, M. (2003). *The organic codes: An introduction to semantic biology*. Cambridge: Cambridge University Press.

Barbieri, M. (2007a). Is the cell a semiotic system? In Barbieri (Ed.), 2007, pp. 179–207.

Barbieri, M. (Ed.). (2007b). *Introduction to biosemiotics: The New biological synthesis*. Berlin: Springer.

Barbieri, M. (2008a). Biosemiotics: A new understanding of life. *Naturwissenschaften, 95*(7), 577–599.

Barbieri, M. (2008b). The code model of semiosis: The first steps toward a scientific biosemiotics. *American Journal of Semiotics, 24*(1–3), 23–37.

Barbieri, M. (2010). On the origin of language: A bridge between biolinguistics and biosemiotics. *Biosemiotics, 3*(2), 201–223.

Barbieri, M. (2012). What is information? *Biosemiotics, 5*(2), 147–152.

Bichakjian, B. H. (1995). Essentialism in language: A convenient but fallacious premise. In Puppel (Ed.), 1995, pp. 33–60.

Bierwisch, M. (1992). Probleme der biologischen Erklärung natürlicher Sprache. In Suchsland (Ed.), 1992, pp. 7–45.

Boeckx, C., & Grohmann, K. K. (2007). The biolinguistics manifesto. *Biolinguistics, 1*, 1–8.

Boeckx, C., & Grohmann, K. K. (Eds.). (2013). *The Cambridge handbook of biolinguistics*. Cambridge: University Press.

Boeckx, C., & Piatelli-Palmarini, M. (2005). Language as a natural object – Linguistics as a natural science. *The Linguistic Review, 22*, 362–379.

Brier, S. (1999). Biosemiotics and the foundation of cybersemiotics. *Semiotica, 127*(1–4), 169–198.

Bruni, L. E. (2007). Cellular semiosis and signal transduction. In Barbieri (Ed.), 2007, pp. 365–408.

[111] Peirce 1866–1913 [1931–1958], CP 7.366, 1902.

Chomsky, N. (1966 [2009]). *Cartesian Linguistics: A chapter in the history of rationalist thought* (3rd ed.). Cambridge: Cambridge University Press.
Chomsky, N. (1968 [2006]). *Language and mind* (3rd ed.). Cambridge: University Press.
Chomsky, N. (2005). Three factors in language design. *Linguistic Inquiry, 36*(1), 1–22.
Chomsky, N. (2007). Of minds and language. *Biolinguistics, 1*, 9–27.
Coletta, W. J. (1999). Literary biosemiotics. *Semiotica, 127*(1–4), 239–271.
Cowley, S. J. (2008). Meaning in nature: Organic manufacture? *Biosemiotics, 1*(1), 85–98.
Deacon, T. W. (1997). *The symbolic species*. New York: Norton.
Deely, J. N. (2007). *Intentionality and semiotics*. Scranton: University of Scranton Press.
Di Sciullo, A. M. (Ed.). (2012). *Towards a biolinguistics understanding of grammar*. Amsterdam: John Benjamins.
Di Sciullo, A. M., & Boeckx, C. (Eds.). (2011). *The biolinguistic enterprise: New perspectives on the evolution and nature of the human language faculty*. Oxford: Oxford University Press.
Eckardt Klein, B. von (1978). What is the biology of language? In E. Walker (Ed.), *Explorations in the biology of language* (pp. 1–14). Hassocks: Harvester Press.
Eco, U. (1988). On semiotics and immunology. In Sercarz, Celada, Mitchison, Tada (Eds.), 1988, pp. 3–15.
Emmeche, C. (1992). Modeling life: A note on the semiotics of emergence and computation in artificial and natural living systems. In Sebeok, Umiker-Sebeok (Eds.), 1992, pp. 77–99.
Emmeche, C., & Hoffmeyer, J. (1991). From language to nature: The semiotic metaphor in biology. *Semiotica, 84*(1–2), 1–42.
Etxeberria, A., & Ibañez, J. (1999). Semiotics of the artificial: The "self" of self-producing systems in cellular automata. *Semiotica, 127*(1–4), 295–320.
Favareau, D. (Ed.). (2010). *Essential readings in biosemiotics: Anthology and commentary*. Berlin: Springer.
Fitch, W. T. (2009). Prolegomena to a future science of biolinguistics. *Biolinguistics, 3*(4), 283–320.
Florkin, M. (1974). Concepts of molecular biosemiotics and of molecular evolution. In M. Florkin & E. H. Stotz (Eds.), *Biochemistry* (Comprehensive Biochemistry, Vol. 29A, pp. 1–124). Amsterdam: Elsevier.
Givón, T. (2002). *Bio-linguistics: The Santa Barbara lectures*. Amsterdam: John Benjamins.
Goudsmit, L. A. (2009). Sense and self-referentiality in living beings. *Biosemiotics, 2*(1), 39–46.
Győri, G. (1995). Animal communication and human language: Searching for their evolutionary relationship. In Puppel (Ed.), 1995, pp. 99–126.
Hauser, M. D. (1996). *The evolution of communication*. Cambridge, MA: MIT Press.
Hoffmeyer, J. (1993 [1996]). *Signs of meaning in the Universe*. Bloomington: Indiana University Press.
Hoffmeyer, J. (2000). The biology of signification. *Perspectives in Biology and Medicine, 43*(2), 253–268.
Hoffmeyer, J. (2005). Code-duality and the semiotics of nature. *Journal of Biosemiotics, 1*(1), 27–64.
Hoffmeyer, J. (2008). *Biosemiotics: An examination of the signs of life and the life of signs*. Scranton: Scranton University Press.
Hoffmeyer, J. (2010). Semiotics of nature. In P. Cobley (Ed.), *The Routledge companion to semiotics* (pp. 29–42). London: Routledge.
Hoffmeyer, J., & Emmeche, C. (Eds.). (1999). *Biosemiotica II*. Berlin: Mouton de Gruyter [*Semiotica*, 127.1–4, pp. 131–655].
Jander, R. (1981). General semiotics and biosemiotics. In R. T. de George (Ed.), *Semiotic themes* (pp. 225–250). Lawrence: University of Kansas Publications.
Jenkins, L. (2000). *Biolinguistics: Exploring the biology of language*. Cambridge: University Press.
Katz, G. (2008). The hypothesis of a genetic protolanguage and epistemological investigation. *Biosemiotics, 1*(1), 57–73.
Kawade, Y. (1996). Molecular biosemiotics. *Semiotica, 111*, 195–215.

Koch, W. A. (1974). Semiotik und Sprachgenese. In W. A. Koch (Ed.), *Perspektiven der Linguistik II* (pp. 312–346). Stuttgart: Kröner.
Koch, W. A. (1983). *Poetry and science: Semiogenetical twins*. Tübingen: Narr.
Koch, W. A. (1986a). *Genes vs. Memes*. Bochum: Brockmeyer.
Koch, W. A. (1986b). *Evolutionary cultural semiotics*. Bochum: Brockmeyer.
Koch, W. A. (Ed.). (1989). *Evolution of culture*. Bochum: Brockmeyer.
Koch, W. A. (1991). *Language in the upper Pleistocene*. Bochum: Brockmeyer.
Koch, W. A. (1993). *The biology of literature*. Bochum: Brockmeyer.
Krampen, M. (1981). Phytosemiotics. *Semiotica, 36*, 187–209.
Krampen, M. (1992). Phytosemiotics revisited. In Sebeok, Umiker-Sebeok (Eds.), 1992, pp. 213–220.
Kull, K. (1992). Evolution and semiosis. In Sebeok, Umiker-Sebeok (Eds.), 1992, pp. 221–234.
Kull, K. (1999). Biosemiotics in the twentieth century. *Semiotica, 127*(1–4), 385–414.
Kull, K. (2001). Jakob von Uexküll: An introduction. *Semiotica, 134*(1–4), 1–50.
Kull, K., Emmeche, C., & Favareau, D. (2008). Biosemiotic questions. *Biosemiotics, 1*(1), 41–55.
Kull, K., Deacon, T., Emmeche, C., Hoffmeyer, J., & Stjernfelt, F. (2009). Theses on biosemiotics: Prolegomena to a theoretical biology. *Biological Theory, 4*(2), 167–173.
Larson, R. K., Déprez, V., & Yamakido, H. (Eds.). (2010). *The evolution of human language: Biolinguistic perspectives*. Cambridge: Cambridge University Press.
Lenneberg, E. H. (1967). *Biological foundations of language*. New York: Wiley.
Lieberman, P. (1984). *The biology and evolution of language*. Cambridge, MA: Harvard University Press.
López-Garcia, Á. (2005). *The grammar of genes: How the genetic code resembles the linguistic code*. Frankfurt am Main: Peter Lang.
Meader, C. L., & Muyskens, J. H. (1950). *Handbook of biolinguistics*. Toledo: H.C. Weller.
Messer, D. J. (1995). Language acquisition and the essentialist-evolutionist debate. In Puppel (Ed.), 1995, pp. 173–194.
Morris, C. (1946 [1971]). Signs, language, and behavior. In C. Morris, *Writings on the General Theory of Signs* (pp. 73–398). The Hague: Mouton.
Nielsen, S. N. (2007). Towards an ecosystem semiotics: Some basic aspects for a new research program. *Ecological Complexity, 4*, 93–101.
Ninio, J. (1990). Petit inventaire biolinguistique. *Cahiers de Praxématique, 15*, 77–86.
Nöth, W. (Ed.). (1994). *Origins of semiosis: Sign evolution in nature and culture*. Berlin: Mouton de Gruyter.
Nöth, W. (1998). Ecosemiotics. *Sign Systems Studies, 26*, 332–343.
Nöth, W. (2000). *Handbuch der Semiotik* (2nd rev. ed.). Stuttgart: Metzler.
Nöth, W. (2008). Natural signs from a synechist perspective. In K.-M. Hingst & M. Liatsi (Eds.), *Pragmata. Festschrift für Klaus Oehler zum 80. Geburtstag* (pp. 130–140). Tübingen: Narr.
Nöth, W. (2009). On the instrumentality and semiotic agency of signs, tools, and intelligent machines. *Cybernetics and Human Knowing, 16*(3–4), 11–36.
Nöth, W. (2011). Semiotic foundations of pragmatics. In W. Bublitz & N. R. Norrick (Eds.), *Foundations of pragmatics* (pp. 167–202). Berlin: Mouton de Gruyter.
Nöth, W. (2012a). Charles S. Peirce's theory of information: A theory of the growth of symbols and of knowledge. *Cybernetics and Human Knowing, 19*(1–2), 99–123.
Nöth, W. (2012b). Signs from the life of organisms, species, languages, and the media. In T. Maran, K. Lindström, R. Magnus, & M. Tõnnessen (Eds.), *Semiotics in the wild: Essays in honour of Kalevi Kull on the occasion of his 60th birthday* (pp. 123–130). Tartu: Tartu University Press.
Nöth, W. (2014a). The life of symbols and other legisigns: More than a mere metaphor? In V. Romanini & E. Fernández (Eds.), *Peirce and biosemiotics* (pp. 171–182). Heidelberg: Springer.
Nöth, W. (2014b). Signs as educators: Peircean insights. In I. Semetsky & A. Stables (Eds.), *Pedagogy and edusemiotics* (pp. 7–18). Rotterdam: Sense.
Nöth, W., & Kull, K. (Eds.). (2001). Semiotics of nature. *Sign Systems Studies, 29*(1), 1–376.

Pateman, T. (2003). What I tell my students about Noam Chomsky and Seymour Papert (www.selectedworks.co.uk/chomskypapert.html; website consulted in March 2013).
Pattee, H. H. (1997). The physics of symbols and the evolution of semiotic controls. In M. Coombs & M. Sulcoski (Eds.), *Control mechanisms for complex systems* (pp. 9–25). Albuquerque: University of New Mexico Press.
Pattee, H. H. (2001). The physics of symbols: Bridging the epistemic cut. *BioSystems, 60*, 5–21.
Peirce, C. S. (1866–1913 [1931–1958]). *The collected papers of Charles Sanders Peirce* (Ch. Hartshorne, P. Weiss & A. W. Burks [Eds.]), 8 vols. Cambridge, MA: Harvard University Press.
Peirce, C. S. (1982). In M. H. Fisch (Ed.), *Writings of Charles S. Peirce: A chronological edition* (Vol. 1). Bloomington: Indiana University Press.
Piattelli-Palmarini, M. (1974). A debate on bio-linguistics, *Centre Royaumont pour une science de l'homme report* [Conference held at Endicott House. Dedham, 20–21 May 1974].
Pollack, R. (1994). *Signs of life: The language and meanings of DNA*. London: Viking.
Prodi, G. (1988a). Material bases for signification. *Semiotica, 69*, 191–214.
Prodi, G. (1988b). Signs and codes in immunology. In Sercarz, Celada, Mitchison, Tada (Eds.), 1988, pp. 53–56.
Puppel, S. (1995). *The biology of language*. Amsterdam: John Benjamins.
Roepstorff, A. (2004). Cellular neurosemiotics. In J. Schult (Ed.), *Biosemiotik – Praktische Anwendungen und Konsequenzen für die Einzelwissenschaften* (pp. 133–154). Berlin: Verlag für Wissenschaft und Bildung.
Rothschild, F. S. (1962). Laws of symbolic mediation in the dynamics of self and personality. *Annals of New York Academy of Sciences, 96*, 774–784.
Rothschild, F. S. (1968). Concepts and methods of biosemiotic. *Scripta Hierosolymitana, 20*, 163–194.
Salthe, S. N. (2007). What is the scope of biosemiotics? Information in living systems. In Barbieri (Ed.), 2007, pp. 133–148.
Santaella, L. (1994). Peirce's broad concept of mind. *S: European Journal for Semiotic Studies, 6*, 399–411.
Santaella, L. (1999). A new causality for the understanding of the living. *Semiotica, 127*(1–4), 497–518.
Scharf, J.-H. (1975). Bemerkenswertes zur Geschichte der Biolinguistik und des sogenannten Sprach-Darwinismus. *Nova Acta Leopoldina, 43*(218), 323–341.
Schult, J. (1992). Species, signs, and intentionality. In Sebeok, Umiker-Sebeok (Eds.), 1992, pp. 317–332.
Schult, J. (Ed.). (2004). *Biosemiotik – Praktische Anwendungen und Konsequenzen für die Einzelwissenschaften*. Berlin: Verlag für Wissenschaft und Bildung.
Sebeok, T. A. (1968a). Zoosemiotics. *American Speech, 43*(2), 142–144.
Sebeok, T. A. (Ed.). (1968b). *Animal communication*. Bloomington: Indiana University Press.
Sebeok, T. A. (1972). *Perspectives in zoosemiotics*. The Hague: Mouton.
Sebeok, T. A. (1985). *Contributions to the doctrine of signs*. Lanham: University Press of America.
Sebeok, T. A. (1986). *I think I am a verb*. New York: Plenum.
Sebeok, T. A. (1994). *Signs: An introduction to semiotics*. Toronto: University of Toronto Press.
Sebeok, T. A. (Ed.). (1999). *Biosemiotica I*. Berlin: Mouton de Gruyter [*Semiotica*, 127.1–4, pp. 1–132].
Sebeok, T. A. (2001). *Global semiotics*. Bloomington: Indiana University Press.
Sebeok, T. A., & Danesi, M. (2000). *The forms of meaning: Modeling systems theory and semiotic analysis*. Berlin: Mouton de Gruyter.
Sebeok, T. A., & Umiker-Sebeok, J. (Eds.). (1992). *Biosemiotics*. Berlin: Mouton de Gruyter.
Sercarz, E. E., Franco, C., Mitchison, N. A., & Tada, T. (Eds.). (1988). *The semiotics of cellular communication in the immune system*. Berlin: Springer.
Sonea, S. (1992). Half of the living world was unable to communicate for about one billion years. In Sebeok, Umiker-Sebeok (Eds.), 1992, pp. 375–392.
Staiano-Ross, K. (2012). The symptom. *Biosemiotics, 5*(1), 33–45.

Stepanov, J. S. (1971). *Semiotika* Moskva: Nauka. [Semiotics].
Suchsland, P. (Ed.). (1992). *Biologische und soziale Grundlagen der Sprache*. Tübingen: Niemeyer.
Swan, L. S. (2011). Signs pointing in a new direction: A biosemiotic framework for biolinguistics. *Biolinguistics, 5*(4), 366–369.
Tembrock, G. (1971). *Biokommunikation*, 2 vols. Berlin: Akademie-Verlag.
Uexküll, J. von (1928 [1973]). *Theoretische Biologie*. Frankfurt am Main: Suhrkamp.
Uexküll, J. von (1940). *Bedeutungslehre*. Leipzig: Barth.
Uexküll, T. von, Geigges, W., & Hermann, J. (1993). Endosemiosis. *Semiotica, 96*(1–2), 5–51.
Vijver, G. van de (1999). Psychic closure: A prerequisite for the recognition of the sign-function? *Semiotica, 127*(1–4), 613–630.
Walker, E. (Ed.). (1978). *Explorations in the biology of language*. Hassocks: Harvester Press.
Weber, B. (2009). On the emergence of living systems. *Biosemiotics, 2*(3), 343–359.
Weber, A. (2011). The book of desire: Toward a biological poetics. *Biosemiotics, 4*(2), 39–46.
Wirth, U. (1993). Die abduktive Wende in der Linguistik. *Kodikas/Code, 16*, 289–301.
Witzany, G., & Baluška, F. (2012). *Biocommunication of plants*. Heidelberg: Springer.

Biology, Linguistics, and the Semiotic Perspective on Language

Prisca Augustyn

Abstract This paper explores the relationship between biology and linguistics by tracing the corresponding parallel developments of phylogenetic thinking in the nineteenth century. The conception of *languages* and *species* as historical entities developed from a philosophical current that originated with philosophies of nature deriving predominantly from Kant, Goethe and Schelling. Following the epistemological and metaphysical trajectory of German *Naturphilosophie*, this paper explains how J. von Uexküll carried this biosemiotic approach to biology and language into the twentieth century while linguistics aligned its methods with psychology and other social sciences. Sebeok's contributions to linguistics and semiotics throughout the twentieth century were characterized by his commitment to biosemiotics, maintaining a close connection to biology and the anti-psychologism associated with the semiotic perspective on language. In several key aspects, Sebeok's views are shown to be compatible with Chomsky's biolinguistics.

Keywords Linguistics • Biology • J. von Uexküll • Th.A. Sebeok • Biosemiotics • N. Chomsky • Biolinguistics

The Nineteenth Century Construction of *Languages* and *Species* as Historical Entities and the Philosophical Origin of Biosemiotics

Most linguists today assume that the tree-diagram as a diagrammatic representation of descent relationships between languages grew in the field of biology with Charles Darwin's theory of evolution.[1] Many would be surprised to learn that it was actually the other way around. Nineteenth century linguists and biologists were connected by a shared attitude towards the living world that influenced the methods they chose to study it. Darwin noted in 1871 that the "formation of different languages and of

[1] Darwin 1859.

P. Augustyn (✉)
Florida Atlantic University, Boca Raton, FL, USA
e-mail: augustyn@fau.edu

E. Velmezova et al. (eds.), *Biosemiotic Perspectives on Language and Linguistics*, Biosemiotics 13, DOI 10.1007/978-3-319-20663-9_9

distinct species, are *curiously parallel*… We find in distinct languages striking homologies due to community of descent, and analogies due to a similar process of formation".[2] By then tree diagrams had been used in linguistics and philology for over half a century.[3]

Before tree-diagrams were introduced in biology, a hierarchical system dominated in the field of animal classification, represented in the Aristotelian *scala naturae.* Up until the seventeenth century the belief persisted that organisms could arise through spontaneous generation from nonliving matter. Similar beliefs existed about linguistic diversity through myths such as the *Tower of Babel*, where the diversification of languages is represented as a spontaneous divine intervention in human affairs. As a result, questions about species lineages and historical relationships didn't arise for a long time. But even Ancient Greek philosophers, upon taking a closer look at historical records of their language, noticed *growth* and *change*. Socrates (469 BC–399 BC) was worried about the *decline* of Greek since Homer (730 BC) in the same way the eighteenth century English grammarians Robert Lowth[4] and Lindley Murray[5] were worried about the *degeneration* of English since William Shakespeare (1564–1616).[6] Today there is a similar concern about the influence of digital media on the "low standards" in grammar and style among younger generations and the negative effects on the assumed "integrity" of our languages.

While language purists and popular belief still maintain illusions of fixed standards today, nineteenth century linguists focused on *growth* and *change* and devoted their energy to documenting and analyzing the historical evolution of languages. Ancient texts became the *fossil record* of dead languages (like Latin, Greek, Sanskrit, or Gothic) that were unequivocally understood to be the *progenitors* of modern languages such as German or English. The idea that one language evolved from another (problematic as it may be upon closer examination) was acceptable and for the most part uncontroversial.

It was understandably more acceptable to think of languages as *evolving* from one into another instead of being *created* **ex nihilo** than it was for living organisms. After all, the eighteenth and nineteenth centuries (at least in Germany) were a time when professors were regularly expelled from their university appointments for atheism. However, those who are familiar with the *natural organicism* of Johann Wolfgang von Goethe know that evolutionary thought existed long before Darwin. Goethe's incessant search for the *Urpflanze* as a precursor to Darwin's work on plant evolution in *The Origin of Species* (1859) attests to the acceptability of phylogenetic thinking about plants preceding phylogenetic thinking about animals. Phylogenetic thinking about languages was uncontroversial even in the nineteenth century.

---

[2] Darwin 1871, pp. 89–90.

[3] Where several manuscripts existed that were copies of an older manuscript, the *stemma* diagrams helped philologists establish a record of which manuscript came first and provided the basis of other, later manuscripts.

[4] Lowth 1762.

[5] Murray 1795.

[6] Atkinson and Gray 2005.

The pioneer of linguistic typology, Wilhelm von Humboldt (1767–1835), had proposed a predominantly hierarchical classification of languages. The perception of inflectional languages (such as the Germanic and Semitic languages) as the most "perfect" was going to last throughout the nineteenth century; and, unfortunately, parallel hypotheses were made about the peoples who spoke them.

The intellectual profiles of Wilhelm von Humboldt, the linguist, and his younger brother, Alexander von Humboldt (1769–1859), the biologist/naturalist exemplify the close relationship between the study of language and other phenomena of living things. Their attitudes towards *life* and the scientific study of *all that lives* unequivocally derived from the philosophical current that is the bedrock of biosemiotic thought. Like many of their contemporary biologists and linguists, their metaphysics, scientific attitudes and methods were anchored in the *Romantic Biology* or *natural organicism* of Immanuel Kant, Johann Wolfgang von Goethe, and Friedrich Wilhelm Joseph von Schelling that sees nature as a creative force and creation at once, where *perfect form* is found in plants and animals as in poetry or art. The nineteenth century biologist was an artist as much as the artist was a naturalist. Nature was seen as artful as much as art was considered a part of nature. Historian of science Robert J. Richards explains that

> [Alexander von Humboldt] believed that the unity of form underlying the diverse profusion of life, [...], could be expressed in biogeographical calculations, with which even his casual essays bulged. Fat numbers alone, though, could not adequately portray the face of nature – only the art of narrative, the poetry of description, could convey to discriminating sensibilities her active, vital features. Behind Humboldt's declarations about the obligation of the naturalist to convey a certain feeling for nature lay the epistemological and metaphysical structures erected by Kant, Schelling, and Goethe.[7]

The complexity of Nature was believed to reveal itself only to the scientist/artist or scientist/poet whose subjective experience is capable of discovering, articulating and representing its underlying principles and capture its form through his own creativity. The ability to perceive and appreciate nature was integral to its scientific discovery as well as its artistic representation. The Humboldt brothers were not the only close relationship among important practitioners of linguistics and biology whose *Naturphilosophie* goes back directly to the metaphysics of Kant, Goethe, and Schelling; but they were celebrity practitioners of a *Romantic biology* that influenced linguists and biologists throughout the nineteenth century and, as we shall see, constitutes the prehistory of the semiotic perspective on language.

It is well known that Charles Darwin and Ernst Haeckel, an important articulator of Darwin's theory of evolution in the German-speaking world, were dedicated readers of the work of A. von Humboldt.[8] That all nineteenth century intellectuals read Kant and Goethe, at least in the German-speaking world, is uncontroversial. But Darwin also exchanged ideas directly with no lesser than the author of the *Stammbaumtheorie*, the Jena linguist August Schleicher (1821–1868). Schleicher also happened to be – not surprisingly for a nineteenth century intellectual – a passionate

[7] Richards 2004, p. 32.

[8] Cf. Richards 2002 and 2008.

botanist, and also a close friend of Haeckel in Jena.[9] That Schleicher considered *languages to be living organisms* is not a mere analogy or metaphor, but instead represents a deep conviction that languages are part of the evolution of life.

It is well known that Schleicher and Haeckel, fellow progressive thinkers and naturalists, enjoyed exercising together at the Jena *Turnverein* and compared notes on how to best represent their intellectual work. More importantly, their theories also had a shared philosophical integrity that characterized a *Romantic biology* and a *Romantic linguistics*.

This integrated view of the living world is expressed in a famous quote attributed to Goethe that served as the epilogue for Haeckel's *Generelle Morphologie der Organismen* (1866). It might have served equally well for Schleicher's *Über die Bedeutung der Sprache für die Naturgeschichte des Menschen* (1865): "There is in nature an eternal life, becoming, and movement. [Nature] alters herself eternally, and is never still. [Nature] has no conception of stasis, and can only curse it. [Nature] is strong, her step is measured, her laws unalterable. [Nature] has thought and constantly reflects – not as a human being, but as nature. [Nature] appears to everyone in a particular form. [Nature] hides herself in a thousand names and terms, and is always the same".[10] Jakob von Uexküll (1864–1944) carried this approach to biology into the twentieth century, when he described *nature as a composer listening to her own composition.*[11] In fact, Uexküll's terminology borrowed from musicology that can also be found in the work of Karl Ernst von Baer and many others, characterizes the scientist/artist as uniquely capable of the profound perception and representation or articulation of the living world. Uexküll used the term *Merkling*[12] for a person of high perceptive and articulatory capabilities. The equivalent of the biologist/artist/poet just like the Romantic ideal that nineteenth century intellectuals saw personified in Goethe.

This holistic view of natural phenomena derived from Goethe captures the spirit of nineteenth century biology and linguistics. Uexküll unequivocally understood linguistics to be part of biology when he wrote to a linguist friend that he was on the right path "towards making [linguistics] a biological science".[13] The distinct metaphysics that connected Haeckel and Schleicher, and also informed Uexküll's biology in the early twentieth century, approached natural phenomena neither from a vitalistic nor a mechanistic perspective. This approach is characteristic of the philosophy of nature that sprang from the early Romanticism of these Jena intellectuals. Upon reading Darwin's *Origin of Species*, Schleicher declared in his *Die Darwinsche Theorie und ihre Bedeutung für die Sprachwissenschaft*:

> The dualism, which one conceives as the opposition of mind and nature, content and form, being and appearance, or however one wishes to indicate it – this dualism is for the natural scientific perspective of our day a completely unacceptable position. For the natural scien-

[9] Cf. Richards 2008.

[10] Goethe, quoted in Richards 2008, p. 111.

[11] Cf. Uexküll 1992.

[12] Apparently it was Fedi Ditmar who invented the term according to Uexküll (1957, p. 61).

[13] Cf. Kull 2001, p. 3.

tific perspective there is no matter without mind [Geist] (that is, without the necessary power determining the matter), nor any mind without matter. Rather there is neither mind nor matter in the usual sense. There is only one thing that is both simultaneously. To accuse this opinion, which rests on observation, of materialism is as perverse as charging it of spiritualism.[14]

The comparative method practiced by Schleicher and his fellow linguists has direct parallels in the work of nineteenth century biologists identifying homologies in the physiologies of sponges, siphonophores and other organisms that exist in great diversity to establish their relatedness. Indeed, the comparative physiology of sponges contributed to Haeckel's progress in providing evidence for Darwin's theory. Just as Haeckel compared the body structures of radiolarians and siphonophores, Schleicher and his fellow linguists compared texts in related languages. Manuscripts that were copies of the same text in different but related language variants served as the *fossil fragments* of language.[15]

Schleicher believed that the developmental history of languages was a main feature of the development of human beings. More particularly, he was convinced that because there are considerably more linguistic fossils than geological fossils, historical language data could provide valuable evidence for the theory of evolution in general. Schleicher was certain that the same processes of competition of languages, the extinction of forms, and the development of more complex languages out of simpler roots suggested mutual confirmation of the basic processes governing such historical entities as species and languages. Finally, since the various language groups were believed to have descended from more primitive forms, Schleicher suggested that language provides analogous evidence that more advanced species descended from simpler organisms. Schleicher intended that these contributions of linguistics to biological theory support an assumption that the pattern of language descent perfectly reflects the pattern of human descent. The monistic point of view (which Schleicher assumed in his commentary on Darwin's *Origin*) held that language was simply the material side of mind and thought.[16]

Haeckel believed that Darwin had advanced powerful evidence (embryology, biogeography, systematics) but he agreed with Darwin's translator into German, Heinrich Bronn, that analytic evidence was desirable. Schleicher thought linguistics could furnish such evidence. Language descent, he proclaimed, was an empirically well-established phenomenon; and he considered the linguist's genealogical tree a perfect model for depicting the evolution of plant and animal species.

Schleicher in his *Zur vergleichenden Sprachgeschichte* (1848) employed a morphological classification of languages that goes back to W. von Humboldt's typology of isolating, agglutinating, and flexional languages. Schleicher, however, did not believe that these types evolved from one another; rather, he thought that they were

[14] Schleicher 1863, quoted in Richards 2008, p. 105.

[15] While it is obvious that the analogy between texts and fossils is more problematic than this comparison allows for, a critique of the comparative methods in linguistics and biology would certainly lead beyond the objectives of this paper.

[16] Cf. Richards 2008, p. 257; Schleicher 1863.

indicative of different groups of human beings. He therefore classified the Germanic and Semitic languages (as flexional languages) as corresponding to the most highly evolved groups of languages indicative of the most highly evolved mental capacities. Haeckel used this line of thinking to argue for the polyphyletic human origin. He believed that languages probably developed only after the species of speechless *Urmenschen* had already split into several separate species or kinds. Within each human species, Haeckel thought, language evolved independently. Haeckel relied on and referred to Schleicher explicitly for this analysis.[17]

Schleicher also maintained, that "the formation of language is for us comparable to the evolution of the brain and the organs of speech".[18] The idea that the human brain evolved with language is popular among evolutionary theorists today and possibly also among many linguists. Both Darwin and Schleicher would have unequivocally agreed with Terrence W. Deacon's theory of the *coevolution of language and the brain.*[19] In *The Descent of Man*, Darwin wrote: "A great stride in the development of the intellect will have followed, as soon as, through a previous considerable advance, the half-art and half-instinct of language came into use; for the continued use of language will have reacted on the brain, and produced an inherited effect; and this again will have reaction on the improvement of language".[20] Schleicher used the bifurcation of the lines in his tree diagram (*Stammbaum*) to signify both the period of time that separates the linguistic data whose phylogenetic relationship he was describing and the degree of separation from the assumed *progenitor*. The general principle of establishing relationships between languages based on shared innovations from here on, as with all poorly defined entities, initiated a long succession of debates between *lumpers* and *splitters*. Historical linguists today know that while some phenomena can be described by a bifurcation in a tree-diagram, many do not. Schleicher's student Johannes Schmidt proposed the wave diagram (similar to a Venn diagram in mathematics) as an alternative explanatory model for change. Without going further into the problematic definition of languages as discrete historical entities, the clear lines and bifurcations of Schleicher's tree-diagrams stand in stark contrast with John McWhorter's recent definition of languages as "jerry-rigged splotches doing the best they can despite countless millennia of slow-but-sure kaleidoscopic distortion".[21]

At least from a diachronic perspective, the concept of species in biology is apparently equally problematic. The philosopher John Wilkins who recently explored the definitions of the concept *species* came to the conclusion that "there are $n+1$ definitions of 'species' in a room of n biologists".[22] One could likewise state that there are $n+1$ definitions of "language" in a room of n linguists.

[17] Richards 2008, p. 259.

[18] Schleicher 1848, p. 258.

[19] Deacon 1997.

[20] Darwin, quoted in Richards 2008, p. 262.

[21] McWhorter 2011, p. 12.

[22] Wilkins 2010.

The *Romantic* biology and *Romantic* linguistics derived from Kant, Goethe, and Schelling unequivocally represent the metaphysics of Uexküll's *Umwelt* theory that inspired the biosemiotics of Thomas A. Sebeok, Jesper Hoffmeyer[23] and others whose thought traverses equally well the phenomena of living things.[24] Their work is representative of a holistic view that preserves a Kantian metaphysics characteristic of nineteenth century biology and linguistics for a semiotic perspective on life.

Linguistics in the Twentieth Century and the Prehistory of Biosemiotics

Jakob von Uexküll preserved in the twentieth century a semiotic approach to all natural phenomena that was increasingly perceived as an anachronism when in the middle of the twentieth century biology was revolutionized by the emerging field of genetics and linguists aligned their methods and approaches with the social sciences (in particular sociology and psychology). The fundamental differences between a semiotic perspective and the emerging mainstream methods in the social sciences, and psychology in particular, become especially clear in light of the debates about *psychologism* around the turn of the twentieth century. The historian of philosophy Robert Lanier Anderson explains the connections between anti-psychologism and neo-Kantianism at the beginning of the twentieth century[25] in the German context as a struggle between diverging schools of thought. Uexküll, the Kantian biologist, represents an anti-psychologism that is characterized by Frederik Stjernfelt[26] as a fundamental necessity for any semiotic perspective on the natural world. Even though the debates among philosophers at the beginning of the twentieth century were admittedly more convoluted than can be addressed in the context of this chapter, "[o]n one significant construal, psychologism is the fallacy of reducing a normative rule of reasoning to an exceptionless, descriptive psychological law".[27] Stjernfelt[28] explains the fallacy of psychologism with a caricature:

> What is this "psychologism" that anti-psychologism takes as its critical target? [Generally], it is the idea that the content and structure of thought and signs form part of the domain of psychology – so that the study of minds and brains forms the primary or even the only way of accessing these issues. A basic problem in psychologism is that it immediately allows for relativism. If one mind holds one thing to be true while another prefers another, who are we

[23] Hoffmeyer 1993[1996] and 2008.

[24] For instance, Hoffmeyer (2007) illustrates concepts such as *semiotic causation*, *semiotic emergence*, and *semiotic scaffolding* in evolution with the movement of an *Escherichia coli* cell, a reproductive disorder in amphibians, and the development of the word *spam* in English respectively.

[25] Lanier Anderson 2005, p. 288.

[26] Stjernfelt 2013.

[27] Lanier Anderson 2005, p. 292.

[28] Stjernfelt 2013, p. 77.

> to judge, when psychology is taken to be the deepest or even the only access to those claims? Psychology studies psychic processes in general with no distinction as to whether particular claims made by those psyches are true or false, and the truth or falsity of a claim may not be decided from investigating the psychological process bringing forth that claim. To make a caricature: If mathematical entities were really of a psychological nature, then the truth about them should be attained by means of psychological investigations. The upshot of psychologism would thus be that a proper way of deciding the truth of the claim that 2+2=4 would be to make an empirical investigation of a large number of individual, psychological assessments of that claim. So, if we amass data of, say, 100,000 individual records of calculating 2+2, we might find that a small but significant amount of persons take the result to be 3, which would give us an average measure of around 3.999 as the result. This might now be celebrated as the most exact and scientific investigation yet of the troubling issue of 2+2, far more precise than the traditional, metaphysical claims of the result being 4, which must now be seen as merely the coarse and approximate result of centuries of dogmatic mathematicians indulging in mere armchair philosophy and folk theories, not caring to investigate psychological reality empirically.

During the twentieth century, questions about language were increasingly understood as psychological entities that should be studied empirically, rather than by making "metaphysical claims" about language as a human sign system from a semiotic perspective. The affinities of the field of linguistics with psychology were famously articulated in a *Course in General linguistics* (*CGL*)[29] between 1911 and 1914. This attempt to define and determine a new linguistics grew out of a sense of unease and dissatisfaction that the discipline concerned with human language was focusing too much on the historical development during the nineteenth century, and lacked the proper units of systematic description other scientific fields had established:

> From a practical point of view, it would be of interest to begin with units; to determine units, and recognize the various kinds of units by providing a classification. It would be necessary to examine what the basis is for division into words. For the word, in spite of being so difficult to define, is a unit that compels recognition by the mind. It has a central role in the linguistic mechanism. (But a discussion of that topic alone would fill a book.) Then one would proceed to classify smaller units, larger units, and so on. By determining in that way the elements to be dealt with, a science of linguistics would fully achieve its goals, having related all relevant phenomena in its domain to one first principle. It cannot be said that this problem has ever been tackled, or that the scope and difficulty of it have been realized. Where languages are concerned, people have always been satisfied to work with poorly defined units.[30]

The *word* as an imprecise and awkward unit was unsuitable for a "serious" scientific analysis of language that steered steadfastly away from philology, the study of texts in historical languages that dominated linguistics throughout the nineteenth century. Linguists looked with envy and hope to the periodic table of elements in chemistry. And anyone familiar with the organization of the IPA (International Phonetic Alphabet) will appreciate the similarities in the visual representation of the

---

[29] Saussure 1916.

[30] Saussure 1916 [1986, p. 109].

periodic table of elements and what Chomsky called *the phonetic capabilities of man*[31] (more on that later).

It is interesting to note that at the time the *CGL* failed to characterize an ideal unit of description (the soon to be invented *phoneme*), an American anthropologist apparently had no difficulty in identifying and cataloguing the speech sounds of indigenous languages of North America.[32] Ironically, the man who failed at characterizing this ideal linguistic abstraction we call the *phoneme*, was later also blamed for the abstractions associated with so-called *structuralism* in spite of his visionary understanding of linguistics as part of a larger "social psychology" he called *semiology* as "a science that studies the role of signs as part of social life. It would form a part of social psychology, and hence of general psychology. We shall call it *semiology* (from the Greek *semeion* 'sign'). It would investigate the nature of signs and the laws governing them".[33] This alignment of linguistics with the methods of psychology and its ensuing unavoidable *psychologism* actually contributed to the marginalization of the semiotic perspective on language that inherently requires anti-psychologism.

The careful reader of the *CGL* will notice the author's apparent despair over the inadequate theoretical concepts in linguistics and hopeful speculation of future semiology within psychology. It was precisely the affinities between linguistics and psychology that alienated the study of language from biology and from the semiotic perspective throughout the twentieth century. While some see psychology as a bridge between biology and linguistics through fields like evolutionary psychology, neurolinguistics, and brain science, the persistent psychologism only intensified in the context of cognitive science during the last decades of the twentieth century and continues well into the twenty-first century.

But first, it was the ensuing productive critique of the *CGL*[34] that brought about the invention of the *phoneme* and other abstractions in phonology and phonetics (such as the *distinctive feature*) that by mid-century became the envy of the social sciences. Following Nikolai Trubetzkoy, Roman Jakobson carried the torch of phonology from Prague to New York. Linguists and anthropologists experimented with phonology-inspired studies in different fields; and there was a great flurry of enthusiasm about all that linguistics had to offer. In the meantime, psychology and behaviorism had become the dominant frameworks in the American context, and linguists had followed right along.[35] Linguistics further flourished and linguistics departments were founded and funded profusely in the wake of WWII and the successes of code-breakers and other promising military applications of linguistic insights. It was a steady continuation of linguistic specialization that moved linguistic inquiry

[31] Chomsky and Halle 1968.

[32] Boas 1911.

[33] Saussure 1916 [1986, p. 15].

[34] E.g., Trubetzkoy 1939.

[35] E.g., Bloomfield 1933.

away from the semiotic perspective towards the empirical methods of psychology and the social sciences.[36]

As noted by Stjernfelt, "[a]nother implication of psychologism may be that signs and their meaning are nothing more than the individual psychic or neuronal phenomena supporting them or associated with them".[37] This assumption is the basis of most psycholinguistic and neurolinguistic research since the 1980s. To illustrate the dominant psychologism in psycholinguistics, consider the work on the question "Why are abstract concepts hard to understand?".[38] The basis of this type of research is the gathering of behavioral evidence by using psycholinguistic methods of measuring reaction times when reading sentences on a computer screen. In these experiments, subjects read sentences like "*All rifles are shot guns*" versus sentences like "*All ideas are thoughts*". They then have to press one of two buttons; one for agreement, another one for disagreement; the pertinent empirical data, of course, being the reaction time. These types of psycholinguistic studies have become the norm in terms of funded research in linguistics and continue to become more and more sophisticated in terms of the technologies and equipment used to measure reaction time and other behavioral and physiological evidence.[39] Anyone familiar with semiotic theory would agree that these are not the type of questions that would come from a semiotic perspective on language, nor would a semiotic perspective seek the behavioral or physiological evidence to answer any questions pertaining to the differences between abstract and concrete concepts.

Stjernfelt considers "[anti-psychologism] [as] basic for semiotics as such. During the founding period of modern semiotics in the decades around 1900, the refusal to take signs to be reducible to psychological phenomena was crucial for the establishment of logical and semiotic phenomena and structures as autonomous objects of research".[40]

Here two American linguists, whose work is characterized by the anti-psychologism that constitutes the semiotic perspective on language, stand out: Sebeok and Chomsky.

---

[36] Some of the articulations in the *CGL* also became the target of other types of criticism. Jacques Derrida (Derrira 1967) could have articulated his *Grammatology* in a positive way based on the semiotic perspective he gleaned from Peirce, but he chose to couch his work in a critique of the "linguist from Geneva", thereby denying the Saussurean legacy of semiology.

[37] Stjernfelt 2013, p. 77.

[38] Schwanenflugel 1991.

[39] E.g., Barber et al. 2013.

[40] Stjernfelt 2013, p. 77. It should be noted that in the twentieth century, some linguists became uncomfortable with linguistic abstractions and critiques came from inside the field. For example, John Rupert Firth criticized linguistics for its exclusion of the *context*. Michael Halliday formulated a social semiotics. The most irreverent and far-reaching criticism of linguistic abstractions is probably Roy Harris' *integrationist* linguistics that takes into consideration all the aspects of linguistic exchanges that *phonemes, morphemes* or *syntagms* cannot capture. Firth's *context*, Halliday's *social semiotics*, Harris' *integrationism*, and Gunther Kress' *multimodality* are all reactions to a linguistics estranged from a semiotic perspective on language.

Sebeok, Chomsky, and the Semiotic Perspective

Sebeok and Chomsky share an intellectual trajectory that began with a re-evaluation of Uexküllian *Umwelt* theory and Peircean semiotics in the 1950s. Their biosemiotic/biolinguistic perspective on language anchored an anti-psychologistic linguistics firmly in biology and ethology.

While the scholarly agendas of biolinguistics and biosemiotics are quite different, they share a common interest in human language as a species-specific cognitive tool. They also share a philosophical core in the Peircean *abduction* and the Uexküllian *Umwelt*[41] that connects them with a view of the living world that is characteristic of the *Naturphilosophie* of Kant, Goethe, and Schelling. Uexküll's concept of *Umwelt* – the subjective species-specific world created by an organism – is central to this approach to human language. Uexküll's son presented his father's *Umweltlehre* as an undogmatic, empirical type of biology by translating the following passage into twentieth century English:

> [...] da die Tätigkeit unseres Gemüts das einzige uns unmittelbar bekannte Stück Natur ist, sind seine Gesetze die einzigen, die mit Recht den Namen Naturgesetze tragen dürfen.[42]
>
> [A]s the activity of the mind is the only aspect of nature immediately known to us, its laws are the only ones which may rightly be called laws of nature.[43]

Chomsky's interest in Uexküll and ethology goes back to his time as a graduate fellow at Harvard working with Morris Halle and Eric Lenneberg in the 1950s.[44] The biolinguistic program, therefore, derives its general approach to human language from ethology; and Konrad Lorenz played an important role in its evolution.[45] Especially Lenneberg's *Biological Foundations of Language* (1964) "anticipated many themes of the coming decades"[46]; and Chomsky concluded in a famous interview that "[linguistics] is really a theoretical biology".[47]

While the cognitive revolution of the mid-twentieth century is generally associated with Chomsky's progress in the understanding of language as a generative system, Chomsky points out that "another influential factor in the renewal of the cognitive revolution was the work of ethologists".[48] In the preface to the third edition of *Language and Mind* (2006), Chomsky writes: "[The framework of ethology] could be adapted to the study of human cognitive organs and their genetically determined nature, which constructs experience – the organism's *Umwelt*, in ethological terminology – and guides the general path of development, just as in all other

---

[41] Cf. Augustyn 2009.

[42] Uexküll 1928, p. 40.

[43] Uexküll 1981 [1987, p. 149].

[44] Cf. Jenkins 2000, p. 1.

[45] *Ibid.*, p. 10.

[46] *Ibid.*, p. 3.

[47] Sklar 1968, p. 213.

[48] Chomsky 2006, p. x.

aspects of growth of organisms".[49] This is also the point of view from which Sebeok's biosemiotics approaches human language. Sebeok moved from the field of Finno-Ugric studies to semiotics to explore the signifying abilities of all organisms – via zoosemiotics to biosemiotics – to promote the view that all life depends on semiosis. For Sebeok, Uexküll was the "chief architect"[50] of biosemiotics, whose origin was "rooted in no antecedent semiotic theory or practice at all; it was, rather, connected to the thought of Plato, Leibniz, especially Kant, Goethe, and a handful of biologists, such as Johannes Müller and Karl Ernst von Baer".[51] To understand the importance of Uexküll's *Umweltlehre* for Sebeok's biosemiotics, it is worth quoting Sebeok's personal account of his first encounter with Uexküll's *Theoretical Biology* whose problematic translation he had already leafed through as a teenager in 1936:

> In the mid 1960s, when at last I read the authentic German version, I came to believe that Ogden, the very animator of Anglo semiotics in the twentieth century, had either known little or no German or, with all his polymathic gifts, had failed to understand what *Theoretische Biologie* was really about: not biology, not psychology, not physiology, but semiotics. What's more, it unfolded a wholly unprecedented, innovative theory of signs, the scope of which was nothing less than semiosis in life processes in their entirety. It created and established the basis for a comprehensive new domain: we now call it *Biosemiotics*.[52]

Sebeok attributed the fact that the notion of *Umwelt* did not reach the Anglo-American and international intellectual community much earlier to the inadequate translation of Uexküll's *Theoretische Biologie* (1920).[53] When Sebeok read the German original, he found it "if not pellucid, nonetheless electrifying"[54] and hereafter recognized in Uexküll the originator of biosemiotic theory in the twentieth century. *Umwelt*, in Sebeok's working definition, "is a model generated by the organism"[55] to which language adds a secondary, cognitive dimension. Based on the affinities between Sebeok's and Chomsky's approach to linguistics, their semiotic perspective on language can be characterized by the following basic assumptions:

(a) *The cognitive capacities of humans are species-specific (as are the semiotic capacities of all organisms)*

Chomsky and Sebeok share the view that an analysis of human language begins with ethology and the Uexküllian principle that all organisms create their own *Umwelt* based on their species-specific capacities. This determines what questions should be asked about language and what are considered permissible hypotheses. Chomsky explained the role ethology played in the articulation of his biolinguistic program:

---

49 *Ibid.*

50 Sebeok 2001, p. 70.

51 Cf. Sebeok 1998, p. 32.

52 *Ibid.*

53 Cf. Uexküll 1928.

54 Sebeok 1998, pp. 32–34.

55 Sebeok 2001, p. vii.

> [It] seems that most complex organisms have highly specific forms of sensory and perceptual organization that are associated with the *Umwelt* and the manner of life of the organism. There is little reason to doubt that what is true of lower organisms is true of humans as well. Particularly in the case of language, it is natural to expect a close relation between innate properties of the mind and features of linguistic structure; for language, after all, has no existence apart from its mental representation. Whatever properties it has must be those that are given to it by the innate mental processes of the organism that has invented it and that invents it anew with each succeeding generation, along with whatever properties are associated with the conditions for its use. Once again, it seems that language should be, for this reason, a most illuminating probe with which to explore the organization of mental processes.[56]

His fellow Harvard graduate Lenneberg, in his *Biological Foundations of Language*, had referred to Uexküll's *Umwelt und Innenwelt der Tiere* to address the species-specificities of all behavior:

> The interaction of integrated patterns of all these different potentialities produces the cognitive specificities that have induced von Uexkuell [**sic**], the forerunner of modern ethology, to propose that every species has its own world-view. The phenomenological implications of this formulation may sound old-fashioned today, but students of animal behavior cannot ignore the fact that the differences in cognitive processes (1) are empirically demonstrable and (2) are the correlates of species-specific behavior.[57]

There is some irony in Lenneberg apologizing for his *Biological Foundations of Language* sounding "old-fashioned" in the 1960s when the "phenomenological implications" quite obviously reflect the principles of the *Naturphilosophie* that Uexküll had tried so hard to preserve in biology.

It is representative both of Chomsky's fame and Sebeok's ambitions to unify a vast variety of semiotic perspectives, that one can find frequent references to Chomsky in Sebeok's work,[58] but not vice versa. Their only documentable "collaboration" is an essay on primate studies in an anthology edited by Sebeok and his wife Jean Umiker-Sebeok.[59] Both Sebeok and Chomsky challenged prominent primate studies of the 1970s, because they both believed that no valuable insights about human language or primate cognition would be gleaned from teaching sign-language to a chimpanzee or a bonobo. It took many unhappy primates practicing abstract symbol recognition and ASL before most linguists and psychologists came to the conclusion that chimps and bonobos have *their own* communication systems that are specific to their species; and that scientific efforts to understand primate cognition had to be refocused on those *species-specific sign systems*.

(b) *Language is primarily a cognitive tool (rather than a communication system)*

The importance of this fundamental idea about human language shared by Chomsky and Sebeok cannot be emphasized enough. They both see human language foremost as a tool of thought, because the species was capable of

[56] Chomsky 2006, p. 83.

[57] Lenneberg 1964, p. 372.

[58] E.g., Sebeok 1977, p. 181 and 2001, pp. xix, 22.

[59] Sebeok and Umiker-Sebeok (eds.), 1980.

communication before it emerged. Sebeok put it like this: "[L]anguage – consisting of a set of features that promotes fitness – had best be thought of as having been built by selection for the cognitive function of modeling, and, as the philosopher Popper as well as the linguist Chomsky have likewise insisted [...], not at all for the message swapping function of communication. The latter was routinely carried on by nonverbal means, as in all animals, and as it continues to be in the context of most human interactions today".[60] The implications for what linguistics ought to be concerned with are far from trivial; and the importance of this basic assumption cannot be emphasized strongly enough. Chomsky readily admits that this view is considered "idiosyncratic" by most linguists, but is perfectly compatible with Sebeok's definition of language as a *secondary modeling system* that allows the species to create models of reality in addition to the species-specific perceptual system (the *primary modeling system*).[61]

(c) *Language is an exaptation*

For Chomsky, as for Sebeok, language is a tool of thought that is based on principles that are *not* specific to language. Chomsky confidently relates these "principles not specific to the faculty of language" to the Galilean intuition that "nature is perfect, from the tides to the flight of birds, and that it is the task of the scientist to discover in just what sense that is true".[62]

Sebeok and Chomsky consequently share the view that language is an *exaptation*[63]; and they both see organism-environment-interaction (i.e. species-specific *Umwelt*) as a crucial component of the growth of language in the individual. This is a view that separates them from a strong evolutionary psychology of language.[64]

While questions of evolution were never central to Chomsky's theoretical work, he considers the diversity of the roughly 6,000 languages on Earth to be superficial as his work focuses on the abstract principles that underlie their grammars.

The notion of optimal design in the Minimalist approach, exemplified by the analogy between language and a *snowflake* within biolinguistics can therefore be understood as the central unifying principle that sees language as a natural object.[65] Darwin and Schleicher could not have agreed more.

(d) *Linguistics is theoretical biology (and habit-taking/abduction are real processes)*

Chomsky outlined his preferred path in linguistics in his review of Burrhus Frederic Skinner's *Verbal Behavior* (1959). In this, he exposed the inadequacy of the predominant behaviorist approaches to issues of mind in general, and the learning of language in particular. He stressed, albeit in a footnote, concerning the

[60] Sebeok 1991, p. 53.

[61] Cf. Andersen and Merrell 1991; Sebeok and Danesi 2000.

[62] Chomsky 2006, p. 178.

[63] Sebeok 2001, p. 29; Boeckx and Piatelli-Palmarini 2005, p. 460.

[64] E.g., Pinker 1994 and 2003.

[65] Boeckx and Piatelli-Palmarini 2005, p. 461.

"unknown character and complexity" of the human "hypothesis formulating ability" – a notion that he later clearly articulated as Peircean *abduction* – "the necessity for carefully analyzing the strategies available to the organism as a complex information-processing system".[66]

The laws and principles of this *philosophical grammar*, he wrote, "are not formulable in terms of even the most elaborate extension of the concepts proper to the analysis of behavior and interaction of physical bodies, and they are not realizable by even the most complex automaton".[67] Chomsky envisioned "a psychology that begins with the problems of characterizing various systems of human knowledge and belief, the concepts in terms of which they are organized and the principles that underlie them, and that only then turns to the study of how these systems might have developed through some combination of innate structure and *organism-environment interaction*".[68] He cautioned psychologists already in 1967 not "to relate the postulated mental structures and processes to any physiological mechanisms or to interpret mental function in terms of 'physical causes'"[69] but, instead, to explore the creative/generative principles of language use. With this fundamentally anti-psychologistic perspective, he regarded the segmentation and classification techniques practiced by the structural linguists of his time as "at best limited to the phenomena of surface structure [that] cannot reveal the mechanisms that underlie the creative aspect of language use and the expression of semantic content".[70]

The important aspect of ethology for Chomsky's philosophical grammar "is its attempt to explore the innate properties that determine how knowledge is acquired and the character of that knowledge".[71] Chomsky, like Sebeok, looked to Peirce in order to explain the problem of development "rather like that of explaining successful abduction".[72] He clarifies his view concerning the acquisition of language as an ideal example of the human *hypothesis-formulating ability*:

> The way in which I have been describing acquisition of knowledge of language calls to mind a very interesting and rather neglected lecture given by Charles Sanders Peirce more than 50 years ago, in which he developed some rather similar notions about acquisition of knowledge in general. Peirce argued that the general limits of human intelligence are much more narrow than might be suggested by romantic assumptions about the limitless perfectibility of man [...]. He held that innate limitations on admissible hypotheses are a precondition for successful theory construction, and that the "guessing instinct" that provides hypotheses makes use of inductive procedures only for "corrective action." [...] To understand how knowledge is acquired, in the rationalist view that Peirce outlined, we must penetrate the mysteries of what he called "abduction".[73]

---

[66] Chomsky 1959, p. 57.

[67] Chomsky 2006, p. 6.

[68] *Ibid.*; italics mine. – *P.A.*

[69] *Ibid.*, p. 12.

[70] *Ibid.*, p. 20.

[71] *Ibid.*

[72] *Ibid.*, p. 84.

[73] *Ibid.*, pp. 79–80.

Chomsky outlines the tasks for the biolinguistic framework, first, "to construct generative grammars for particular languages that yield the facts about sound and meaning", and second, "to account for the acquisition of language".[74]

What has been a constant throughout the 50 years of the biolinguistic approach is its anchoring in the concept that language depends on a unique interplay of innate faculties and organism-environment interaction, and a "genetically determined instinct" of formulating hypotheses that Chomsky sees explained in Peircean abduction.

Sebeok's linguistics likewise begins with the idea that at the core of this secondary modeling system are abstract principles that can only be explained through semiotic analysis. His work is openly grounded in Peircean semiotics and he shared what Chomsky considered "the preferred path" in linguistics to be theoretical biology. Peircean abduction, likewise, is at the heart of the biosemiotic enterprise. This core principle that defines the biosemiotic perspective outlined by Sebeok, resonates in Hoffmeyer's assertion that "[i]t lies at the heart of biosemiotics and of Peircean cosmological philosophy that 'habit taking' or interpretation are real processes in the world, and therefore that belief in the law of necessity is unfounded".[75]

(e) *Language is a natural object*

Sebeok and Chomsky refute the common distinction between nature and culture. They are *hybrids* in the sense of Bruno Latour's analysis of what he calls the Modern Constitution. In his essay *We Have Never Been Modern* (1991),[76] Latour lays out the Modern Constitution that separates "three regions of being",[77] *nature – politics* – and *discourse* through the processes he calls *purification* and *mediation*.

The paradox of the Modern Constitution is that the separation of nature and society (= *purification*) makes *mediation* possible, but marginalizes it and renders it invisible at the same time. But only *hybrids*, says Latour, "can change the future".[78] Mainstream linguists and mainstream biologists who suffer from the illusions of the Modern Constitution practice purification so that nature and society must remain distinct. This includes the illusion (1) that even though we construct nature, nature is as if we did not construct it, and another (2) that even though we do not construct society, it is as if we construct it.[79] More importantly, Latour shows us that the Modern Constitution entails, besides the dichotomy between *purification* and *mediation*, the separation between non-humans (as nature) and humans (as culture).

Hybrids who reject the Modern Constitution, because they practice *mediation* (such as, for instance, anthropologists who study non-Western cultures or ethologists who study the physiological and cognitive capacities of other species) are seen as outsiders of the purified disciplines of the mainstream. This becomes especially

[74] Chomsky 2007, p. 14.

[75] Hoffmeyer 2004, p. 73.

[76] Latour 1991 [1993].

[77] *Ibid.*, p. 39.

[78] *Ibid.*, p. 11.

[79] Latour 1991 [1993].

apparent when anthropologists study cultures in the West, or when ethologists, biologists, linguists, or semioticians study the cognitive capacities of humans.

Chomsky and Sebeok's grounding in Peircean *semeiotic* and Uexküllian *Umwelt* theory clearly makes them *hybrids.*[80] The difficulty of their position within the field of linguistics (or semiotics, even though *purification* is much less of an issue there) is that their work is prone to gross misinterpretation, precisely because the mainstream lives by the illusions that uphold the Modern Constitution. As Latour explains, "[t]he essential point of this Constitution is that it renders the work of mediation that assembles hybrids invisible, unthinkable, unrepresentable".[81]

This can be explained with the predominant folk-definition of *Universal Grammar* (*UG*), an unfortunate misinterpretation that can be attributed to the artificial dichotomies that are the result of the disciplinary purification that wants to see the field of linguistics in the social sciences or the humanities (culture) rather than, as Chomsky and Sebeok would have it, as a domain of biology that approaches the study of human language as a phenomenon of nature. The folk-definition of *UG* is something like an equivalent of linguistic universals or the things that are shared by all languages, a definition that does not depend on the ethological perspective and is not in contradiction with the laws of the Modern Constitution.

For most students of linguistics, it is difficult to accept Chomsky's definition of *UG* as the properties of the initial state of the human faculty of language that are specific to the species. For those who live by the Modern Constitution, the *hybrid* character of this concept remains nebulous, "unthinkable, unrepresentable",[82] because they want to ground everything in the Modern Constitution, keep language in the domain of culture, and the field of linguistics separate from biology. This is also because most of biology follows the Modern Constitution in the form of evolutionary psychology. For those who understand the philosophical background behind the faculty of language as a combination of (1) innate capacities, (2) organism-environment interaction (*Umwelt*), and (3) abstract principles *not* specific to the faculty of language,[83] the *hybrid* character of this concept is quite uncontroversial.

Modernity has made it impossible for some to take the ethologist's perspective on our species, to mediate instead of separating nature and culture. Chomsky's *Cartesian Linguistics* (1966)[84] likewise defies the paradoxes of the Modern Constitution, because it begins with the unresolved questions of the seventeenth century. Because the very title of Chomsky's *Chapter in the History of Rationalist Thought* is perpetually mischaracterized and misinterpreted, especially by those who don't care to read it and prematurely associate its title with a folk definition of the Cartesian mind/body dualism. The Introduction to the 2009 edition explains that Descartes "was among the first to recognize the importance of this 'ordinary' form

[80] *Sensu* Latour 1991 [1993].

[81] Latour 1991 [1993, p. 34].

[82] *Ibid.*

[83] Cf. Chomsky 2005, p. 6.

[84] Chomsky 1966 [2009].

of linguistic creativity [...] for the study of the human mind".[85] Connecting biolinguistics to the questions Descartes addressed at the end of the sixteenth century declares this perspective on language and mind scientifically pre-Modern.

Chomsky's *cognitive revolution* of the mid-twentieth century is a renewal and further development of the cognitive revolution of the seventeenth century, while another influential factor in the renewal of the cognitive revolution was the work of ethologists, ethology being a field that defies the principles of the Modern Constitution. Sebeok, the linguist whose life work was to turn semiotics into a science of all life, obviously *has never been modern*. He would certainly agree that the fundamental questions of biolinguistics articulated by Chomsky[86] have yet to be answered:

1. What constitutes knowledge of language? (Plato's problem)
2. How is this knowledge acquired? (Humboldt's problem)
3. How is this knowledge put to use? (Descartes' problem)

To these three fundamental questions, the following two have been added cautiously:

4. What are the related brain mechanisms?
5. How did language evolve in the species?

Chomsky's preferred path in linguistics steered away from physiological and behavioral evidence for a long time, slowly and cautiously considering such evidence for what are considered permissible hypotheses within biolinguistics. In particular, Chomsky has been critical of the many confident pronouncements coming from neuroscience about how the "brain produces language". Chomsky's collaborator Tecumseh Fitch recently expressed this kind of skepticism towards physiological evidence when he accused neuroscientists for "a decade or so of somewhat self-indulgent neo-phrenology".[87] Like in all "academic tribal societies", biolinguistics is plagued by challenges "concerning terminology, disciplinary turf wars, and struggles for dominance".[88] The same is true for biosemiotics.

Among the real challenges, not sociological but intellectual in nature, Fitch points to the theoretical shortcomings in neuroscience and the lack of good collaboration with theoretical linguists because neuroscientists still "do not understand how brains generate minds" and "principles underlying brain development and evolution remain only dimly understood".[89] Likewise, neuroscientists do not know how brains generate language, and there is very little collaboration between neurolinguists and theoretical linguists.[90]

---

[85] *Ibid.*, p. 1.

[86] Jenkins 2000.

[87] Fitch 2009, p. 284.

[88] *Ibid.*, p. 285.

[89] *Ibid.*

[90] Cf. Andrews 2011.

An important issue for biolinguists, according to Fitch, are "questions of meaning" and what he calls "unresolved semiotic challenges [that] pose problems for any aspect of cognition".[91] Maybe Fitch and those who agree with him would find more satisfying theories of meaning in the foundational literature associated with biosemiotics? When Fitch writes "[we] have a good theory of information (Shannon information theory), but we lack anything even approaching a good theory of meaning",[92] he is looking for an alternative to "many currently popular models and metaphors for understanding genes, brain and language [that] need to be abandoned if [biolinguists] hope to make any substantial progress"[93] that many biosemioticians see in mainstream biology.

Most biosemioticians would see eye to eye with Fitch on that central challenge, although they may not all agree on how to best connect biolinguistics and biosemiotics.[94] Hoffmeyer, who turned to philosophy to address these issues in biology would agree that it is precisely the vagueness of concepts such as *information* or *signal* in biology that drove biologists to philosophy and semiotics, fueled the biosemiotic movement and helped crystalize its central theses.[95] According to Hoffmeyer, "[biosemiotics] does not turn experimental biology to metaphysics but instead replaces an outdated metaphysics – the thought that life is only chemistry and molecules – with a far better, more contemporary, and more coherent philosophy. Life rather than natural law – and signs rather than atoms – must become natural science's fundamental phenomena".[96] To be sure, even though the "ideas and the personalities who embody and propagate them, are in [Sebeok's] view kept asunder at one's peril",[97] biolinguistics and biosemiotics are what he would have considered to be "complementary domains"[98] because they unequivocally share an anti-psychologistic perspective on language that is rooted in semiotic theory.

When the biologist/philosopher Andreas Weber anticipates a "revolution of the life sciences",[99] it becomes very clear that what Weber is hoping for is that biology (along with other fields) may *return* to a view of living organisms that is in agreement with the monist metaphysics of nineteenth century *Romantic biology* and the anti-psychologism of Kantian biologists like Jakob von Uexküll. That was the intellectual climate that gave rise to the concept of ecology[100] and the idea that the analysis of human language can make a direct contribution to a natural history of the genus *Homo*.

---

[91] Fitch 2009, p. 284.

[92] *Ibid.*, p. 285.

[93] *Ibid.*, p. 286.

[94] E.g., Barbieri 2010.

[95] Cf. Kull et al. 2009.

[96] Hoffmeyer 2008, p. 15.

[97] Sebeok 1998, p. 25.

[98] *Ibid.*, p. 24.

[99] Weber 2008.

[100] Haeckel 1866.

References

Andersen, M., & Merrell, F. (1991). *On semiotic modeling*. Berlin: Mouton de Gruyter.

Andrews, E. (2011). Language and brain. Recasting meaning in the definition of human language. *Semiotica, 184*(1–4), 11–32.

Atkinson, Q., & Gray, R. D. (2005). Curious parallels and curious connections – Phylogenetic thinking in biology and historical linguistics. *Systematic Biology, 54*(4), 513–526.

Augustyn, P. (2009). Uexküll, Peirce, and other affinities between biosemiotics and biolinguistics. *Biosemiotics, 2*(1), 1–17.

Barber, H., Otten, L. J., Kousta, S.-T., & Vigliocco, G. (2013). Concreteness in word processing: ERP and behavioral effects in a lexical decision task. *Brain & Language, 125*, 47–53.

Barbieri, M. (2010). On the origin of language: A bridge between biolinguistics and biosemiotics. *Biosemiotics, 3*(2), 201–223.

Bloomfield, L. (1933). *Language*. Chicago: University of Chicago Press.

Boas, F. (1911). *Handbook of American Indian languages*. Washington, DC: GPO.

Boeckx, C., & Piatelli-Palmarini, M. (2005). Language as a natural object; Linguistics as a natural science. *The Linguistic Review, 22*, 447–466.

Chomsky, N. (1959). A review of B.F. Skinner's verbal behavior. *Language, 35*(1), 26–58.

Chomsky, N. (1966 [2009]). *Cartesian linguistics. A chapter in the history of rationalist thought.* Cambridge: Cambridge University Press.

Chomsky, N. (2005). Three factors in language design. *Linguistic Inquiry, 36*(1), 1–22.

Chomsky, N. (2006). *Language and mind*. Cambridge: Cambridge University Press.

Chomsky, N. (2007). Biolinguistic explorations: Design, development, and evolution. *International Journal of Philosophical Studies, 15*(1), 1–21.

Chomsky, N., & Halle, M. (1968). *The sound patterns of English*. New York: Harper & Row.

Darwin, C. R. (1859). *On the origin of species by means of natural selection or the preservation of favoured races in the struggle for life*. London: John Murray.

Darwin, C. R. (1871). *The descent of man, and selection in relation to sex*. London: John Murray.

Deacon, T. W. (1997). *The symbolic species*. New York: W.W. Norton & Co.

Derrida, J. (1967). *De la Grammatologie*. Paris: Éditions de Minuit.

Fitch, T. (2009). Prolegomena to a future science of biolinguistics. *Biolinguistics, 3*(4), 283–320.

Haeckel, E. (1866). *Generelle Morphologie der Organismen*. Berlin: Georg Reimer.

Hoffmeyer, J. (1993 [1996]). *Signs of meaning in the universe*. Bloomington: Indiana University Press.

Hoffmeyer, J. (2004). Uexküllian Planmäßigkeit. *Sign Systems Studies, 32*(1–2), 73–97.

Hoffmeyer, J. (2007). Semiotic scaffolding of living systems. In M. Barbieri (Ed.), *Introduction to biosemiotics. The new biological synthesis* (pp. 149–166). Dordrecht: Springer.

Hoffmeyer, J. (2008). *Biosemiotics: An examination into the life of signs and the signs of life*. Scranton: University of Scranton Press.

Jenkins, L. (2000). *Biolinguistics. Exploring the biology of language*. Cambridge: Cambridge University Press.

Kull, K. (2001). Jakob von Uexküll. An introduction. *Semiotica, 134*(1–4), 1–50.

Kull, K., Deacon, T., Emmeche, C., Hoffmeyer, J., & Stjernfelt, F. (2009). Theses on biosemiotics: Prolegomena to a theoretical biology. *Biological Theory, 4*(2), 167–173.

Lanier Anderson, R. (2005). Neo-Kantianism and the roots of anti-psychologism. *British Journal for the History of Philosophy, 13*(2), 287–323.

Latour, B. (1991 [1993]). *We have never been modern.* Cambridge, MA: Harvard University Press.

Lenneberg, E. H. (1964). *Biological foundations of language*. New York: Wiley.

Lowth, R. (1762). *A short introduction to English grammar with critical notes*. London: Dodsley and Cadell.

McWhorter, J. (2011). *What language is (and what It Isn't and what It could be).* New York: Gotham Books.

Murray, L. (1795). *English grammar adapted to the different classes of learners*. Philadelphia: J. and W. Paul.
Pinker, S. (1994). *The language instinct. How the mind creates language*. New York: Harper Perennial.
Pinker, S. (2003). *The blank slate*. New York: Penguin.
Richards, R. J. (2002). The linguistic creation of man: Charles Darwin, August Schleicher, Ernst Haeckel and the missing link in evolutionary theory. In M. Dörries (Ed.), *Experimenting in tongues: Studies in science and language* (pp. 21–48). Stanford: Stanford University Press.
Richards, R. J. (2004). *The romantic conception of life: Science and philosophy in the Age of Goethe*. Chicago: University of Chicago Press.
Richards, R. J. (2008). *The tragic sense of life. Ernst Haeckel and the struggle over evolutionary thought*. Chicago: University of Chicago Press.
Saussure, F. de (1916). *Cours de linguistique générale*. Lausanne/Paris: Payot.
Saussure, F. de (1916 [1986]). *Course in general linguistics*. La Salle: Open Court.
Schleicher, A. (1848). *Zur vergleichenden Sprachgeschichte*. Bonn: H.B. König.
Schleicher, A. (1863). *Die Darwinsche Theorie und die Sprachwissenschaft*. Weimar: Böhlau.
Schleicher, A. (1865). *Über die Bedeutung der Sprache für die Naturgeschichte des Menschen*. Weimar: Böhlau.
Schwanenflugel, P. J. (1991). Why are abstract concepts hard to understand? In P. J. Schwanenflugel (Ed.), *The psychology of word meanings* (pp. 223–250). Hillsdale: Lawrence Erlbaum Associates.
Sebeok, T. A. (1977). *A perfusion of signs*. Bloomington: Indiana University Press.
Sebeok, T. A. (1991). In what sense is language a primary modeling system? In T. A. Sebeok, *A sign is just a sign* (pp. 49–58). Bloomington: Indiana University Press.
Sebeok, T. A. (1998). The Estonian connection. *Sign Systems Studies, 26*, 20–41.
Sebeok, T. A. (2001). *Global semiotics*. Bloomington: Indiana University Press.
Sebeok, T. A., & Danesi, M. (2000). *The forms of meaning*. Berlin: Mouton de Gruyter.
Sebeok, T. A., & Umiker-Sebeok, J. (Eds.). (1980). *Speaking of apes. A critical anthology of two-way communication with man*. New York: Plenum.
Sklar, R. (1968). Chomsky's revolution in linguistics. *The Nation*, 9 September 1968, (pp. 213–217).
Stjernfelt, F. (2013). The generality of signs: The actual relevance of anti-psychologism. *Semiotica, 194*, 77–98.
Trubetzkoy, N. (1939). *Grundzüge der Phonologie* [Travaux du Cercle Linguistique de Prague 7].
Uexküll, J. von (1928). *Theoretische Biologie* (2nd Rev. ed). Berlin: Verlag von Julius Springer.
Uexküll, J. von (1957). *Niegeschaute Welten*. München: Paul List Verlag.
Uexküll, T. von (1981 [1987]). The sign theory of Jakob von Uexküll. In M. Krampen, K. Oehler, R. Posner, T. A. Sebeok & T. von Uexküll (Eds.), *Classics of semiotics* (pp. 147–179). New York: Plenum.
Uexküll, T. von (1992). Preface to *A stroll through the worlds of animals and men: A picture book of invisible worlds*. *Semiotica, 89*(4), 319–391.
Weber, A. (2008). *Alles Fühlt. Mensch, Natur, und die Revolution der Lebenswissenschaften*. Berlin: Berlin Verlag.
Wilkins, J. (2010). What is a species? Essences and generation. *Theory in Biosciences, 129*(2–3), 141–148.

Before Babel: The Evolutionary Roots of Human Language

Piera Filippi

Abstract The aim of the present work is to identify the evolutionary origins of the ability to speak and understand a natural language. I will adopt Botha's "Windows Approach" (*Language and Communication*, 2006, 26, pp. 129–143) in order to justify the following two assumptions, which concern the evolutionary continuity between human language and animals' communication systems: (a) despite the uniqueness of human language in sharing and conveying utterances with an open-ended structure, some isolated components of our linguistic competence are shared with non-human primates, grounding a line of evolutionary continuity; (b) the very first "linguistic" utterances were holistic, that is, whole bunches of sounds able to convey information despite their lack of modern syntax. I will address such suppositions through the comparative analysis of three constitutive features of human language: syntax, the semantic value of utterances, and the ability to attribute mental states to conspecifics, i.e. the theory of mind.

Keywords Language evolution • Non-human primates • Holistic protolanguage • Semantics • Syntax • Theory of mind

The aspects of things that are most important for us are hidden because of their simplicity and familiarity. (One is unable to notice something – because it is always before one's eyes.) The real foundations of their inquiry do not strike people at all. Unless that fact has at some time struck them. – And this means: we fail to be struck by what, once seen is most striking and most powerful.

(Wittgenstein 1953 [2010, § 129]).

P. Filippi (✉)
University of Vienna, Vienna, Austria

Vrije Universiteit Brussel, Ixelles, Belgium
e-mail: pie.filippi@gmail.com

E. Velmezova et al. (eds.), *Biosemiotic Perspectives on Language and Linguistics*, Biosemiotics 13, DOI 10.1007/978-3-319-20663-9_10

Homo loquens

Words, thoughts, and reasoning are all constitutive parts of human *natural* history. Humans' ordinary life is so permeated by them that, as fate would ironically have it, they are one of the most mysterious topics of studies accessible to the human mind; mysterious and difficult, for sure, but nonetheless extremely fascinating. I believe that one of the most efficient ways to explore the nature of such a complex phenomenon as language is studying its origins' dynamics, which can shed light on those features that distinguish it from other animals' systems of communication; in short, what makes human communication unique. Indeed, for many centuries theorists of language sciences have speculated on the evolution of language, but the impossibility to find direct evidence has repeatedly led to a state of impasse. In fact, unlike other phenomena addressed by evolutionary research, language cannot be studied through paleontological data, as it has never fossilized on rocky stratifications able to indicate an evolutionary path towards *species* and time. Back in 1866, the lack of any scientific progress in the study of this topic led the *Societé linguistique de Paris* to publish an edit banning any communication related to the origins of language or to the existence of a universal language that all modern languages share. However, in the last century this topic has seen a considerable revival due to the emergence of a new fertile methodology of research, in which multiple disciplines related to language and biology interact with each other (cf. Fig. 1). Within this methodological frame, the aim of the present work is to explore the origins of language bridging research on linguistics and philosophy of language to the comparative investigation of animal communication.

In order to avoid any conceptual misinterpretation, I wish to initially point out a terminological distinction which is missing in numerous spoken languages: (1) the semantic distinction between the faculty of *language*, meant in a broad sense, as a general biological tool that allows communication, (2) the ability to speak and understand a natural language. The first meaning refers to the ability to produce a visual and/or acoustic sign in association with a specific referential object. Thus,

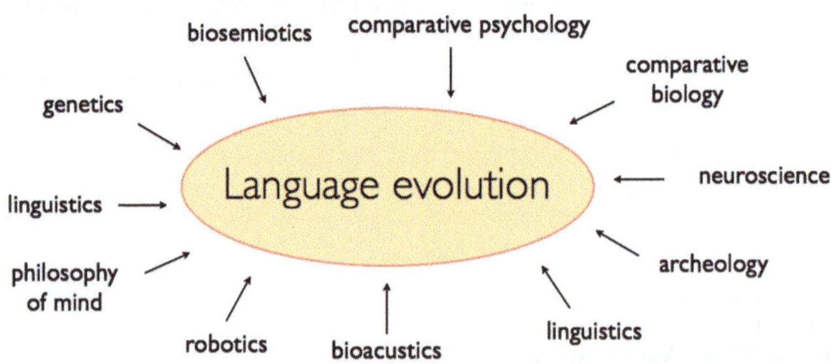

Fig. 1 The interdisciplinary approach on language evolution. In order to pinpoint the evolutionary dynamics of language, a coalition of multiple types of expertise is required. Observations from different fields are now encouraged to be integrated (modified from Christiansen and Kirby 2003)

non-human animals' cognitive and communication systems are part of this broad biological set. Notably, the faculty of language (broad sense) includes the second type of ability, i.e. to speak (or sign) a natural language such as Hindi, Chinese, or Italian following specific combinatorial and morfo-syntactical rules. This latter order of language is specific to humans, which typically employ it in a social group, conveying information or influencing behaviours. In order to clarify the distinction between these two orders of the faculty of language, it is worth taking into consideration Terrence W. Deacon's observations.[1] His hypothesis is that in animal communication systems, each sound or sequence of sounds relates to one referential object (indexical association). In contrast, what makes human language unique is that "the relationship that a word has to an object is a *function* of the relationship that word has to other linguistic units within the sentence".[2] This means that in human language the propositional system of linguistic units (be they morpho-syntactic elements, words, or entire sentences), which are ruled by combinatorial rules, guides the act of reference. Thus, the combinatorial dimension is one feature that makes human language unique. In fact, unlike animal referential calls, propositional languages have specific morpho-syntactic organizations. Specifically, a set of language-specific rules governs the combination of morphological and syntactic units, generating a potentially infinite set of utterances, and thus enabling much of the generative power specific to human language.[3]

In the present article, I will argue that in animals' communication systems, one can identify general language-related cognitive traits that were critical for the evolutionary path of propositional language. Thus, the underlying assumption is that although human language includes a set of intertwined morpho-syntactic and conceptual-intentional operations, some components of our linguistic competence, taken in isolation, are shared with other animals. Indeed, although much research has been dedicated to the individuation of one (monogenesis) or more (polygenesis) natural languages as a common root of modern spoken languages, here I wish to adopt an interdisciplinary, comparative approach with the aim to identify the biological constraints underlying the evolution of language in non-human animals' communication systems. In fact, comparative research on different species can help shed light on the biological constraints underlying the emergence of human language. I will apply a comparative framework with the aim to analyze the evolution of three constitutive components that are highly intertwined in human language, but that I will keep separate merely for methodological reasons: semantics, syntax and the ability to attribute mental states to conspecifics. Ultimately, this approach could help us grasp a better understanding of the communicative abilities shared across animals, thus shedding light on the cognitive features that make human verbal communication species-specific.

[1] Deacon 1998.

[2] *Ibid.*, p. 79.

[3] Chomsky 1980; Yip 2006.

Methodology

In order to provide a scientifically valid contribution to the study of the evolution of language – a topic which itself tends to be an object of speculation – it is necessary to adopt a precise empirical methodology. First, it is opportune to adopt an interdisciplinary approach to the topic, linking the different theoretical observations and the empirical data within a coherent frame of concepts, which could lead to an increasing understanding of language and its evolutionary dynamics.

In this direction, the "Windows Approach to language evolution", proposed by Rudolf P. Botha[4] is a valid methodology, which in my opinion is able to root the research on language evolution on informative, empirically grounded theory. According to Botha, we should explore the evolution of language putting together empirical data from multiple research areas that are linked to this broad topic, for instance animal communication or archaeology. This methodological strategy enables the investigation of a phenomenon that is not directly observable empirically, as it is the case with the origins of language. To be scientifically valid, such an empirically informed theory – which the author refers to as the "window theory" – should be characterized by three basic features. First, it should be grounded on phenomena about which there is direct evidence.[5] Second, it should be warranted, in the sense that it "has to take an empirical form which gives a systematic account of how properties of present forms of language and (properties of) stages in the emergence of language are interlinked".[6] Finally, the window theory should be pertinent: "[Window] inferences can be pertinent – that is, about the evolution of the 'right entity', namely language – only if they are underpinned by a restrictive theory of what language is".[7]

The Nature of the First Human Utterances

Given such methodological assumptions, I will take the studies concerning monkeys and apes' communicative systems as a conceptual window through which one can observe the phylogenetic path of human language. More specifically, I will use this conceptual frame of research in order to justify the evolutionary thesis according to which the very first "linguistic" utterances were holistic, that is to say, whole bunches of sounds able to convey information despite their lack of modern syntax.

A fertile question one could start the exploration of such issues with could be the following: shall we refer to the first *Homo* vocal units as mere representational

[4] Botha 2006.

[5] *Ibid.*, p. 134.

[6] *Ibid.*, p. 137.

[7] *Ibid.*, p. 139.

labels attached to objects in the surrounding environment, or is it not more correct to conceive of them rather as functionally referential units? There are two opposite theses that follow this later approach in the present debate about language evolution. On one hand, Dereck Bickerton's analytical model of explanation asserts that names were labels (mostly referring to environmental objects such as food or aggressors), whose increasing number and complexity consequentially gave rise to syntax.[8]

This idea collides with the holistic model of explanation, first proposed by Otto Jespersen in the early twentieth century,[9] and recently revived by Alison Wray, according to which the first meaningful units were not mere labels, but had a complex intrinsic internal meaning: "In this holistic protolanguage the messages are semantically complex and agrammatical. [...] Simply, the whole thing means the whole thing".[10] In particular, Wray's idea is that the first hominids may have communicated by means of random sequences of sound, to which they associated functionally referential meanings relying on the pragmatic context of use. The first expressions were, according to her ideas, formulaic and internally amorphous, though efficient in their performative, *manipulative* purposes.[11]

Let us imagine a situation in which the protagonists are the very first hominids who become aware of an imminent attack from a dangerous predator, e.g. a leopard. Most likely, our very first ancestor would have given an alarm call, similar to that of the great apes. In this situation, would we translate such a vocalization not as a simple name, but rather as a more complex message with an intrinsic emotional connotation, which could lead to an appropriate reaction somehow achievable by the utterance: "I've just seen a leopard... Behave accordingly!", or "Warning, ground danger!"?

In order to address this question, I will review relevant research on the communication system of our non-human primate ancestor, with whom we share genetic traits inherited by a common ancestor. The idea is to examine three core abilities that might have grounded a line of phylogenetic continuity (and discontinuity at the same time) between monkeys' communication system and the human language: syntax, the semantic value of utterances, and the ability to attribute mental states to conspecifics, i.e. the theory of mind.

---

[8] Bickerton 2002.

[9] Jespersen 1922 [2013, Chapter XXI].

[10] Wray 2002, p. 118. Cf. Fitch 2010, Chapter 14.

[11] Cf. Wray 1998, p. 51: "Protolanguage would, then, be a phonetically sophisticated set of formulaic utterances, with agreed function-specific meanings, that were a direct development from the earlier noises and gestures, and which had, like them, no internal structure. Each would be phonetically arbitrary, unrelated in sound to even those utterances that meant similar things".

Semantics

Regarding the semantic level, the meaning value of primate alarm calls refers to several different domains. For instance, eminent researchers on monkeys' communication system such as Robert M. Seyfarth and Dorothy L. Cheney have addressed this by defending the thesis according to which their signals are highly informative, given the agreed meaning of "information" as the reduction of uncertainty in the recipient.[12] Indeed, as they observe, the signal can be used by listeners to extrapolate information concerning the presence of food, the caller's identity, the kind of predator and the urgency of the danger. Concerning the signalling of the presence of food, it is worth noting that recent research conducted by Zanna Clay and Klaus Zuberbühler on bonobos has revealed that: "Captive bonobos at two locations produced five acoustically distinct call types when interacting with food: barks, peeps, peep-yelps, yelps and grunts. The production and distribution of these call types within a sequence was not random but was significantly associated with the preference score of the food".[13]

Similarly, alarm calls can indicate the presence of specific types of predators, and the related level of danger, eliciting the most appropriate behavioural response.[14] Furthermore, by hearing the signals exchanged by two or more monkeys, the listeners can infer the kind of relationship and approach that exists between them, perceiving them as actors predisposed to behave according to specific social patterns, such as who is supposed to groom or threaten who on the basis of the affiliated dominance rank:

> In groups of long-lived, highly social animals, communication and cognition are linked to fitness. To survive, avoid stress, reproduce, and raise offspring who are themselves successful, individuals need both a system of communication that allows them to influence other animals' behaviour and a system of mental representations that allows them to recognize and understand other animals' relationships. Because these mental representations concern animate creatures and are designed to predict behaviour, they include information (if rudimentary) about other individuals' mental states, and about the causal relations between one social event and another.[15]

These observations suggest that monkeys' vocalizations have a semantic value. At this point, however, we should address the question whether there is a strict link between the sound of the call itself and its meaning, or as it sometimes happens in human language (in the case of synonymy), whether the different calls could convey the same "meaning". Indeed, calls with similar acoustic features might elicit different responses. For instance, an eagle alarm call can lead a monkey placed on a tree to jump into a bush, while a monkey already located in a safe position does not react by moving to a different place. On the other hand, it is also true that calls with

---

[12] Seyfarth et al. 2010.

[13] Clay and Zuberbühler 2009, p. 1392. Cf. Hauser and Marler 1993.

[14] Struhsaker 1967.

[15] Cheney and Seyfarth 2008, p. 270.

different acoustic features elicit similar responses: a leopard growl and a monkey's alarm call elicit the same behavioural response, which is climbing up a tree. As Seyfarth and Cheney observe,[16] this phenomenon tells us that the recipients' response depends either on the physical properties of the signal and on the specific information they acquire from it. Also, Zuberbühler and his colleagues[17] provided evidence that female Diana monkeys do not respond to the shriek of an eagle if they are exposed to an alarm call emitted by a Diana monkey male five minutes earlier, even though these two types of signals are acoustically completely different. This suggests that Diana monkeys do not classify sound merely on the basis of their acoustic features, but also by the semantic meaning they convey. Such considerations support the hypothesis that monkeys are provided with a mental representation of the object linked to the conveyed signal.

Finally, for the purpose of our study, it is necessary to emphasise that one cannot refer to monkeys' vocalizations as to mere automatic innate reflexes:

> Monkeys, then, seem genetically predisposed to give particular contexts. But this is not to say that their vocalizations are entirely reflexive and involuntary. Although their call *repertoire* may be relatively fixed, their choice of whether to call or to remain silent is more flexible. [...] There is no obligatory link between the sight of a predator and the production of an alarm.
>
> [...] Primate vocalizations are not involuntary reflexes, impossible to suppress. They are, instead, much more like the other behaviours in which animals choose to engage. As they go about their daily lives, baboons decide whether or not to vocalize, just as they decide whether or not to groom, play or form alliance. Their behaviour depends on a complex combination of their own motivation, the particular situation at hand, and who else is involved. Primates can control whether they vocalize or not; what they cannot control are the detailed acoustic features of the calls they choose to produce.[18]

Thus, as clearly inferable through field observations, monkeys' vocalizations are linked to a mental representation of the referred object. In fact, it has been reported that vervet monkeys are able to suppress a vocalization, if a conspecific has previously emitted it in response to the same predator encounter. Moreover, acoustically similar vocalizations can lead to different responses, relying on the involved subjects and on the specific situation in which they occur. These data tell us that the potential meanings of monkey alarm calls are not strictly fixed to a mere genetic level, but are, in contrast, bearers of associations learned through experiences and interactions.

In addition, multiple studies have reported the use of informative calls in a wide range of animal species such as birds, frogs, rats, bats, chickens, bees.[19] The pervasive presence of this core communicative feature in widely distant species indicates that the ability to convey information that favors survival in the environment, i.e. calls linked, for instance, to the presence of food, predators, sexual attraction

[16] Seyfarth et al. 2010.

[17] Zuberbühler et al. 1997 and 1999.

[18] Cheney and Seyfarth 2008, pp. 226–227, 233.

[19] Cf. Hauser 1996 for a detailed review.

or emotional state is a pivotal biological constraint shared across phylogenetically distant species.

Syntax

Recently, Peter Marler, a researcher on animal communication, revived Martinet's concept of duality of pattern, and applied it to the overall analysis of animal signals.[20] Specifically, he highlights the distinction between two levels of syntax:

1. The phonological syntax,[21] which consists of the meaningless recombination of sounds into longer sequences. This syntactic level concerns the rules for the combinatorial structure of sounds;
2. The lexical syntax, whose rule of recombination concerns the generation of meaning within the sentence context.

For the purposes of this paper I will address the question whether there is any observable evidence that either of these steps, or at least some crucial aspects of them, are present in monkey communication systems, in order to find some evolutionary precursors of language. In order to avoid terminological confusion, it is worth emphasizing that with the term *syntax* I refer to the meaning modelled on the Greek word *syntaxis*, composed by "*syn*" ('together', 'with') – and "*taxis*" ('order', 'connection', 'coordination of the parts according to structural rules'), which must be kept conceptually distinguished from the definition of the term *syntax* as, intrinsically tied to the semantic values of the lexical units occurring within the sentence context.

As to mere phonological syntax, we can find examples of sound sequences in animal vocal communication. Erroneously, indeed, it has historically been assumed that animal vocalizations are merely an acoustically graded continuum, in contrast to human utterances, which are perceived as differentiated into phonetic discrete units. By this regard, Cheney and Seyfarth assert:

> Given the potential ambiguity inherent in a graded series of calls, and the importance of distinguishing both between different call types and between the call of different individuals, it appears that baboon listeners have been under strong selective pressure to detect subtle distinctions within a graded acoustic continuum and to link these differences in acoustic structure with differences in individual identities, social events, predators and so on.[22]

In other words, monkeys are indeed able to categorize their communicative vocalizations into different acoustic features which convey different meanings, relying on contextual cues linked to the environment (presence of food or predators), to

---

[20] Marler 2000.

[21] I use this term referring to the regulated combination of monkey vocalization sounds, which by themselves are not as differentiated as human languages phonemes.

[22] Cheney and Seyfarth 2008, pp. 232–233.

the social relationship occurring between the vocalizing monkeys (in the case of vocal interactions), or to the emotional state of the caller. The inferred meaning of the vocalization relies either on the acoustic features of the signal, or on the information acquired on the basis of associations experienced in the past.

Concerning the second level of description adopted by Marler, the lexical syntax, recent studies suggest that the levels of syntactical complexity characterizing human verbal propositions are not widespread in animal communication systems.[23] Primate calls cannot be broken into meaningful units, and there are no parts comparable to words which can be combined in any rule-governed structure within a meaningful sentence, conveying a message which would be more than the sum of its parts. Nonetheless, recent field research has revealed the existence of a few important exceptions concerning rudimental cases of "vocal syntax" in non-human animals. Zuberbühler has observed that Campbell monkeys, a species living on the western Ivory coast, emit a pair of low "boom" calls before their alarm calls, in the presence of less dangerous situations such as a falling branch or upon hearing the predator alarm call of a distant group. As the author asserts, it seems that this acoustic component somehow affects the overall meaning of the call:

> ["Boom" vocalization] is given in pairs separated by some seconds of silence and typically precedes an alarm call series by about 25s. Boom-introduced alarm call series are given to a number of disturbances, such as a falling tree or large breaking branch, the far-away alarm calls of a neighbouring group, or a distant predator. Common to these contexts is the lack of direct threat in each, unlike when callers are surprised by a close predator.[24]

In this direction, a study conducted on the potty-nosed monkey reveals that this species uses two types of signals (pyows and hawks) and inverting them generates different meaning effects:

> Series consisting of "pyows" are a common response to leopards, while "hacks" or "hacks" followed by "pyows" are regularly given to crowned eagles. Sometimes, males produce a further sequence, consisting of 1–4 "hacks". These "pyows-hack" (P-H) sequences can occur alone, or they are inserted at or near the beginning of another call series. Regardless of the context, P-H sequences reliably predict forthcoming group progression. […] *We conclude that, contrary to current theory, meaningful combinatorial signals have evolved in primate communication and future work may reveal further examples.*[25]

Although these data confirm the ability, at least in some species of monkeys, to combine a few signals in a very rudimental way generating qualitatively different meanings, they lack the general capacity to apply combinatorial rules to produce an open-ended set of vocal productions, an ability that is typically human.

Importantly, evidence suggests that songbirds and whales also possess the ability of phonological syntax; in fact, a number of studies addressed have shown that these species are able to concatenate the notes of their songs following a hierarchical and non-random transitional structure.[26] Further, it has been shown that in chickadees,

---

[23] Cf. Collier et al. 2014 for a review.

[24] Zuberbühler 2002, p. 294. Cf. Ouattara et al. 2009.

[25] Arnold and Zuberbühler 2008, pp. 202–203; italics mine. – *P.F.*

[26] Okanoya 2004; Suzuki et al. 2006; Clarke et al. 2006; Jansen et al. 2012; Berwick et al. 2011.

experimental change to songs composition, rhythm, or component order tends to interfere with its communicative function.[27] Based on these data, we can identify in the ability to concatenate sounds within an utterance an "analogous" trait, i.e. a biological trait that has evolved independently in phylogenetically distant species, under the same selective forces. Importantly, studies suggest that this ability has evolved under the evolutionary pressures linked to sexual selection,[28] territory defense,[29] or group bonding.[30]

Theory of Mind

A study concerning the evolutionary dynamics of language cannot disregard the research on the precursors of the capacity that had a key role in determining the specificity of human cognition: the ability to attribute mental states to conspecifics within a frame of shared intentions and joint actions.

It is worth asking, then, whether non-human animals are equipped with some equivalent ability. In order to address this question, it is necessary to distinguish the signaler's perspective from the receiver's one. As Seyfarth and Cheney assess, indeed, the formers are not aware of the state of knowledge of the receivers, neither do they communicate on the explicit goal to change it. Nonetheless, on the other hand, the achieved effect is to supply the listeners with useful information, or to cause an emotional and behavioural response:

> [...] the co-evolution of caller and recipient has favored signalers who call strategically and listeners who acquire information from vocalizations, using this information to represent their environment. The inability of animals to recognize the mental states of others places important constraints on their communication and distinguishes animal communication most clearly from human language. With the possible exception of chimpanzees, animals cannot represent the mental state of another. As a result, whereas signalers may vocalize to change a listener's behavior, they do not call with the specific goal of informing others or in response to the perception of ignorance in another. Similarly, whereas listeners extract subtle information from vocalizations, this does not include information about the signaler's knowledge. Listeners acquire information from signalers who do not, in the human sense, intend to provide it.[31]

Interestingly, multiple studies suggest that a wide variety of species (phylogenetically both related and distant from humans) posses the ability to know what other individuals see.[32] This might be considered an evolutionary precursor of the theory of mind. Importantly, although extensive research has been dedicated to

[27] Freeberg and Lucas 2002.

[28] Searcy and Andersson 1986.

[29] Holland et al. 2000.

[30] Boeckle and Bugnyar 2012; Doupe and Kuhl 1999; Treisman 1978.

[31] Seyfarth and Cheney 2003, p. 168.

[32] Bugnyar and Bernd 2006; Hare and Tomasello 1999; Hare et al. 2000 and 2003; Anderson et al. 1996; Ruiz et al. 2008.

animals' ability to infer others' states of mind, no common agreement on the interpretation of the resulting findings was achieved. In fact, much of the observed behaviours might be merely explained in terms of associative learning from previous experience. Thus, we can conclude that although the ability to attribute mental states to other individual (i.e. to understand the other's beliefs and desires in intentional terms and to use this knowledge to trigger specific behaviors) is uniquely human, certain evolutionary constraints underlying this ability are present also in non-human species.

Could We See a Holistic Protolanguage Through Monkeys' Communication System?

The data discussed above can be used as a window through which the evolution of language can be studied. According to the methodological criteria proposed by Botha, this approach satisfies the three conditions of groundedness, warrantedness, and pertinence. Indeed, an overall analysis of non-human animals' vocalization system has provided pivotal empirical data (although further investigations are still necessary). This allows us to recognise that the criterion of groundedness of the theory is satisfied. Moreover, the comparative approach I have adopted is empirically supported by the evolutionary data provided by studies on "homologs" – i.e. structurally similar traits that belong to phylogentically close species and on "analogs" – i.e. functionally similar traits that phylogenetically distant species have acquired independently. Finally, the condition of pertinence is guaranteed by the identification of language with the ability to speak and understand a natural language, where meanings are: (1) syntactically structured, (2) acquired through social practises and (3) ontologically tied to the pragmatic and/or emotional situation in which they occur.

The data reviewed in the present study support the adoption of the holistic model proposed by Wray[33] as more plausible than the analytic one proposed by Bickerton. Indeed, even if the signalers are not able to communicate intentionally (that is, with a conscious, explicit aim to provide other individuals with specific information) – the listeners are nonetheless able to get from such unintentional utterances an arrangement of complex meanings, not reducible to mere lexical labeling. Regarding this last point, it is noteworthy to remark Cheney and Seyfarth's observations about primates' alarm calls: "Baboon alarm calls, like those of other primates, are thus holistic utterances, simultaneously both eventish and objectish because they incorporate both reference to an object and a disposition to behave toward that object in a particular way".[34]

[33] Wray 1998 and 2002.

[34] Cheney and Seyfarth 2008, p. 256.

Conclusions

The assumption that the first human utterances were holistic is an important step in the study of the origin of language, and opens new questions to address. For instance, it would be interesting to study the specific dynamics concerning the evolution of the ability to know what other individuals see into the ability to infer what they know: a faculty that is closely related to the ability to share thoughts, attention targets, and goals. A second question concerns the pragmatic and cognitive process that, within an increasing complex frame of shared attention and actions, grounds the evolution of the holistic messages into syntactically structured sentences. These research questions might pave the way for an increasing understanding of the evolution of propositional language and to its links to animal communication.

Acknowledgments Funding was provided by a European Research Council Advanced Grant SOMACCA (No. 230604) awarded to WTF. The funders had no role in decision to publish, or preparation of the manuscript.

References

Anderson, J. R., Montant, M., & Schmitt, D. (1996). Rhesus monkeys fail to use gaze direction as an experimenter-given cue in an object-choice task. *Behavioural Processes, 37*(1), 47–55.

Arnold, K., & Zuberbühler, K. (2008). Meaningful call combinations in a non-human primate. *Current Biology, 18*(5), 202–203.

Berwick, R. C., Okanoya, K., Beckers, G. J. L., & Bolhuis, J. J. (2011). Songs to syntax: The linguistics of birdsong. *Trends in Cognitive Sciences, 15*(3), 113–121. Doi:10.1016/j.tics.2011.01.002.

Bickerton, D. (2002). Foraging versus social intelligence in the evolution of protolanguage. In A. Wray (Ed.), *The transition to language* (pp. 207–225). Oxford: Oxford University Press.

Boeckle, M., & Bugnyar, T. (2012). Long-term memory for affiliates in ravens. *Current Biology, 22*, 1–6. Doi:10.1016/j.cub.2012.03.023.

Botha, R. P. (2006). On the windows approach to language evolution. *Language and Communication, 26*, 129–143.

Bugnyar, T., & Bernd, H. (2006). Pilfering ravens, Corvus corax, adjust their behaviour to social context and identity of competitors. *Animal Cognition, 9*(4), 369–376.

Cheney, D. L., & Seyfarth, R. M. (2008). *Baboon metaphysics. The evolution of a social mind.* Chicago: The University of Chicago Press.

Chomsky, N. (1980). Rules and representations. *Behavioral and Brain Sciences, 3*(01), 1–15.

Christiansen, M. H., & Kirby, S. (2003). Language evolution: Consensus and controversies. *Trends in Cognitive Sciences, 7*, 300–307.

Clarke, E., Reichard, U. H., & Zuberbühler, K. (2006). The syntax and meaning of wild gibbon songs. *PLoS ONE, 1*(1), e73. Doi:10.1371/journal.pone.0000073.

Clay, Z., & Zuberbühler, K. (2009). Food-associated calling sequences in bonobos. *Animal Behaviour, 77*(6), 1387–1396.

Collier, K., Bickel, B., Schaik, C. P. P. van, Manser, M. B., & Townsend, S. W. (2014). Language evolution: Syntax before phonology? *Proceedings of the Royal Society B: Biological Sciences, 281*(1788). Doi:10.1098/rspb.2014.0263.

Deacon, T. W. (1998). *The symbolic species: The co-evolution of language and the brain*. New York: WW Norton & Company.

Doupe, A. J., & Kuhl, P. K. (1999). Birdsong and human speech: Common themes and mechanisms. *Annual Review of Neuroscience, 22*(1), 567–631.

Fitch, W. T. (2010). *The evolution of language*. Cambridge: Cambridge University Press.

Freeberg, T. M., & Lucas, J. R. (2002). Receivers respond differently to chick-a-dee calls varying in note composition in Carolina chickadees, *Poecile carolinensis*. *Animal Behaviour, 63*(5), 837–845. doi:10.1006/anbe.2001.1981.

Hare, B., & Tomasello, M. (1999). Domestic dogs (*Canis familiaris*) use human and conspecific social cues to locate hidden food. *Journal of Comparative Psychology, 113*(2), 173–177. doi:10.1037/0735-7036.113.2.173.

Hare, B., Call, J., Agnetta, B., & Tomasello, M. (2000). Chimpanzees know what conspecifics do and do not see. *Animal Behaviour, 59*(4), 771–785.

Hare, B., Addessi, E., Call, J., Tomasello, M., & Visalberghi, E. (2003). Do capuchin monkeys, Cebus apella, know what conspecifics do and do not see? *Animal Behaviour, 65*(1), 131–142.

Hauser, M. D. (1996). *The evolution of communication*. Cambridge: MIT Press.

Hauser, M. D., & Marler, P. (1993). Food-associated calls in rhesus macaques (*Macaca mulatta*). 1. Socioecological factors. *Behavioral Ecology, 4*, 194–205.

Holland, J., Dabelsteen, T., & Paris, A. L. (2000). Coding in the song of the wren: Importance of rhythmicity, syntax and element structure. *Animal Behaviour, 60*(4), 463–470.

Jansen, D. A., Cant, M. A., & Manser, M. B. (2012). Segmental concatenation of individual signatures and context cues in banded mongoose (*Mungos mungo*) close calls. *BMC Biology, 10*(1), 97. doi:10.1186/1741-7007-10-97.

Marler, P. (2000). Origins of music and speech: Insights from animals. In N. L. Walli, B. Merker, & S. Brown (Eds.), *The origins of music* (pp. 31–48). Cambridge: MIT Press.

Okanoya, K. (2004). Song syntax in Bengalese finches: Proximate and ultimate analyses. *Advances in the Study of Behavior, 34*, 297–346.

Jespersen, O. (1922 [2013]). *Language. Its nature, development and origin*. Oxford: Routledge.

Ouattara, K., Lemasson, A., & Zuberbuhler, K. (2009). Campbell's monkeys concatenate vocalizations into context-specific call sequences. *Proceedings of the National Academy of Sciences, 106*, 22026–22031.

Ruiz, A., Gómez, J. C., Roeder, J. J., & Byrne, R. W. (2008). Gaze following and gaze priming in lemurs. *Animal Cognition, 12*(3), 427–434. doi:10.1007/s10071-008-0202-z.

Searcy, W. A., & Andersson, M. (1986). Sexual selection and the evolution of song. *Annual Review of Ecology and Systematics, 17*, 507–533.

Seyfarth, R. M., Cheney, D. L., Bergman, T., Fischer, J., Zuberbühler, K., & Hammerschmidt, K. (2010). The central importance of information in studies of animal communication. *Animal Behaviour, 80*(1), 3–8. doi:10.1016/j.anbehav.2010.04.012.

Seyfarth, R. M., & Cheney, D. L. (2003). Signalers and receivers in animal communication. *Annual Review of Psychology, 54*, 145–173.

Struhsaker, T. T. (1967). Auditory communication among vervet monkeys (*Cercopithecus aethiops*). In S. A. Altmann (Ed.), *Social communication among primates* (pp. 281–324). Chicago: University of Chicago Press.

Suzuki, R., Buck, J. R., & Tyack, P. L. (2006). Information entropy of humpback whale songs. *The Journal of the Acoustical Society of America, 119*(3), 1849. doi:10.1121/1.2161827.

Treisman, M. (1978). Bird song dialects, repertoire size, and kin association. *Animal Behaviour, 26*, 814–817. doi:10.1016/0003-3472(78)90146-X.

Wittgenstein, L. (1953 [2010]). *Philosophical investigations*. New York: Wiley.

Wray, A. (1998). Protolanguage as a holistic system for social interaction. *Language and Communication, 18*, 47–67.

Wray, A. (2002). Dual processing in protolanguage: Performance without competence. In A. Wray (Ed.), *The transition to language* (pp. 113–137). Oxford: Oxford University Press.

Yip, M. J. (2006). The search for phonology in other species. *Trends in Cognitive Sciences, 10*(10), 442–446.

Zuberbühler, K. (2002). A syntactic rule in forest monkey communication. *Animal Behaviour, 63*, 293–299.

Zuberbühler, K., Noë, R., & Seyfarth, R. M. (1997). Diana monkey long-distance calls: Messages for conspecifics and predators. *Animal Behaviour, 53*(3), 589–604.

Zuberbühler, K., Cheney, D. L., & Seyfarth, R. M. (1999). Conceptual semantics in nonhuman primates. *Journal of Comparative Psychology, 113*(1), 33–42.

Part IV
History of Ideas

Biosemiotics, Politics and Th.A. Sebeok's Move from Linguistics to Semiotics

Sara Cannizzaro and Paul Cobley

Abstract This paper will focus on the political implications for the language sciences of Sebeok's move from linguistics to a global semiotic perspective, a move that ultimately resulted in biosemiotics. The paper will seek to make more explicit the political bearing of a biosemiotic perspective in the language sciences and the human sciences in general. In particular, it will discuss the definition of *language* inherent in Sebeok's project and the fundamental re-drawing of the grounds of linguistic debate heralded by Sebeok's embrace of the concept of modelling. Thus far, the political co-ordinates of the biosemiotic project have not really been made explicit. This paper will therefore seek to outline

- how biosemiotics enables us to reconfigure our understanding of the role of language in culture;
- how exaptation is central to the evolution of language and communication, rather than adaptation;
- how communication is the key issue in biosphere, rather than language, not just because communication includes language but because the language sciences often refer to language as if it were mere "chatter", "tropes" and "figures of speech";
- how biosemiotics, despite its seeming "neutrality" arising from its transdisciplinarity, is thoroughly political;
- how the failure to see the implications of the move from linguistics to semiotics arises from the fact that biosemiotics is devoid of old style politics, which is based on representation (devoid of experience) and "construction of [everything] in discourse" (which is grounded in linguistics, not communication study).

In contrast to the post-"linguistic turn" idea that the world is "constructed in discourse", we will argue that biosemiotics entails a reconfiguration of the polis and, in particular, offers the chance to completely reconceptualise ideology.

Keywords Th.A. Sebeok • Linguistics • (Bio)semiotics • Politics

S. Cannizzaro (✉)
London Metropolian University, London, UK
e-mail: sblissa@gmail.com

P. Cobley
Middlesex University, London, UK
e-mail: P.Cobley@mdx.ac.uk

E. Velmezova et al. (eds.), *Biosemiotic Perspectives on Language and Linguistics*, Biosemiotics 13, DOI 10.1007/978-3-319-20663-9_11

The Role of Language in Culture

Although Thomas A. Sebeok played a central role in propelling international study of communication in the 1950s, his career before his sojourn studying animal communication at Stanford was focused mainly on linguistics and the ethnographic study of language.[1] His study of language (singular) as a general phenomenon was based on solid fieldwork in immense quantity on languages (plural). Sebeok repeatedly defined himself as "a biologist manqué"[2] and even suggests that he "became a professional linguist and, alas forever, a geneticist manqué".[3] *Perspectives on Zoosemiotics*, the book which collects the post-1964 watershed works on animal communication is dedicated to the geneticist who taught Sebeok at Chicago, Joseph J. Schwab. However, the figure who haunts its earlier pages is Roman Jakobson, Sebeok's one-time mentor who bequeathed the idea of "distinctive features" as "universal building blocks of language".[4] Sebeok, at this stage in his career, clearly considered distinctive features the "most concretely and substantively realized"[5] part of general linguistic theory. In this way, then, Sebeok's thinking on linguistics followed the orthodoxy of the time, in thrall to the "language myth"[6] in which linguistic communication is seen to be embodied in basic coded elements quasi-independent of human interaction.[7] Yet, even in the post-1964 essays re-printed in *Perspectives* he notes that the "phylogeny of distinctive features [...] has clearly not yet progressed beyond mere speculation",[8] thus opening biosemiotic questions even while engaging in the customary closure of communicational questions characteristic of that period in institutional linguistics.

Ultimately, Sebeok's project was to lead to the fundamental re-drawing of the grounds of linguistics through his embrace of the concept of modelling. A small part of this project was inspired by the Chomskyan revolution in language study from mid-century. Yet while this revolution morphed into a further variant of the "language myth", particularly in its spawning of cognitivism,[9] Sebeok embedded language in the much broader frame of semiotics, revealing language to be a modelling process whose origins and ramifications were to be found far beyond the utilizing of coded elements. That modelling was central to Sebeok's semiotics after his rediscovery of Jakob von Uexküll in the mid-1970s and that this effectively forged the field of biosemiotics is well known. What is less discussed but will be considered in what follows is the massive political shift that this development heralded.

---

[1] Sebeok 2001a.

[2] E.g., in Sebeok 1991a and 2011, p. 457.

[3] Sebeok 1972, p. 2.

[4] *Ibid.*, p. 86.

[5] *Ibid.*

[6] For example, the essays in Sebeok 1972.

[7] Cobley 2014.

[8] Sebeok 1972, p. 88.

[9] Harris 2008.

Communication as Adaptation Versus Language as Exaptation

But firstly, let us consider what Sebeok's notion of modelling entails. Taking his cue from the Tartu-Moscow notion of modelling system, and Juri Lotman's model of the semiosphere, Sebeok proposed a reconfiguration[10] of the "Primary Modelling System" that can be argued to constitute a core distinctive paradigmatic feature of biosemiotics. In reconfiguring the pre-existing notion of modelling system Sebeok suggested that (what was once called) "Soviet semiotics"[11] did not sufficiently take into account how humans could communicate and build "cultures" well before mastering externalised verbal signs. Primary modelling, evident in humans since *Homo habilis* circa 300,000 years ago, preceded and is the basis of the verbal encoding and decoding that developed with *Homo sapiens* (around 300,000 years ago). In the previous millennia communication had been carried out among humans by exclusively nonverbal communication; verbal communication, speech and writing – syntax-based linear communication or externalised verbal communication – were exapted[12] as opposed to adapted. Human modelling as such is unique among animals because it features both nonverbal and verbal communication[13] or, as Terrence W. Deacon[14] puts it, we are "apes plus language". Early humans' possession of a mute verbal modelling device featuring a basic capacity for syntax allowed humans to assemble standardised tools but circumstances had not yet arisen whereby it was expeditious or hominids were in agreement to encode communication in articulate linear speech.[15] Thus, for Sebeok, there are sign systems (nonverbal communication) which in terms of evolution are antecedent to, and give rise to, externalised linguistic sign systems. Nonverbal communication is recognised by Sebeok as an adaptive communicational capacity possessed by all living beings.[16] It is, in fact, only hominids across the whole animal kingdom that possess two mutually sustaining repertoires of signs: the zoosemiotic nonverbal and the anthroposemiotic verbal.[17]

The perspectives of the erstwhile "Soviet semiotics", which put verbal language at the basis of all communications and of the organisation of culture, was at risk of both glottocentrism and anthropomorphism. In light of the recognition that there is communication prior to verbal language, Sebeok recast Tartu-Moscow notion of modelling systems and observed that (verbal language) "is the modelling system the Soviet scholars call primary but which, in truth, is phylogenetically as well as

[10] Sebeok 1991b.

[11] Lucid 1977.

[12] Gould and Vrba 1982.

[13] Sebeok 1991b.

[14] Deacon 1997, p. 5.

[15] Sebeok 1991b, p. 55.

[16] Sebeok 1981 and 1991b.

[17] Sebeok 1991b, p. 55.

ontogenetically secondary to the nonverbal".[18] Thus, according to Sebeok, "natural language" or the primary modelling system is not verbal language, but is a cognitive capacity manifested in "nonverbal communication" through chemical, thermal, olfactory, acoustic and visual means. In humans, such primary modelling existed, phylogenetically, alongside the cognitive capacity manifested in the production of externalised verbal signs (secondary modelling system). However, it was not until *Homo sapiens* that such signs (secondary modelling systems) were routinely circulated.

To grasp this point, it is necessary to move back further, as Sebeok does, beyond the period of "Soviet semiotics", to the work of the theoretical biologist, J. von Uexküll. Signs, as well as what makes up signs, constitute what Uexküll[19] has called an *Umwelt*. The theory of *Umwelt* posits that all species live in a "world" that is constructed out of their own signs, the latter being the result of their own sign-making and receiving capacities. A fly, for example, has a much different sensory apparatus for making/receiving signs than does the human. Beyond those capacities of semiosis (sign action) there is a world, the "real" one, in a sense, which cannot be reached. Yet, while it is true that within a species' *Umwelt* there are all manner of possibilities of illusion – through misinterpretation of signs, through overlooking of signs and through signs not being 100 % adequate representations of reality – the testimony that an *Umwelt* is a fairly good guide to reality is offered by the survival of the species within a given *Umwelt*. Semiotics is the study of comparative *Umwelten*[20] and, as such, must be concerned with animal and plant communication whilst principally attending to the human *Umwelt* which is characterised by what Sebeok called "language" – not linguistic communication but the innate and phylogenetically developed "modelling" device mentioned above.

It is in this that Sebeok develops what is probably the core proposition of biosemiotics: that the primordial and overarching form of communication is nonverbal.[21] Nonverbal communication characterizes all life, including a large part of human life. Although humans also utilize verbal communication, nonverbal communication is implicitly overlooked in many realms of human endeavour. In fact, as we signalled above, Sebeok holds that natural language "evolved as an adaptation; whereas speech developed out of language as a derivative exaptation".[22] That is, while the primary modelling system (refigured by Sebeok), sustaining nonverbal communication and driven by the increased brain size and differentiation capacity had a palpable survival function, the development of the secondary modelling system was not a necessary survival mechanism. Primary modelling, argued Sebeok, "has been built by selection for the cognitive function of modelling and, as the philosopher Popper and the linguist Chomsky have likewise insisted, not at all for

[18] *Ibid.*

[19] Uexküll 2001a, b.

[20] Cobley 2001.

[21] Cf., especially, Sebeok 2001b.

[22] Sebeok 1991b, p. 56.

the message-swapping function of communication".[23] Hypothetically, hominids might have continued communication by nonverbal means for many more millennia; yet they began to utilize their capacity for differentiation along with their evolved vocal apparatus to produce verbal communication, little knowing that the much later developments of speech and cheirography would generate oral narratives forging communities and written scripts facilitating agriculture and economics.

Exaptation, here and also as Stephen J. Gould and Elisabeth S. Vrba discussed it, demonstrates that one should not assume that the current utility of a biological phenomenon is a result of natural selection. An exaptation may be desirable and potentially an enhancement of the capacity for survival; but that does not necessarily entail that it is indispensable for survival, nor that the phenomenon in question is the product of natural selection. As Davide Weible shows,[24] *exaptation* has become a useful term for scholars in biosemiotics. Yet, what exaptation demonstrates most strikingly in respect of human evolution is that the phenomenon often central to definitions of humanity – language – is, in the verbal forms that have provided the foundation for communication and culture, only beneficial in evolutionary terms at one remove or more, or even, perhaps, in various cases, not beneficial at all. The communicational forms that are often taken for granted in the human *Umwelt* and, sometimes, have been assumed to be the only portal through which humans can grasp life, are, in this account, merely the veneer of anthroposemiosis.

Transdisciplinarity as Apparent "Neutrality"

Moving the focus on semiosis from the level of signs circulating in the polis to those circulating in comparative *Umwelten* in the way that we propose might be seen as a gesture towards the apolitical, a gesture that is ultimately reactionary since it is a denial of the politicization of signs. Yet nothing could be further from the truth. Contemporary semiotics, in its transdisciplinarity, has no pretentions to "neutrality".

A lesson is offered from history. In addition to his work in cybernetics and communication theory during the 1950s, as well as his inauguration of semiotics for the present era beginning with his editing of the *Approaches to Semiotics* volume in 1964, Sebeok also disseminated the transdisciplinary approach that was characteristic of Tartu-Moscow semiotics as a whole.[25] However, the Tartu-Moscow school was grounded in the interdisciplinary developments of 1950s and 1960s Soviet academia which were, in turn, influenced by cybernetics; thus, it seems that biosemiotics, in building on Tartu-Moscow semiotics' transdisciplinarity, automatically inherits cybernetics' transdisciplinarity.[26] Historically, transdisciplinarity did

[23] *Ibid.*

[24] Weible 2012.

[25] Randviir 2007.

[26] Waldstein 2008, p. 17.

become putatively aligned with "neutrality". Maxim Waldstein claims that due to its closeness to mathematical sciences, cybernetics appealed to Soviet scholars as an "ideology-free" and thus neutral language. This is because cybernetics was believed to aid the clear formulation of problems and thus could have favoured the reception and expansion of structural linguistics. Arguably then, cybernetics was being "marketed" as the "maths of the humanities", particularly in light of it being underwritten, as previously shown, by instances of mathematical modelling. For example, such a "rhetoric of exactness" is found in Warren McCulloch and Walter Pitts' theory of formal neural networks which postulates that "any functioning [of a system] which can be defined in its entirety logically, strictly and unambiguously in a finite number of words, can also be realised by such formal neural networks"[27]; that is, anything that can be put into a question with words can be solved. As Waldstein contends, this impetus towards exactness and the "ideological neutrality" that is indigenous to cybernetics constituted a point of appeal for the semiotics developed during Soviet times in that it promised to be "a recipe for transformation of linguistics and other human sciences into 'true sciences'".[28] This is because such a promise was directly in opposition to Stalinist thinking which impeded scientific discovery. As Laurent Schwartz usefully illustrates: "In physics [...] such fields as quantum physics were sometimes condemned as anti-Marxist, and in biology all progress was rendered practically impossible for twenty-five years because of Lysenko",[29] who was a member of the USSR Academy of Sciences who championed the non-Darwinian theory that within species there is no overpopulation nor struggle for survival. He affirmed that a progressive biological science would be indebted not to Darwin but to Lenin and Stalin.[30] On the other hand, "under Stalin, mathematics was probably more secure than other branches of science, doubtless because it is less accessible".[31] Hence, the reason why academics in the humanities fixated on mathematical models: they were representative of theory that was not accessible to the majority, even intellectually. In fact, in 1964 the term *secondary modelling system* (notably, *modelling* is a mathematical term) was used as an euphemism for *semiotics* because the very term *semiotics* became quasi-prohibited by scientific state officials.[32]

In other words, one may argue that the early "alliance" of Soviet academia with cybernetics can be seen as the beginning of a process of de-Stalinisation of knowledge, which is the core of what was later dubbed Eurocommunism, or "the vast process of change involving the left everywhere in the world – that of de-Stalinisation".[33] In fact, as Carl Boggs and David Plotke argue, Eurocommunism presents itself as a political formation that sets out to transcend the failures of the

[27] McCulloch and Pitts, quoted in Neumann 1948 [1963, p. 309].

[28] Waldstein 2008, p. 18.

[29] Schwartz 1984, p. 179.

[30] *Ibid.*, p. 185.

[31] *Ibid.*, p. 179.

[32] Chernov 1988, p. 12.

[33] Ross 1980, p. 15.

past through, amongst other things, involvement in political struggles that take place within institutions and a principled support of social and political pluralism[34] much like that which "Soviet" academics were trying to achieve. Additionally, Massimo L. Salvadori[35] argues that the core basis of Western European Communist parties (the Eurocommunists) was a desire for autonomy from the [Stalinist] USSR and the adhesion to principles of democracy. Hence one may argue that Soviet academia's desire for autonomy and its pursuit of "scientific neutrality" through mathematical models could be seen as a precedent for Eurocommunism, or its historical context.

Yet, of course, one can see how such a "neutral" view was an ideological – in this case, anti-Stalinist – position in itself. Hence, as a prefigurement of neutrality and transdisciplinary applicability that was in itself fundamentally political, cybernetics is said to have favoured (in Soviet academia, but arguably also in West-European countries) the birth of semiotics as a science aimed at the study of "any sign system in human society".[36] The universal model of applicability proposed by cybernetics, or its transdisciplinary character, was thus assimilated into the "Soviet semiotics" project as illustrated by Daniel Peri Lucid.[37] The recognition that cybernetics had a strong influence on the birth of Tartu-Moscow semiotics is important because this division of semiotics was then co-opted by biosemiotics, through the elaboration of Lotman's work on modelling[38] and semiosphere.[39] In this respect, an awareness of Soviet semiotic interest in cybernetics constitutes the historical and disciplinary basis for understanding, conceiving and relaunching a new biosemiotic and transdisciplinary polis, for however contradictory this expression might sound.

The transdisciplinarity of contemporary semiotics after Sebeok is a curious phenomenon. It stems, in part, from the acutely political attempt to carry out research in a "neutral" frame under a repressive regime. Yet, the broadening of semiotics is also a political move in a much more general sense. Discovering that semiosis is politically charged in the polis is one thing; but conveniently forgetting that semiosis occurs and is built on the development of signs in realms far beyond the polis is considerably more "apolitical" and reactionary than attempting to assume a supposedly "neutral" transdisciplinary vantage point. It is the equivalent of mapping some of the co-ordinates within the dark cupboard under the stairs of a vast mansion and proclaiming "We're now able to know the house".

[34] Boggs and Plotke 1980, p. 7.

[35] Salvadori 1978, p. xxv.

[36] Ivanov, quoted in Waldstein 2008, p. 20.

[37] Lucid 1977.

[38] Sebeok 1988; Sebeok and Danesi 2000.

[39] Kull 1998; Hoffmeyer 1993 [1996]; Brier 2008.

Old Polis: Representation and the Construction [of Everything] in Discourse

Biosemiotics' ranging across the whole of semiosis – animal and plant – has been in distinct contrast with the powerful idea, developed in the last 40 years, that many of the determinants of human life are "constructed in discourse". The "linguistic turn" in social thought, inaugurated by Richard Rorty's 1967 collection,[40] has been influential in areas of knowledge where the volume is seldom if ever cited. More important still, perhaps, and arguably more nebulous, has been the work of structuralism and poststructuralism and their basis in a philosophy of the sign derived from Ferdinand de Saussure that is often critiqued but infrequently rejected altogether. This has been elaborated upon, disseminated through the human sciences in the West and almost naturalized in Francophone academia from the 1950s onwards and from the late 1960s onwards in the Anglophone world. One subject area in which this perspective has held sway is the one in which both authors of the current article work: media, communications and cultural studies.

The idea of the world "constructed in discourse" has underpinned much of the study of the media which is concerned with the key issue of "representation". Introduced in its recognizable form by, among others, Roland Barthes in *Mythologies,*[41] representation has occupied a privileged role in signification, even as its variant of representation through code,[42] generally neglecting the pragmatic/subjective aspect of sign processes. For example Stuart Hall[43] claims that "the meaning is not in the object, person or thing, nor is it in the word… The meaning is constructed by the system of representation". One can immediately see that this approach privileges representation over other aspects of signification, as if construction of meaning excluded emotional, physiological and environmental constraints or its actual context of use. Hence one may argue that Hall's view tends to worry about the "text in principle" rather than the "text in practice". Even the tedious ideological debate[44] about the active or passive status of readers or media audiences which was conceived in the 1980s as a solution to the orthodoxy embedded in approaches to representation, misses the point. It ignores the fact that representation certainly influences the process of signification but it is very far from being the sole player in signification, or the only factor responsible for the construction of meaning.

More redolent, still, of a linguistic perspective, in *Mythologies*[45] Barthes introduces the concept of myth; that is, a linguistic epiphenomenon amounting to a collective representation of reality which, in his view, is not a reflection of reality

[40] Cf. Rorty 1967.

[41] Barthes 1957b [1973].

[42] Cobley 2013.

[43] Hall 1997, p. 21.

[44] Quoted by e.g. Fiske 1989; Bignell 1997.

[45] Barthes 1957b [1973].

itself but a reflection of culture. In other words, myths are responsible for making "culture" pass as "nature" or for turning "the social, the cultural, the ideological, the historical into the 'natural'".[46] For example, in discussing the myth of the Romans in films, Barthes states that "in Mankiewicz's Julius Caesar, all the characters are wearing fringes. Some have them curly, some straggly, some tufted, some oily, all of them well combed, and the bald are not admitted, although there are plenty in Roman history".[47] In short, Barthes underlines the discrepancy between fiction (Romans with a fringe) and reality (Romans who must have suffered hair loss and thus no fringe). In "Myth today", the final theory-based essay of *Mythologies*, Barthes uses Louis Hjelmslev in order to turn this amusing but simple observation into a complex linguistic argument, invoking different levels of form, substance, plane of expression and plane of content in the act of representation.

Barthes claims that "myth hides nothing and flaunts nothing: it distorts; myth is neither a lie nor a confession: it is an inflection".[48] Yet he constantly proposes the idea that the representation of reality as elicited by myths is false: "The […] sign, the fringe of Roman-ness […] reveals a degraded spectacle, which is equally afraid of simple reality and of total artifice. For although it is a good thing if a spectacle is created to make the world more explicit, it is both reprehensible and deceitful to confuse the sign with what is signified".[49]

Barthes' worry is that viewers of Joseph L. Mankiewicz' s movie will inevitably confuse the false Romans (with the fringe) with the real Romans (who may have not had the fringe). Through the concept of myth, Barthes takes complexity away from signification and turns it into a typical formal logic problem in which the analyst's job is to determine the True or False aspect of a final proposition (in this case, the cultural proposition elicited in representation i.e. that Romans with fringes is a false statement). In so doing, Barthes reduces the whole process of signification solely to its representational aspect, a view that subsists merely "at the surface level" of analysis.[50] This perspective, not just prevalent but naturalised in media studies, reduces signification to representation. It singularly fails to address the question of why audiences/readers/human beings willingly and persistently allow themselves to "get fooled" in watching movies that present false Romans or characters or settings that are equally fictitious. Nor can the question simply be answered by quasi-ethnographic audience study. It needs to be addressed by broadening, or even abandoning, the current linguistically-based concept of representation.

It is hardly surprising that the "representational" perspective ultimately finds itself in a cul-de-sac. It is glottocentric and therefore fails to take account of humans as thoroughly semiotic entities within a vast environment of (non-human) semiosis. Based on linguistics, it can only posit a very limited version of the sign, one which is mired in the vicissitudes of linguistic communication, chatter and figures of

46 Barthes 1977, p. 165.

47 Barthes 1957c [1973, p. 26].

48 Barthes 1957a [1973, p. 129].

49 Barthes 1957c [1973, p. 27]; italics ours. – *S.C., P.C.*

50 Cobley 2006, p. 417.

speech which make up the loose, common understanding of "language". When one thinks of the sign in its full complexity – as semiotics does, but other fields do not have the time to do adequately – a different picture emerges. This fact is exemplified especially in the work of the American philosopher, John Deely, whose intellectual lineage can be traced back through the work of Sebeok and Uexküll through the Catholic thinker, Jacques Maritain, Charles Sanders Peirce, and to the *Tractatus* of João Poinsot, Aquinas and, ultimately the Stoics and Epicureans. For Deely, following Poinsot, signs are a matter of "relation" – not, as the representational perspective would have it, some entity standing in for some other entity from which it is different. For Poinsot and, later, for Peirce, the sign needs to be understood as the entire relation of its constituents. What is frequently considered the sign – the "relation" between some ground and some terminus – was discovered by the Latin thinkers to be false because it excluded the very awareness of sign functioning that distinguishes humans from other animals. The real relation that constitutes the sign consists of ground, terminus and "relation" as a triad. Furthermore, Poinsot delineates the functions of signs in relation to objects. As such, the relation of representation must differ from that of signification simply because an object can represent another and also represent itself. A sign is only a sign of something if that something is other than the sign.[51] Lastly, Poinsot emphasized that the relation in a sign is not so much suprasubjective as contextual: in one set of circumstances the relation in a sign could be of the order of *ens reale* (mind independent), in another set it could be of *ens rationis* (mind dependent).[52]

"Representation" assumes that human semiosis is mind-dependent (*ens rationis*), constantly preventing humans from gaining anything other than a tantalising glimpse of the mind-independent (*ens reale*) universe. Yet, as Deely is at pains to stress in the wake of Poinsot, the sign fluctuates between both forms of dependency according to context. One might add that implicit in the contextuality of the sign is the sharing of some parts of signhood across the world of humans, other animals and plants, the variegation of semiosis being so extensive that "representation" does not really come close to capturing it. Deely writes, initially with reference to St. Thomas,

> So the levels of dependency in being are complete, from the most tenuous of pure relations to the fullness of the divine being, with the twist that, according to Aquinas, the inner life of God consists in a community of persons each of which is a pure relation, but now relations themselves subsisting! It is an astonishing picture, much more interesting and intricate, actually, than anything dreamed of in modern philosophy, bogged down as it became in the technical detail necessary to try to maintain at all costs the facade of representations blocking our access to the order of ens reale, our development of knowledge of the things-in-themselves, things in the subjective constitution according to which they exist and interact among themselves and with our bodies.[53]

[51] Deely 2001.

[52] *Ibid.*, p. 729.

[53] Deely 2009, pp. 115–116.

Ultimately, Deely[54] concludes that "the social construction of reality, no doubt, occurs in the political order" and that "reality" "as we experience it is neither purely objective [in the sense of things experienced as objects in an *Umwelt*] nor purely subjective nor purely intersubjective, but rather a constantly shifting mixture and proportion of all three not at all easy (perhaps not even fully possible) to keep complete track of". As a statement about semiosis aligned with the biosemiotic idea of *Umwelt*, this demonstrates how biosemiotics does not really partake of the old style of politics based on linguistics and envisaging power in the masking of reality with illusion. Rather, it proceeds from humans' suspension in a universe of changing relations, sometimes "illusory", sometimes "real"; sometimes reliable enough to preserve members of a species, sometimes not. Such fluctuation and change entail that humans are not forever barred from reality, as the theory of representation insists; nor are they able to easily access the road to reality as adopting the theory of representation seems to imply. Rather, humans are charged with the task of enacting a semiotic awareness appropriate to the vagaries of relation. These changing relations, in the world of humans, have often been investigated by theories of ideology.

New Polis: Ideology as the Lived Biosemiotic Relationship to Existence

As adumbrated above, Barthes' concern was with the discovery of the power relations hidden in texts through representation; in other words, with ideology. The concept was especially taken up in media and cultural studies in the wake of Louis Althusser's "Ideology and ideological state apparatuses".[55] Arguably, though, insufficient attention has been paid to some of the complexities of signhood in this landmark essay and, consequently, approaches in media and cultural studies which advocate the falsity of representation[56] fall victim to one problem in Althusser's statements, specifically that "ideology" "represents the imaginary relationship of individuals to their real conditions of existence".[57] As Kevin McDonnell and Kevin Robins[58] convincingly contend, this aspect of Althusser's argument is vitiated by the idea of falsity implied in ideology: "It reduces ideology to mere false consciousness. [...] Ideology is no false consciousness, because it duplicates a concrete reality, one that really does exist, one that imposes itself on the texture of everyday life. [...] Nor is ideology, in this conception, immaterial, a mere epiphenomenon; it is an

[54] *Ibid.*, p. 116.

[55] Althusser 1971.

[56] E.g., Hall 1980; Dyer 1982 [1999]; Fiske 1989 and 1990; Goffman 1979; Jhally 1990; Vestergaard and Schrøder 1985; Williamson 1995 [2002]; Bignell 1997.

[57] Althusser 1971, p. 162.

[58] McDonnell and Robins 1980, p. 222.

illusion, but one that is 'the most efficacious reality, the spell that holds the world bewitched' (Adorno)".

This statement of the importance of the "imaginary", with its Lacanian overtones, has been seized too readily by many from a representational perspective. For example, in media and cultural studies John Fiske argues[59] that when confronted with popular texts, which are supposedly high in ideological content, one can choose whether to produce "a preferred reading according to the dominant code", a "negotiated reading", or a "radically opposed reading".[60] Clearly, it is important, as Fiske argues, that through the act of reading one can rebel against the repression exercised through ideology by (what Althusser calls) the ideological state apparatuses, in the very fact that readership implies "subjectivity" and activity (rather than passivity). Yet, the choices offered are rather too mechanical: very ideological, quite ideological, anti-ideological. It is easy to see why the renewed idea of "ideology as false consciousness" was readily taken up in media studies with Althusser as a cover – it provided the means to theorise a heroic reader who was not only capable of "resisting" or opposing ideologies (like Barthes' reader) but was also totally in control of his cognitive capacities and was free to choose whether to accept or resist ideological propositions.

However, a more careful reading of Althusser's essay, reveals a fact that rather undermines this position: that is, that the subjectivity implied in readership, does not exist prior to ideology, but is constituted by it. Famously, Althusser explains this point by positing a hypothetical situation in which, a policeman (representing the ideological state apparatus) shouts at a passer-by: "You, for whom I have shed this drop of my blood"/"Hey, you there!" The passer-by is then compelled to pay attention to, and reply, upon turning around: "Yes, it's me!" This vignette illustrates how the ideological state apparatus (the policeman) constitutes the subject, the individual whose identity (it's me! – self-recognition) has emerged at the same time in which the ideological act (the shout) was perpetrated. Impinging on the same example, Fiske argued – possibly following Michel Pêcheux[61] – that one can "resist" ideology in that "if you hear in the street a shout 'Hey You!', you can either turn in the belief that you are being addressed or you can ignore it… you thus reject the relationship implicit in the call".[62] However, conceiving the reader as a form of active audience that is active by the very means of being capable of resisting ideology is a view flawed from the start, because it presupposes that the subject exists and is as such (i.e. an "active" reader) before its encounter with ideology. This is a contradiction, despite its pretention to be an exploitation of an Althusserian loophole. In Althusserian terms, ideology cannot be resisted in that it is constitutive. That is, there is no such thing as an "I" before the very call "You", a perspective which is fully semiotic (subjectivity emerges out of relations of meaning) and that

[59] Following Hall 1973.

[60] Cf. in Fiske 1990.

[61] Pêcheux 1982.

[62] Fiske 1990, p. 175.

puts a heavy burden on the workings of "culture", "nurture" and "ideology" to sustain selfhood.

Neglecting this fact amounts to a desire to take the most "convenient" aspect of Althusser's Marxism (that authorities are repressive – the convenient aspect of this statement serving as a rationale for "response", including "reader response"), and neglecting the less convenient, that is, that the human being's subjectivity is not as unconstrained as such approaches would like to think. In contemporary semiotic terms, humans do not pre-exist semiosis and then struggle when they are somehow "inserted" into it. Nor are humans the conscious creators of semioses by which they can exercise control and power. In an *Umwelt*, as has been noted, humans inhabit from the start the very signs that their sensorium allows them to promulgate. Humans cannot "get outside" semiosis and control it; along with other living creatures, they are semiosis. This corresponds with the other plank of Althusser's work on ideology: concrete reality as a lived relation. For Althusser,[63] the imaginary and the lived are in a complex interplay: ideology

> is a matter of the lived relation between men [**sic**] and their world. This relation, that only appears as "conscious" on condition that it is unconscious, in the same way only seems to be simple on condition that it is complex, that it is not a simple relation but a relation between relations, a second degree relation. In ideology men do indeed express, not the relation between them and their conditions of existence, but the way they live the relation between them and their conditions of existence: this presupposes both a real relation and an "imaginary", "lived" relation. Ideology, then, is the expression of the relation between men and their "world", that is, the (overdetermined) unity of the real relation and the imaginary relation between them and their real conditions of existence. In ideology the real relation is inevitably invested in the imaginary relation, a relation that expresses a will (conservative, conformist, reformist or revolutionary), a hope or a nostalgia, rather than describing a reality.

In this formulation there is an indication of the consonance of Althusser's conception of ideology with the insistence of contemporary semiotics on the sign as always relation, but a relation oscillating between mind-dependent reality and mind-independent reality. Ultimately, Althusser's "imaginary", an idea that sustains the representational perspective, indicates a falling back on extraneous and confused speculations from Lacanian psychoanalysis in the hope that it will shore up a theory of subjectivity that will then complement the theory of ideology. Clearly, Althusser's insight into ideology as both "lived" and a "relation" was groundbreaking, more so than the representational paradigm which grew out of the "imaginary" view of ideology; but a more consistent approach would focus on ideology, its instruments and its effects, in terms of that which constitutes them: human semiosis.

This bears upon the issue of representation and resistance. Ideology, like "information" cannot be "resisted" because it is not something that is transferred or forced upon humans; it is instead the relation of meaning that emerges when humans interact with real objects in a cultural, physiological and environmental context. These three contextual levels, and not just the cultural-linguistic one, all play a part in framing the way in which ideology is constituted. Ideology frequently showcases

[63] Althusser 1969, p. 233.

untruths, to be sure; but, from a semiotic standpoint, it is no more "false" than shouting out or laughing uncontrollably are "false". The "linguistic turn", along with the representational paradigm, has fostered the seemingly ineluctable impression that, for humans, ideology supervenes on a realm of mendacity and a realm of reality. Upheld by "language" as a representational medium, the realm of mendacity suffuses the polis, holding it in a firm grip which refracts all perception and only very occasionally gets broken in such a way that it allows humans to glimpse the real – i.e. social – relations that obtain within the polis alone. In semiotics, particularly after biosemiotics, humans inhabit a synthesis of their sensoria and their cognition, constantly negotiating mind-dependent and mind-independent relations.

Relying on linguistics as the basis of an understanding of how semiosis occurs, as well as for an assessment of sociality and what to do with problems that arise from sociality, not only occludes humans' consanguinity with non-human inhabitants of this planet but also fails to address the complex edifice of human communication. Biosemiotics has had this edifice in its sights since being founded by Sebeok. Biosemiotics has sought to proceed, in a transdisciplinary mode, from a concept of semiosis as "global" and with its own contextual effectivities sustaining *Umwelten*, rather than assuming that signification can be graded according to measures of truth and falsity derived from cultural taxonomies. In short, biosemiotics' reconfiguration of the polis consists of having bigger fish to fry than traditional political approaches that signal the tyrannies of language and pursue the representational paradigm. This is not a matter of biosemiotics simply drawing back and stating that local political struggles are somehow less significant than the bigger picture, as some advocates of environmental politics have done. Rather, it is a global view recognizing that every semiosis, local and quotidian, is subject to relation and is therefore the object of politics. Central to the representational view and, for Deely,[64] the key impediment of modern thought, is the inability to arrive at a coherent distinction between mind-dependent and mind-independent being. Relations create a public sphere in which there is room for freedom, but there is also the possibility of reaching an understanding of nature, likewise through relations. The task for science and philosophy, then, is to sort out what belongs to the mind and what belongs to nature,[65] an advanced act of modelling that falls to the human alone. Sebeok espoused through biosemiotics a new semiotics driven by the idea of modelling; whether he contemplated, in the terms outlined above, that he was inaugurating a radically new understanding of the polis, is not known. What is clear, however, is that the transdisciplinary project of biosemiotics heralds an opportunity to completely recapitulate politics, avoiding, this time, the blinkered representational stalemate born of linguistic approaches' parochialism.

---

[64] Deely 2009, p. 172.

[65] *Ibid.*

References

Althusser, L. (1969). *For Marx*. London: New Left Books.

Althusser, L. (1971). Ideology and ideological state apparatuses. In L. Althusser, *Lenin and philosophy and other essays* (pp. 121–176). London: New Left Books.

Barthes, R. (1957a [1973]). Myth today. In Barthes 1957b [1973], pp. 109–129.

Barthes, R. (1957b [1973]). *Mythologies*. London: Paladin.

Barthes, R. (1957c [1973]). The Romans in films. In Barthes 1957b [1973], pp. 26–28.

Barthes, R. (1977). Change the object itself. In R. Barthes, *Image – Music – Text* (pp. 165–169). London: Fontana.

Bignell, J. (1997). *Media semiotics: An introduction*. Manchester: Manchester University Press.

Boggs, C., & Plotke, D. (1980). Introduction. In C. Boggs & D. Plotke (Eds.), *The politics of eurocommunism* (pp. 5–12). Montreal: Black Rose.

Brier, S. (2008). *Cybersemiotics: Why information is not enough!* Toronto: University of Toronto Press.

Chernov, I. (1988). Historical survey of Tartu-Moscow semiotic school. In H. Broms & R. Kaufmann (Eds.), *Semiotics of culture. Proceedings of the 25th Symposium of the Tartu-Moscow School of Semiotics, Imatra, Finland, 27th – 29th July 1987* (Vol. 1987, pp. 7–16). Helsinki: Arator.

Cobley, P. (2001). Introduction. In P. Cobley (Ed.), *The Routledge companion to semiotics and linguistics* (pp. 3–12). London: Routledge.

Cobley, P. (2006). Myth. In K. Brown (Ed.), *Encyclopedia of language and linguistics* (Vol. 8, pp. 415–418). Amsterdam/London: Elsevier.

Cobley, P. (2013). Semiotic models of communication. In P. Cobley & P. J. Schulz (Eds.), *Theories and models of communication* (Handbooks of communication science, Vol. 1, pp. 223–240). Berlin: Mouton de Gruyter.

Cobley, P. (2014). Codes and coding: Sebeok's zoosemiotics and the dismantling of the fixed-code fallacy. *Semiotica, 198*, 32–45.

Deacon, T. W. (1997). *The symbolic species*. Harmondsworth: Penguin.

Deely, J. (2001). *The four ages of understanding: The first postmodern survey of philosophy from ancient times to the turn of the 21st century*. Toronto: University of Toronto Press.

Deely, J. (2009). *Purely objective reality*. Berlin: Mouton de Gruyter.

Dyer, G. (1982 [1999]). Semiotics and ideology. In G. Dyer (Ed.), *Advertising as communication* (pp. 91–109). London: Routledge.

Fiske, J. (1989). Popular texts. In J. Fiske (Ed.), *Understanding popular culture* (pp. 83–101). London: Unwin Hyman.

Fiske, J. (1990). Ideology and meanings. In J. Fiske (Ed.), *Introduction to communication studies* (pp. 156–179). London: Routledge.

Goffman, E. (1979). *Gender advertisements*. New York: Harper and Row.

Gould, J., & Vrba, E. S. (1982). Exaptation – A missing term in the science of form. *Paleobiology, 8*(1), 4–15.

Hall, S. (1973). Encoding and decoding in the television discourse. *Centre for contemporary cultural studies stencilled occasional Paper No. 7*. Birmingham: Centre for Contemporary Cultural Studies (http://www.birmingham.ac.uk/Documents/college-artslaw/history/cccs/stencilled-occasional-papers/1to8and11to24and38to48/SOP07.pdf; website consulted on 19 February 2015).

Hall, S. (1980). Encoding/decoding. In S. Hall, D. Hobson, A. Lowe, & P. Willis (Eds.), *Culture, media, language* (pp. 128–138). London: Unwin Hyman.

Hall, S. (1997). The work of representation. In S. Hall (Ed.), *Representation: Cultural representations and signifying practices* (pp. 13–74). Thousand Oaks/London: Sage.

Harris, R. (2008). *Mindboggling*. Luton: The Pantaneto Press.

Hoffmeyer, J. (1993 [1996]). *Signs of meaning in the universe*. Bloomington: Indiana University Press.

Jhally, S. (1990). *The codes of advertising. Fetishism and the political economy of meaning in the consumer society*. New York/London: Routledge.
Kull, K. (1998). On semiosis, Umwelt and semiosphere. *Semiotica, 120*(3–4), 299–310.
Lucid, D. P. (1977). *Soviet semiotics: An anthology*. Baltimore/London: Johns Hopkins University Press.
McDonnell, K., & Robins, K. (1980). Marxist cultural theory: The Althusserian smokescreen. In S. Clarke, T. Lovell, K. McDonnell, K. Robins, & V. Jeleniewsk (Eds.), *One-dimensional Marxism: Althusser and the politics of culture* (pp. 157–231). London: Allison and Busby.
Neumann, J. von (1948 [1963]). The general and logical theory of automata. In A. H. Taub (Ed.), *Collected works* (Vol. 5, pp. 288–318). New York/Paris: Pergamon Press.
Pêcheux, M. (1982). *Language, semantics and ideology*. London: Macmillan.
Randviir, A. (2007). On spatiality in Tartu-Moscow cultural semiotics: The semiotic subject. *Signs Systems Studies, 35*(1–2), 137–158.
Rorty, R. M. (Ed.). (1967). *The linguistic turn: Essays in philosophical method with two retrospective essays*. Chicago/London: University of Chicago Press.
Ross, G. (1980). The PCF and the end of the Bolshevik dream. In C. Boggs & D. Plotke (Eds.), *The politics of eurocommunism* (pp. 15–48). Montreal: Black Rose.
Salvadori, M. L. (1978). Introduzione. In M. L. Salvadori (Ed.), *Eurocomunismo*. Torino: Giulio Einaudi.
Schwartz, L. (1984). Soviet anti-semitism and Jewish scientists. In T. Freedman (Ed.), *Anti-semitism in the Soviet Union: Its roots and consequences* (pp. 179–185). New York: Freedom Library Press of the Anti-defamation League of B' nai B'rith.
Sebeok, T. A. (1972). *Perspectives in zoosemiotics*. The Hague: Mouton.
Sebeok, T. A. (1981). The ultimate enigma of "Clever Hans": The union of nature and culture. In T. A. Sebeok & R. Rosenthal (Eds.), *The clever Hans phenomenon: Communication with horses, whales, dolphins, apes and people* (pp. 199–205). New York: New York Academy of Sciences.
Sebeok, T. A. (1991a). Communication. In T. A. Sebeok (Ed.), *A sign is just a sign. Advances in semiotics* (pp. 22–35). Bloomington/London: Indiana University Press.
Sebeok, T. A. (1991b). In what sense is language a "Primary Modeling System"? In T. A. Sebeok (Ed.), *A sign is just a sign. Advances in semiotics* (pp. 49–58). Bloomington: Indiana University Press.
Sebeok, T. A. (2001a). Introduction. In T. A. Sebeok, *Global semiotics* (pp. 1–7). Bloomington/London: Indiana University Press.
Sebeok, T. A. (2001b). Nonverbal communication. In P. Cobley (Ed.), *The Routledge companion to semiotics and linguistics* (pp. 14–27). London: Routledge.
Sebeok, T. A. (2011). Summing up: In Lieu of an introduction. In P. Cobley, J. Deely, K. Kull, & S. Petrilli (Eds.), *Semiotics continues to astonish: Thomas A. Sebeok and the doctrine of signs* (pp. 451–459). Berlin: Mouton de Gruyter.
Sebeok, T. A., & Danesi, M. (2000). *The forms of meaning: Modeling systems theory and semiotic analysis*. Berlin: Mouton de Gruyter.
Uexküll, J. von (2001a). An introduction to Umwelt. *Semiotica, 134*(1–4), 111–123.
Uexküll, J. von (2001b). The new concept of Umwelt: A link between science and the humanities. *Semiotica, 134*(1–4), 107–110.
Vestergaard, T., & Schrøder, K. (1985). *The language of advertising*. London: Basil Blackwell.
Waldstein, M. (2008). Soviet science and academic autonomy: The structuralist *sturm und drang*. In M. Waldstein, *Soviet empire of signs: A history of the Tartu school of semiotics* (pp. 15–38). Saarbrücken: VDM Verlag.
Weible, D. (2012). The concept of exaptation between biology and semiotics. *International Journal of Signs and Semiotic Systems, 2*(1), 72–87.
Williamson, J. (1995 [2002]). *Decoding advertisements: Ideology and meaning in advertising*. London: Marion Boyars.

How Useful Is *état de langue* for Biosemiotics? An Exploration of Linguistic Consciousness and Evolution in F. de Saussure's Works

Jui-Pi Chien

Abstract In the field of biosemiotics in our time, Saussure's theory of semiology has been dismissed for its glottocentric, anthropocentric, and dyadic characteristics and as such unsuitable for the said field. Such accusation is symptomatic of a narrow view of Saussure, which ignores the efforts he made in tackling problems concerning the unification of biology (natural sciences) and semiotics (human sciences). A broader view of Saussure, emerging from the newly-discovered orangery manuscripts along with his thought-provoking lectures, reveals that his epistemology is actually grounded upon evolutionary differences and the concept of uniformitarianism. This study points out how the network of differences, which Saussure proposes in his manuscripts, blurs disciplinary or systematic boundaries between language and nonverbal systems, and how it might serve as a framework for appreciating true analogies between natural sciences and the science of language. Moreover, Saussure's concept of *état de langue* is made comprehensive in relation to appropriations of the Darwinian model and Neo-Darwinian ideas. His model of evolution is seen to have amplified the phenomenon of symbiogenesis, which is non-linear, non-adaptive, non-restrictive as regards localities, yet claims certain truths about nature and culture. All in all, this study draws attention to the implications of conceptualizing non-linear evolution within and across systems.

Keywords Biosemiotics • *état de langue* • Evolutionary differences • Uniformitarianism • Symbiogenesis • F. de Saussure

---

This contribution is a slightly modified part of an already published article (cf. in *Semiotica*, 2011, 185[1–4], pp. 51–77).

J.-P. Chien (✉)
Department and Institute of Foreign Languages and Literatures,
National Taiwan University, Taipei, Taiwan
e-mail: jpchien@ntu.edu.tw

E. Velmezova et al. (eds.), *Biosemiotic Perspectives on Language and Linguistics*, Biosemiotics 13, DOI 10.1007/978-3-319-20663-9_12

Introduction

Discursive Problematics: Saussure Versus Peirce in Biosemiotics

How useful is the semiology of Ferdinand de Saussure for the study of biosemiotics? Since the 1970s, with the organized efforts of certain advocates, the study of biosemiotics has been divided into two strands: biological semiotics and semiotic biology. While advocates of the latter have been pursuing a model for the interpretation of codes and meanings within organic mechanisms, they have also looked back to history in order to ensure the continuity and validity of their scope. Several linguists and biologists in history have been credited with practicing or envisioning biosemiotics, but such efforts have exposed more stumbling blocks than sound connections. Saussure has turned out to be one of the main obstacles for such efforts: his theory has been considered to be glottocentric, anthropocentric as well as dyadic and as such unsuitable for the study of biological phenomena in general.

In comparison with Charles S. Peirce's triadic model of sign formation and interpretation, Saussure's theory is criticized for failing on three points: (1) the absence of thirdness, i.e. the interpretant; (2) the arbitrary relation between signifier and signified; (3) the priority of synchronic states over diachronic changes. Another reason why Saussure has been thought to be unsuitable for the study of biosemiotics is that he appeared not to take into account environmental factors.[1] However inadequate his theory appeared to the demand for a far-reaching model of the living, it should not be so easily dismissed.[2] Since the 1990s, the "meaning" mechanism has been redefined, certain "boundaries" have been extended, and concepts such as *group properties* have been introduced to improve and enhance biosemiotics.[3] Drawing on Saussure's critical thoughts in his orangery manuscripts may shed light on these areas of research.

Textual Problematics: The Orangery Manuscripts in Relation to Other Materials

The standard version of Saussure's ideas derives from the *Course in General Linguistics* (*CGL*) in which his ideas were presented by his editors. A new kind of Saussure, inspiring to biosemiotics, has emerged with the publication of his own *Writings in General Linguistics* (*WGL*), the main part of which constitutes the so-called "orangery manuscripts". This postponed publication gives rise to the question of whether the ideas that have emerged from standard readings of the *CGL* really do justice to Saussure's innovative – even revolutionary – theory. For Saussure

[1] Sebeok 1976, p. 152; Hoffmeyer 1993 [1996, p. 17–18]; Nöth 1998, p. 337.

[2] Bouissac 2004, p. 241, 256; Barbieri 2008, p. 594–596.

[3] Barbieri 2003, p. 236–237; Kull 1998, p. 348–349 and 2003, p. 56; Chebanov 1998, p. 418.

scholars like Roy Harris and Simon Bouquet, the immediate light that these manuscripts brings to the world is to prove that Saussure's colleagues, Charles Bally and Albert Sechehaye, have misrepresented his ideas and imposed their own on him.[4] It is quite natural to make such a hasty judgment since there is indeed a discrepancy between the *CGL* where Saussure's ideas were put into succinct and systematic categories and the manuscripts where Saussure questions himself and his contemporaries. This reflects a standard problem in the discovery of new materials: there is a temptation to discover hidden aspects ignored by established schools of thought. Thus, before glorifying the presumed greater authenticity of those recently found in the manuscripts, we should consider the actual discrepancies between these two groups of materials.

Saussure wrote the major portion of the orangery manuscripts which he entitled the "Dual Essence of Language" shortly after his three inaugural lectures in 1891.[5] Bally and Sechehaye on the other hand edited the *CGL* on the basis of notes students attending Saussure's courses on general linguistics had made and published the work in 1916, 3 years after Saussure's death. Bally and Sechehaye ignored inconsistencies between students' notes and Saussure's writings and concentrated on what they believed were the most definitive distinguishing features of the Geneva school.[6] Actually, Saussure insisted that one should never publish writings before they have reached their definitive form.[7] As he was constantly experimenting with his thoughts, it is hardly surprising to find constant questionings and self-criticism in the manuscripts – which eventually caused the manuscripts to be left unfinished when he suddenly passed away in 1913. Nevertheless, the orangery manuscripts should be thought as more original material than the *CGL*.

Until the discovery of the manuscripts, scholars relied on the critical edition in four volumes in which the editor Rudolf Engler lays out corresponding passages from editions of the *CGL*, students' notes, and Saussure's personal notes on six parallel columns. After examining their textual variations closely, Engler does not criticize Bally and Sechehaye for professional misconduct, but instead draws attention to the nature of Saussure's writings. First, he indicates that Saussure constantly renewed (or rephrased) his ideas and had the capacity to develop his ideas in all sorts of directions without contradicting himself. Secondly, it is only human that one's words in written form and in oral presentation sometimes contradict each other, which certainly has happened to Saussure.[8] Based on these observations, we can argue that there has actually been only one Saussure: we always find the same group of key concepts whatever source material we start from. Comparing the texts with one another, we can cast light on Saussure's rephrasings, questionings, and criticisms. The multiple sources may well be synthesized into a more or less unified whole where key concepts can be related to relevant frames of reference. Therefore,

[4] Harris 2002; Bouquet 1997; Saussure 2006, p. xvi.

[5] Engler 2004, p. 48.

[6] Amsterdamska 1987.

[7] Engler 2004, p. 47–48.

[8] Saussure 1967, "Preface".

though this study introduces newly-discovered material into biosemiotics, it does not mean that there is a completely new Saussure awaiting us – all materials which have been made available up to the present are equally valid when discussing his method. However, redefining the theoretical framework of biosemiotics in the light of the newly available material may help us appreciate the effectiveness of his conceptual tools.

Methodological Problematics: The Significance of Genuine and Overall Change Within a System

Many of the misunderstandings of Saussure's ideas can be attributed to the misuse of the word *état* ('state') and its adjectival form ('static') in translations and commentaries. The way these words have been used gives the impression that Saussure thought language should be studied as if it were in a permanent state, not subject to change. However, according to the manuscripts, the diachronic dimension of language is every bit as important as the synchronic, and the study of the latter in its own right, manifested in the form of the *état de langue* (abbreviated as *ÉDL* afterwards), facilitates exploring the former.[9]

On top of the false impression of the *ÉDL*, the Saussurean notion of arbitrariness has been much exploited: it is either criticized as a lack of natural links between signifier and signified or overgeneralized to the whole of his system.[10] In his third course lecture, a major portion of which touches upon *la langue* (the rule or law summarized from different languages), he called his students' attention immediately to the matter of choice and *a posteriori* connections between signifier and signified. For him, they not only shape the learning and speaking of a language, but also explain the fact that every language (or system) is different in its own right. By employing the notion of arbitrariness to define his idea of signs, he was able to defend his system against the philosophy that there is a shared and predetermined origin of languages. Furthermore, within a specific system, he saw that arbitrariness functions in a solitary and limited fashion: it gives rise to the making of terms which cannot be related to (or associated with) others [*il fait appel à rien*].[11] The outreaching and sociable force *en vivant* goes to relative arbitrariness (analogy) which is able to reproduce shared units in different terms. Arbitrary and relatively arbitrary states of mind make up a system – the former, though idealized, serves as a starting point, while the latter (analogy) explains the status quo. Although it has been generally conceived that the latter sets limits to the former, they both are governed by the law of generating *a posteriori*, artificial, and reworkable connections (*ÉDL*'s) within one sign entity and across many others within a system.

---

9 Saussure 2006, p. 7.

10 Jakobson 1966 [1990].

11 Saussure 1967, p. 86.

Fig. 1 The schema of sign (Saussure 1993, p. 139)

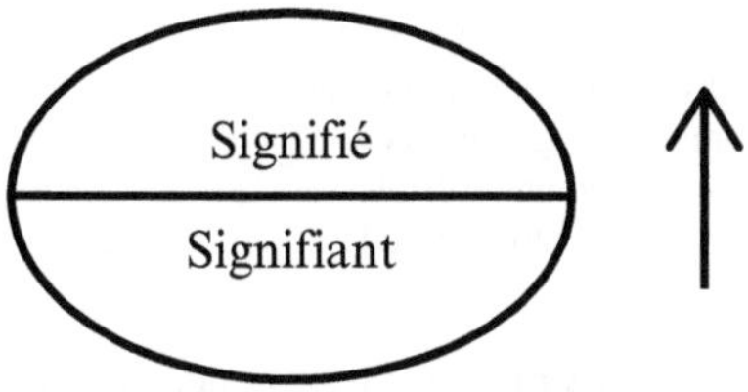

Such seemingly contradictory but actually compatible states can be made comprehensive by looking into Émile Constantin's notes (taken for the third course lecture) along with Saussure's diagrams and writings. At the beginning of his lecture on the nature of linguistic sign (dated 2 May 1911), Saussure declared that he would reject approaches to *la langue* which do not start from the two-term structure, i.e. form and meaning.[12] However, as he gradually elaborated on it together with his critical tools, he indicated at one point that such structure (or schema) is not the "starting point" [*n'est donc pas initial*] in the study of *la langue* at all – the schema should be made into play and address phenomena of all kinds (beyond the formation of vocabulary only) [*ce schéma peut entrer en jeu; il s'agit de n'importe quoi*] (dated 4 July 1911).[13] Looking further back in time, as revealed in his orangery manuscripts (1891), we find that already at this early stage he has criticized the structure (Fig. 1) as something that is "entirely rough" and shows a "profound misconception" of *la langue.*[14] He then proposed "a range of possible forms and possible meanings", which interact through their distinctions but in no way correspond to each other, as a better representation of signs (Fig. 2).

For unknown reasons, Saussure propagated the self-criticized structure (and schema) (Fig. 1) to his students but stored most of his ramifications on the network of differences for his own. The "true thoughts" [*pensée intime*] (Fig. 2) which he cherished and used as the starting point of his manuscripts appear only in elliptical lines towards the end of his third course lecture.[15] Moreover, in his orangery manuscripts, he had already conceptualized the functioning of a system in accordance with the idea of a "complicated game leading to a final balance" [*les différences qui résultent du jeu compliqué et de l'équilibre final*].[16] Such a game generates networks among signs without limit, starting point, or any fixed point. It not only enables signs to absorb and locate new ideas at any time, but also serves to change the boundary of general meaning from time to time.[17] Despite having introduced such prospect into the actual functioning of his system, he still affirmed that the whole thing is the "result" of the schema which he has defined and criticized.[18] With such

[12] Saussure 1967, p. 150 and 1993, p. 75a; Fig. 1.

[13] Saussure 1967, p. 264 and 1993, p. 140a.

[14] Saussure 2006, pp. 22, 24.

[15] *Ibid.*, p. 59; Saussure 1993, pp. 140a–143a.

[16] Saussure 2006, p. 43.

[17] *Ibid.*, pp. 22, 50–51, 60.

[18] *Ibid.*, p. 43.

General difference of meanings
(only exists in relation to forms)

General difference of forms
(only exists in relation to meanings)

Fig. 2 Saussure's proposed approach to the study of signs in his orangery manuscripts (Saussure 2006, p. 24)

a model of mind that generates and learns from endless differentiations, Saussure annuls binary oppositions generally made between literal and figurative meanings, physical words and spiritual meanings, signs inside and outside consciousness, and above all, the philosophy (or metaphysics) of language since the eighteenth century.[19]

Before having access to the manuscripts, one could even suspect that the *ÉDL* is empty of precise content.[20] This, however, is mistaken. Already in his inaugural lectures Saussure warned about defining the *ÉDL* in terms of concrete but "loose markers", such as historical period, language, dialect, the borders of a country, or geographical area.[21] The manuscripts reveal how he then sought to define the *ÉDL* in more abstract terms in an attempt to develop it into a working tool for theorizing differentiations, multiplications, fragmentations, and discontinuities in the continuous currents of languages. He was planning a drastic move away from contemporary received notions about language. Examining his inaugural lectures (1891), orangery manuscripts (1891), students' notes (1907–1911), and *CGL* (1916) together allows us to argue that the *ÉDL* is as vigorous and sophisticated a term as an analogical creation, negative categorization, and general difference. With this group of terms, Saussure has explored the changing state of human consciousness as it develops new forms of verbs and expands semantic networks. The kind of change he conceptualizes is not to be measured by the variation of forms on the surface but should rather be considered as a system within which the boundary of general meaning functions across and beyond several *ÉDL*'s.

ÉDL*(s) as the "Central Object" of Linguistics*

ÉDL can, in its plural form, indicate explicitly different stages of *la langue* in history; it can also, in singular form, manifest certain mental capacity (such as analogical thinking) which constantly generates new relationships between forms and meanings. The two relate reciprocally to each other with regard to their occurrences in time, but Saussure insists on keeping them conceptually apart, as he reclaims his object of study towards the end of his orangery manuscripts. He wrote: "The examination of *a sequence of states* offers the linguist one central object. This object does not enter into a straightforward, marked opposition with the preceding one, but into

[19] *Ibid.*, pp. 47–51, 56, 59, 64.

[20] Camara 1995, p. 128.

[21] Saussure 2006, p. 111.

a relationship of radical disparity, which necessarily precludes all comparison, opening up a new order of ideas which can in no way exist in relation to a given [*ÉDL*]".[22] As he continued to elaborate on this point in his first course lecture, he made the point that *ÉDL* is in essence social and psychological.[23] Meanwhile, he would like to take this notion to put forward a different sense of history and evolution – *ÉDL* is what speaking subjects have "immediate sense and control of," "a movement in time without worrying about whether it is a development, [a movement] forward or backward".[24] He valued the power of creating new orders to the extent that he denounced the need of *ÉDL* in its plural form: speaking subjects can have forgotten or ignored *ÉDL*s in history, but they are still able to coin new forms in their environments. On the other hand, *ÉDL* is also the power to negate (involving interpreting and decomposing entities or units), with which speaking subjects not only collocate *ÉDL*s, but also restructure the links between forms and meanings across *ÉDL*s.[25] The benefit of such mental work is to build up multiplying networks without the limitation of one's immediate condition or environment:

> [U]nderstanding the purely negative, purely *differential*, essence of each of the elements of language that we hastily assume to exist is a never-ending task. Not one of them, however ordered, possesses this supposed existence – although admittedly we may have to recognize that without this invention *(cette fiction)* the mind would be literally incapable of dealing with such a mass of differences, with no positive, solid reference point at any place or time… these differences which make up the whole of [*la langue*] would represent nothing.[26]

ÉDL in its most abstract form is not any kind of tangible entity but an abstraction which Saussure formed and utilized for introducing the value of "new combinations" or "sudden constructions" into the evolution of linguistic (or any other living and semiological) system.[27] These non-ceasing changes and growing differences which have no hierarchical orders contribute to the increase of novel relationships between signs and the extension of boundaries between grammatical categories.[28] Such deep and overall change of a system is not measured by the presence (appearance) or absence (disappearance) of specific signs since they can recur at any time in history. Signs in general absorb and elaborate on the value which has been defined by certain stages or categories; however, a complete change within a system is not to continue the elaboration – it simply alters, ignores, or abolishes some such previously-defined value.

[22] *Ibid.*, p. 59.

[23] Saussure 1996, p. 2.

[24] *Ibid.*, p. 27–28.

[25] *Ibid.*, p. 90–91.

[26] Saussure 2006, pp. 42–43.

[27] Saussure 1996, pp. 63, 90–91.

[28] Saussure 2006, p. 60.

Saussure's Epistemology of Evolution

The validity of a tool for the observation of life within a long period has been a matter of concern to scientists and philosophers. This explains why the Darwinian term *species* is still much favoured in our time: it not only serves as an analytical unit for taxonomy, but also facilitates cross-scientific diachronic studies at multiple levels.[29] Somewhat in the manner of how the notion of species serves scientific and philosophical purposes, *ÉDL* is a concept which is reduced but capable of extending its validity to as many language phenomena as possible. Recent studies on gene-language co-evolution (including the hypothetical forms of protolanguage), palaeolithic continuity theory, and glossogenetic process point to the importance of culture and society in language change: (1) languages evolve as humans within their communities produce a range of more or less structured and lasting combinations of linguistic units; (2) languages spread and change together with the migrations of *Homo sapiens.*[30] They show that we cannot draw an encompassing picture of language evolution if we simply adopt a linear viewpoint of the development of languages. Part of the problem is that – apart from the misappropriation of certain metaphors and thereby confused epistemologies – Darwinism did not really have a sufficiently revolutionary impact upon linguists in the nineteenth century.[31] Saussure's notion of the *ÉDL* was overshadowed by the efforts of mapping the idea of species, organs, and natural selection unto language variations. Thus, our task is to explore: (1) Saussure's criticism of the fallacious applications of Darwinian assumptions to linguistics; (2) how Saussure's uncompromising approach is compatible with the idea of "symbiogenesis", i.e. non-linear and non-adaptive evolution[32]; (3) how *ÉDL* helps reveal greater truths about nature and culture.

The Unity and Universality of Object

Discussing the problems sciences faced in the nineteenth century, August Schleicher suggested that the study of languages was in a more advantageous position than that of the development of species: it is rich in written records from ancient to contemporary which can be used to confirm linguists' observations, whereas natural scientists base their work on less well preserved and randomly surviving specimens or remains of animals and plants.[33] Schleicher only considered the quantitative differences of objects, i.e. their degrees of completeness, and did not explore their

[29] McCauley 2007.

[30] Hurford 1992 and 2007; Danesi 2004; Alinei 2006; Katz 2008.

[31] Nerlich 1989; Auroux 2007; Klippi 2007.

[32] Gontier 2006b.

[33] Koerner 1983, pp. 42–44.

qualitative or "specific" differences – which he thought are not really essential.[34] He affirmed that parallels can be found between the gradual variations of languages and species: (1) "languages are organisms of nature; they have never been directed by the will of man [...] they [grow] old and [die] out"[35]; (2) "for the terms species, subspecies, variety, we substitute the words language, dialect, patois".[36] Most of all, he believed that viewing species and languages as concrete entities can make a significant breakthrough in conceptualizing modern science. He argued for the prospect as follows:

> The tendency of modern thought is undeniably towards *monism. Dualism* [binary oppositions] [...] is no longer a firm ground to stand upon, if we wish to survey the field of modern science [...]. It is now more than ever necessary to occupy oneself with the most minute special study of the object, without thinking at all of a systematic upbuilding of the whole [...]. The importance which the observation of facts has acquired for science in general, but more especially for natural sciences, is the unavoidable result of the monistic principle, which does not look for anything behind the things, but looks upon the object as identical with its form or appearance. Observation is the foundation of modern knowledge; nothing else is acceptable but the necessary conclusions arrived at through that channel.[37]

Schleicher shifted the focus of scientific research from its previous paradigm governed by Carl von Linné's *Systema Nature* to the minute details of objects. Nevertheless, he was still constrained by the idea that languages as well as living beings should be specified and put into categories according to their appearances. Without looking into how Darwinian terms serve to implicate the observation of nuances which are likely to divide languages from species, Schleicher simply kept on repeating the Darwinian jargon and making baffling statements – he failed to justify how "widely different" speech is from the animal and vegetable kingdoms.[38] He illustrated the origin and divergences of language families in accordance with Charles Darwin's tree diagram, but it is questionable how his bringing together of terms used in different domains serves to devise a new approach to the conceptualization of language diversities. This problem was tackled in a cynical tone in Saussure's manuscripts and notes. He pointed up the dilemma as follows:

> If we were invited first to determine the chemical classification of a sheet of iron, gold, or copper, and then the zoological species of a horse, cow, or sheep, these would be two easy tasks. But if we were asked to determine what "species" is represented by the odd combination of an iron plaque attached to a horse, a gold plate on a cow, or a sheep adorned with something copper, we would exclaim that the task was absurd. The linguist has to realize that it is precisely this absurd task that faces him right from the very outset.[39]

Saussure's criticism reveals that a new object of study is not likely to emerge if linguists simply divide and combine singular and isolated substances – such efforts

---

[34] *Ibid.*, p. 45.

[35] *Ibid.*, pp. 20–21.

[36] *Ibid.*, pp. 47–48.

[37] *Ibid.*, pp. 23–24, 25–26.

[38] *Ibid.*, pp. 64–65.

[39] Saussure 2006, p. 3.

trivialize their observations rather than bringing unity to what is still waiting to be defined. What needs to be unified is their viewpoint: a perspective which forces them to look straight into the actual functioning of languages and to form a holistic and comprehensive view of diversities. Furthermore, it is futile to explore whether languages are like species or chemical elements, or how many functions (or capabilities) they have in parallel with functions in organs or compounds. To break up languages in the same way in which anatomists, physiologists, and chemists have done in their respective fields of study simply shies away from exploring what actually governs the production of languages. According to Saussure, unlike in the field of natural history, death has no dominion in the development of languages. In his notes, he pinpointed the sharp distinction between his viewpoints and those of naturalists':

> In an organized being a function can die without the organ dying. Even a corpse still has its organs, which are the material of anatomical science. In the word, there is absolutely nothing anatomical, i.e. no difference in parts based on a relationship between the function and the part which carried out this function. There is only a sequence of acoustic productions which are perfectly *similar* to one another, in that lung [*poumon*] and foot [*pied*] are all the same in the word. Principle of *Identical capacity*.[40]

Saussure takes it as an urgent task to unify and universalize the study of the functions for the production of words and parts of speech: it should not just pertain to certain language families or specific members of a language family but all languages. In contrast, naturalists observe various specific functions when they shift from organ to organ. Linguists who try to adapt such an approach not only become confused about the right path of linguistics, but also eliminate its metacritical power which is expected to explicate non-linguistic systems as well. In order to avoid the indiscriminate adoption of naturalists' terms and the implications of the analogy with bodily organic functions, Saussure offers another term, *an instance of pure consciousness* [*un fait de conscience pure*], to distinguish the evolution of languages from that among animals and plants.[41] As a challenge to pseudo Darwinian attempts at identifying origins, permanent features, functions, and stages of development, the Saussurean idea of looking into human consciousness initiates a new linguistic epistemology which is a step forward from indiscriminate adaptations of natural history.[42] It is a way of observing phenomena in a way which gives rise to the *ÉDL* as the abstract yet complete object of linguistics:

> This is not a trivial observation: the phenomenon under study exists only in the presence of those phenomena that are in contrast to it [...] [F]or any linguistic term [...] cannot exist, even fleetingly, in its own right and independently of its contrast with others, and which cannot be anything greater than an approximate encapsulation of the sum of differences at work. Only these differences exist; which means that the whole object of the science of language finds itself in the realm of relativity.[43]

As a new object to be observed in linguistics and evolution (rather than the evolution of languages in naturalists' terms), the *ÉDL*'s aim is the discovering of contrasts

[40] *Ibid.*, p. 77.

[41] *Ibid.*, p. 4.

[42] *Ibid.*, p. 110.

[43] *Ibid.*, pp. 42–43.

between entities: it does not aim at reconstructing specific needs, functions, organs, and origins on the basis of pieces of evidence – as in models inspired by the supposed analogies between anatomy and philology – rather, it is highly charged with the interest in establishing ever more links and differences between such pieces of evidence. Furthermore, the *ÉDL* overcomes a problem which approaches on the lines of the Darwinian tree diagram face regarding linguistic evidence: while in the Darwinian approaches, the notion of languages evolving on certain teleological lines, in which the continuation or disappearance of certain features is a matter of life and death, often leads into ignoring pieces of evidence which do not fit this model,[44] in the study of the *ÉDL* there is no need to ignore undesirable facts and construct perfected entities of language development. It constantly jumps between at times seemingly unrelated linguistic entities in its study of hybridizations and mutations which emerge from its repertoire of differences. Moreover, the *ÉDL* as an analytical tool pays equal attention to ontogenesis and phylogenesis: children can make errors while they are learning conjugations; different natural languages cross-fertilize by exchanging units of words. On the basis of such cycling and recycling of pieces, the *ÉDL* also shuns the strict distinctions made between genotypes and phenotypes in the Modern Synthesis, in which the former are taken as the groundwork while the latter are given more attention as regards their variations in phylogenic environments.[45] In comparison with the two standard evolutionary models, the unity and universality of *ÉDL* lies in the fact that it brings together pieces of evidence found from the past and the present – the bottom and the surface, the near and the distant – for our scrutiny here and now.

Intercourses, Innovating Waves, and Concordances

Saussure's ambition of challenging natural history and the notion of linear evolution is manifested in his third course lecture as he extended his discussion on the unity of linguistic object to dialects and their distributions. He saw that drawing topographical boundaries for the sake of discussing language variations is as misleading as putting them into species and families. Already in his third inaugural lecture he suggests studying *la langue* or the *ÉDL* as a way of transcending strict localities and single points in time. He reveals to his audience that he prefers to study each dialect as well as language as an "intermediate link between its two neighbours to the East and the West".[46] However, a more radical idea presented in his third course lecture is to look into the life of the *ÉDL* which changes with time but is not constricted within any space or direction [*le développement libre*].[47] He informed his students of the biological and evolutionary agenda he was considering for the *ÉDL*:

[44] Gontier 2006a, pp. 211–213.

[45] Gontier 2006b, pp. 11–12.

[46] Saussure 2006, pp. 115–116.

[47] Saussure 1993, pp. 21a–23a.

> To some extent one could also say that [...] *la langue* could be expanded to read the life of *la langue* [...] that [...] would contain things of importance for the characterization of *la langue*, and that these things are all part of a life, a biology. But there are other things that would not be included: among others, the whole logical side of *la langue*, involving invariables unaffected by time or geographical boundaries. [...] Instead of geographical differences we have evolutionary differences. The evolution of *la langue* is one of the major concerns of linguistics. Geographical differentiation is only *particular* application of the facts of evolution; it has to be completely contained therein.[48]

First and foremost, such agenda elaborates on the *ÉDL* in its abstract form: it discusses (1) why language users change their discourse as they constantly migrate from one place to another; (2) how linguists can still unify their observations of these changes which approximately accord with inhabited regions. Unlike species which – under the naturalists' scope – are supposed to change their needs and develop new organic functions when they move from place to place, human consciousness shifts wherever and however humans change their places of living. In emphasizing the autonomy of the *ÉDL* which is distinguished from geographical localities, Saussure problematized the subtle scheme of time required for the observation of the life of the *ÉDL*. Such a scheme goes beyond conventional notions of diachrony and history as it is about neither the programmed development nor the adaptation of species.[49] Secondly, in the context of species evolution, theoretical interests have been invested on isolated areas or places on the borders where new species are supposed to appear faster and more abundantly than in the mainlands. In developing the *ÉDL*, Saussure deconstructed such distinctions by pointing out that both contiguity and separation of areas end up with the same result of differentiations of languages – it does not really take an island to prove the working of the *ÉDL*.[50] Even when there is concentration on a certain geographical area, the purpose of assuming a continuum or aggregate of regions is not to prove the rate (or speed) of evolution. The aim is rather to illustrate the uneven distribution or propagation of the *ÉDL* throughout that area.[51] The actual linguistic forces which crisscross an area are of more interest than the geographical area itself. Therefore, studying cases of extremely closed and provincial states where people adhere to a limited number of tongues cannot shed much light on the complexities of the *ÉDL*. It is rather within areas where people are open to the intercourse of various tongues that we can avoid coming to a standstill and meanwhile expand our understanding of the *ÉDL* in time.[52] Saussure relates the force of intercourse to his scientific principle of unity as follows:

> [T]he force of intercourse [...] will be the force promoting linguistic unification [*le principle unifiant pour la langue*]. [...] The influence of intercourse may appear in two forms: sometimes the new feature emerging at one point *will be* opposed and suppressed by the influence of intercourse. But you cannot say which among such innovations will be stifled. That is a process of conservation, of resistance. In other cases, an innovation introduced in

---

[48] *Ibid.*, pp. 11a, 22a.

[49] *Ibid.*, pp. xxi.

[50] *Ibid.*, pp. 39a–40a; Saussure 1959, p. 210.

[51] Saussure 1993, p. 23a.

[52] Saussure 1959, pp. 206–208.

one place is propagated, transmitted by this influence. Here again the result makes for unity, levelling, but by an active, positive process. It is this propagative form of the influence of intercourse that I propose to examine.[53]

His idea of intercourse enhances the fact that new features or innovations are composites (rather than modifications of previous ones) due to the coming together of different tongues and that there is no ladder or specific direction for their emergence and travelling: they emerge in tune with the conditions of the *ÉDL* which are rather unpredictable as well. Furthermore, whether the conditions of the *ÉDL* are favourable or not, innovations always keep their fundamental tone of life – they can hold their breath in harsh conditions and wait for the chances to resume their lives again. Just as much as geographical forces have shaped the appearances of the earth, the force of intercourse has been configuring the *ÉDL* to such an extent that there is no benefit of distinguishing the past from the present, the present from the future, one area from the next. What is observed about the *ÉDL* here and now could have happened before and might still act out in the future. The notion of *ÉDL* thus puts forward an alternative concept to catastrophism – instead of presuming disparities between the past and the present, uniformitarianism not only appreciates the preservation of changes, but also affirms the unity, continuity, and contiguity of processes in time.[54] Saussure's discourse on uniformitarianism revised vulgar appropriations of Darwinism and geology which stick to the tree diagram and atlases showing natural languages, nations or races bordering against each other. In his third course lecture, he offered to visualize the intercourse of languages as a succession of waves and their overlaps in the form of loops:

A glance at a linguistic atlas will sometimes reveal two or three waves that almost coincide or even overlap in one zone. [Figure 3] [The invasion of a territory by a number of features may be compared to waves. These linguistic waves or innovations sometimes coincide over a certain stretch] [Figure 4]. The two points A and B, which are separated by such a zone, obviously have some divergences and constitute two rather clearly differentiated forms of speech. These concordances, instead of being partial, may characterize the whole perimeter of two or more zones [Figure 5]. A dialect is defined, roughly speaking, by a sufficient accumulation of such concordances. Their foundations are social, political, religious, etc., matters which do not concern us at the moment but which veil, without ever erasing completely, the basic and natural fact of differentiation from zone to zone.[55]

Fig. 3 A zone crisscrossed by several waves (Saussure 1993, p. 26)

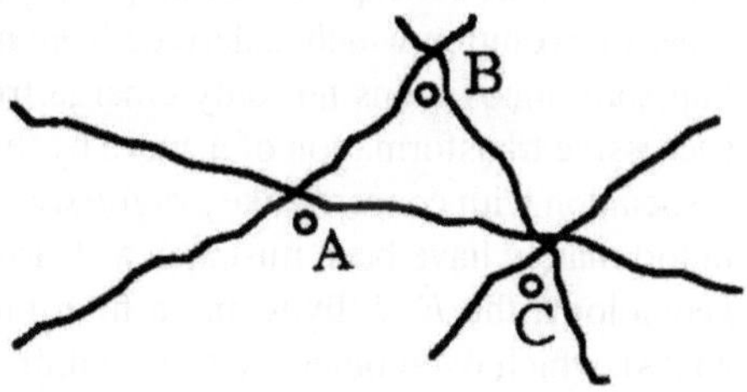

[53] Saussure 1993, pp. 33a–34a.

[54] Gould 1977 [2007, pp. 150–152].

[55] Saussure 1959, pp. 202–203 (lines within parentheses: Saussure 1993, p. 27a).

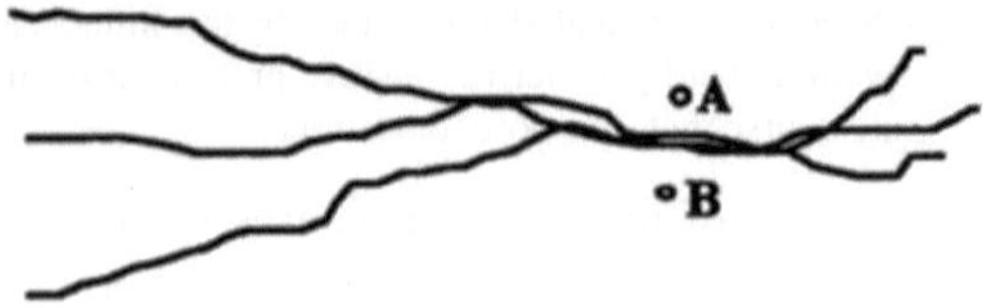

Fig. 4 Several waves coincide over a certain stretch (Saussure 1993, p. 27)

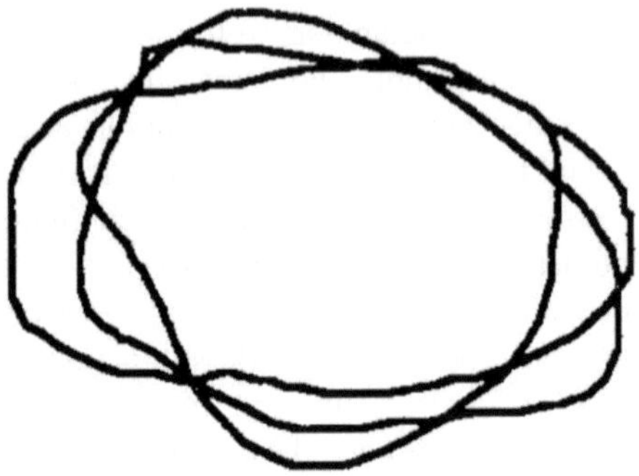

Fig. 5 Loops showing the convergence and divergence of waves (Saussure 1993, p. 28)

These innovating waves vividly picture the propagation of the *ÉDL* across zones and areas – they are not adaptations to specific natural environments. Following the waves, we can find converged points which help refer to a couple of concordances (A, B, C, etc.). Although these concordances are segregated by the waves – they are by all means distinct from one another – they somehow share some features which make them look similar. Within the same zone of converged waves, the closer they are, the more features they share with each other; further away, these waves run parallel for a while, and it takes other conditions to find some other concordances, and so on and so forth. Since humans have been migrating, the *ÉDL* can find its life wherever and however – we never know how far away it will come to a stop. In addition, there is a host of other conditions outside linguistics – cultural and social situations, etc. – which participate in shaping the waves and their degrees of convergence. Most of all, the concept of unending flowing and converging waves reveals a hidden sense of evolution which is derived from mathematics and geometry – according to Saussure, innovations not only emerge from differentiations, but also follow "the successive transformation of a curve by the alteration of conditions".[56] It avoids the association with concepts like *progressive development* and *perfected species* which unfortunately have been mistaken as Darwin's genuine contribution. In the light of geobiology, the *ÉDL* lives on "a high ratio of surface to volume" (two or more zones), which overcomes fixations within geographical, national or racial boundaries and assists in flowing, jumping, and travelling as much as can be.[57]

[56] Simpson and Weiner 1989, s.v. "evolution".

[57] Gould 1977 [2007, pp. 34–38, 197–198]; Saussure 1959, pp. 192–193, 204–205.

Concluding Remarks

Symbiogenesis in Nature and Culture

We gain from Saussure's formulations of analogy and the *ÉDL* a vista of the nature and intensity of evolution: his model is summarized from linguistic and cultural phenomena far and wide and – for its metacritical power accumulated from criticism of misappropriations of Darwinism – may well be considered as an alternative to Neo-Darwinian approaches in our time which base their studies of animals and humans on the tree of lineages, on the classification of species and subspecies on a purely genetic basis, and, in particular, on the determination of ontogeny, phylogeny, and reproduction by genes. Saussure's model illustrates evolution as a continuous process, which alternates between convergence and divergence of (new) features and properties, and implies the fact that living beings of different ontogenetic and phylogenetic traits have been cross-fertilizing in accordance with multiple conditions in time. Neo-Darwinism with its genetic reductionism has empirically traced the seat of the variation of species back to genes and chromosomes; however, it is more the differentiation and combination than the specific functions of genes that explicate the biological potentiality of merging and blending across species. Such happenings in the horizontal dimension have been underestimated in Neo-Darwinian approaches which focus on the heredity of genes and the ways it encodes or determines characteristics and behaviours in a linear fashion. My study proposes to take the diagram of general difference in Saussure's orangery manuscripts (Fig. 2) as a framework to observe symbiogenesis in order to achieve greater truths about nature and culture. This framework enables us to perceive that both ontogeny and phylogeny have been constantly restructuring – in the words of Stephen J. Gould, it takes "multivariate analysis" to go beyond taxonomy, isolated traits, and gene determinism so as to discover inspiring patterns of (new) life based on the observation of a composite of factors or conditions simultaneously.

Theoretical Framework(s) of Biosemiotics

In his retrospective review of Thomas A. Sebeok's career, John Deely highlights the fact that his master's thinking about linguistics and biology in parallel became a converged and enlarging theme after the coinage of key terms like *zoosemiotics* (in 1963) and *anthroposemiotics* (in 1968) – both terms were aimed at analyzing species-specific communications among animals and humans respectively.[58] Meanwhile, in the 1970s Sebeok came to read the original and first version of Jakob von Uexküll's *Theoretische Biologie* (1920), which led him to widen his scope of

[58] Deely 2004, pp. 2–5.

semiotics in the rather encompassing domain of biosemiotics.[59] He came up with a preliminary definition of biosemiotics as follows:

> It seems likely that a full-fledged synthesis will be achieved before long, offering both a new paradigm and a methodology for the comparative analysis of semiosis in its full diversity, ranging from the two vast linked polymer languages [genetic codes] at one end of the scale to the thousands of natural languages at the other, with a host of singular information coding and transmission devices, inside and outside the body of every organism, in between. Semiosis, independent of form or substance, is thus seen as a universal, criterial property of animate existence.[60]

His idea was paraphrased and included in a glossary entry, defining semiosis as follows: "Also, semiose [semiosis] is understood in *biosemiotics* as a transfer of *signs*, e.g., an *interpretation* through which signs are replicated and further developed. Examples of such dynamic reproductions extend, for Sebeok, from the genetic *code* through all stages of transmission of information up to human speech, action and thought".[61] As a new approach, envisaged to shed light on the animate, biosemiotics synthesizes the studies of animals with those of human beings, the coding and transmission of signals inside organic bodies with those outside, and the symbolic codes used among scientists with the communication of languages in our daily lives. However, without making any clear distinctions between these diversities, biosemiotics has taken all physical and biological processes as a matter of semiosis and semiotics. Genetic codes which define the biological properties of living beings are – according to Darwinian and Neo-Darwinian conceptualizations – rigid and not subject to be modified in the regular process of reproduction and transmission. They are quasi-automatic and self-sufficient so that each code can find a stereotypical counterpart, just as a key fits a keyhole. The logic in an enclosed genetic system is to obey the necessities and to bridge the equivalences among codes. Nevertheless, the biological nature of consciousness- and language-based semiotics is about making options from available codes – a subject picks up a favored code, interprets, and complicates the meanings of the code as infinitely as possible.

Furthermore, in the light of Peirce, semiosis is a precondition of semiotics, and it is a growing and enlarging circle of interpretation that gradually deviates from an original sign and involves other objects in the world. The opaqueness in such a process lies in the fact that it is difficult to tell how the interpreting subject is going to shift his frameworks of references. Semiotics in this sense is not far-fetched from Saussure's idea of consciousness and the *ÉDL*, the nature of which is supposed to be making comparisons and associations as well. Even though Peirce and Saussure theorize consciousness as a process of internal mental reconstruction, they would agree that outsiders, heterogeneous beings, or individuals of different kinds are required in appreciating the ambiguity and flexibility of semiotic processes. Unfortunately, Sebeok in his preliminary definition imposes a kind of opaque, unfaithful and untransparent quality of semiosis upon the presumably stable and

[59] Sebeok 1998, p. 32.

[60] Sebeok 1973, p. 1189 and 1998, p. 32; Krampen et al. 1981 [1987, p. 214].

[61] Krampen et al. 1981 [1987, p. 244–245].

transparent quality of biological codes and processes in nature. The eclectic attitude of Sebeok in bringing these two conflicting states under the same paradigm of biosemiotics has committed the fallacy of a confused epistemology – biosemiotics is torn between approaches in natural sciences and those in cultural and social semiotics which actually prefer to avoid certain Darwinian doctrines. We should draw on Saussure's underestimated discourse on *ÉDL* and his evolutionary epistemology if we take it a task to unify diverse approaches which have been made available to biosemiotics.

References

Alinei, M. (2006). Darwinism, traditional linguistics and the new Palaeolithic Continuity Theory of language evolution. In Gontier, van Bendegem, Aerts (Eds.), 2006, pp. 121–147.

Amsterdamska, O. (1987). *Schools of thought: The development of linguistics from Bopp to Saussure*. Dordrecht/Norwell: D. Reidel/Kluwer.

Auroux, S. (2007). Introduction: Le paradigme naturaliste. *Histoire Épistémologie Langage, 29*(2), 5–15.

Barbieri, M. (2003). *The organic codes: An introduction to semantic biology*. Cambridge: Cambridge University Press.

Barbieri, M. (2008). Biosemiotics: A new understanding of life. *Naturwissenschaften, 95*(7), 577–599.

Bouissac, P. (2004). Saussure's legacy in semiotics. In C. Sanders (Ed.), *The Cambridge companion to Saussure* (pp. 240–260). Cambridge: Cambridge University Press.

Bouquet, S. (1997). *Introductions à la lecture de Saussure*. Paris: Éditions Payot/Rivages.

Camara, J. B. de (1995). *Saussure, chess and time: The role of an analogy in a scientific revolution*. Lisboa: Instituto Superior De Ciencias Socias E Politicas/Universidade Tecnica De Lisboa.

Chebanov, S. V. (1998). Totality of semiosphere. *Sign Systems Studies, 26*, 417–424.

Danesi, M. (2004). *A basic course in anthropological linguistics*. Toronto: Canadian Scholars' Press.

Deely, J. (2004). Thomas Albert Sebeok, "Biologist Manqué": Memorial essay for 2004 Congress of International Association for Semiotic Studies in Lyon. Unpublished Paper, 10 pages.

Engler, R. (2004). The making of the *Cours de linguistique générale*. In C. Sanders (Ed.), *The Cambridge companion to Saussure* (pp. 47–58). Cambridge: Cambridge University Press.

Gontier, N. (2006a). Evolutionary epistemology and the origin and evolution of language: Taking symbiogenesis seriously. In Gontier, van Bendegem, Aerts (Eds.), 2006, pp. 195–226.

Gontier, N. (2006b). Introduction to evolutionary epistemology, language and culture. In Gontier, van Bendegem, Aerts (Eds.), 2006, pp. 1–32.

Gontier, N., Bendegem, J. P. van, & Diederik A. (Eds.) (2006). *Evolutionary epistemology, language and culture: A non-adaptationist, systems theoretical approach*. Dordrecht: Springer.

Gould, S. J. (1977 [2007]). *Ever since Darwin: Reflections on natural history*. New York/London: W.W. Norton.

Harris, R. (2002). Ferdinand de Saussure, "Écrits de linguistique générale". *Times Literary Supplement*, 5182 (26 July 2002), (p. 30).

Hoffmeyer, J. (1993 [1996]). *Signs of meaning in the universe*. Bloomington: Indiana University Press.

Hurford, J. R. (1992). An approach to the phylogeny of the language faculty. In J. A. Hawkins & M. Gell-Mann (Eds.), *The evolution of human languages* (SFI studies in the sciences of complexity. Proceedings, Vol. XI, pp. 273–303). Reading: Addison-Wesley.

Hurford, J. R. (2007). Why communicate? Squaring with evolutionary theory. In J. R. Hurford (Ed.), *The origins of meaning: Language in the light of evolution* (pp. 243–306). New York: Oxford University Press.

Jakobson, R. (1966 [1990]). Quest for the essence of language. In L. R. Waugh & M. Monville-Burston (Eds.). *On language* (pp. 407–421). Cambridge, MA/London: Harvard University Press, 1990.

Katz, G. (2008). The hypothesis of a genetic protolanguage: An epistemological investigation. *Biosemiotics, 1*, 57–73.

Klippi, C. (2007). La première biolinguistique. *Histoire Épistémologie Langage, 29*(2), 17–40.

Koerner, K. (Ed.). (1983). *Linguistics and evolutionary theory: Three essays by August Schleicher, Ernst Haeckel and Wilhelm Bleek*. (Amsderdam classics in linguistics, Vol. 6). Amsderdam: John Benjamins.

Krampen, M., Oehler, K., Posner, R., Sebeok, T. A. & Uexküll, T. von (Eds.). (1981 [1987]). *Classics of semiotics*. New York: Plenum Press.

Kull, K. (1998). Semiotic ecology: Different natures in the semiosphere. *Sign Systems Studies, 26*, 344–371.

Kull, K. (2003). Thomas A. Sebeok and biology: Building biosemiotics. *Cybernetics and Human Knowing, 10*(1), 47–60.

McCauley, R. N. (2007). Enriching philosophical models of cross-scientific relations: Incorporating diachronic theories. In M. Schouten & H. Looren de Jong (Eds.), *The matter of the mind: Philosophical essays on psychology, neuroscience, and reduction* (pp. 199–223). Oxford: Blackwell.

Nerlich, B. (1989). The evolution of the concept of "linguistic evolution" in the 19th and 20th century. *Lingua, 77*(2), 101–112.

Nöth, W. (1998). Ecosemiotics. *Sign Systems Studies, 26*, 332–343.

Saussure, F. de (1959). *Course in general linguistics*. New York: McGraw-Hill.

Saussure, F. de (1967). *Cours de linguistique générale*, fasc. 1 & 2. Wiesbaden: Otto Harrassowitz.

Saussure, F. de (1993). *Troisième cours de linguistique générale (1910–1911) d'après les cahiers d'Émile Constantin*/Saussure's third course of lectures on general linguistics (1910–1911) from the notebooks of Émile Constantin. Oxford: Pergamon.

Saussure, F. de (1996). *Premier cours de linguistique générale (1907) d'après les cahiers d'Albert* Riedlinger/Saussure's first course of lectures on general linguistics (1907) from the notebooks of Albert Riedlinger. Oxford: Pergamon.

Saussure, F. de (2006). *Writings in general linguistics*. Oxford/New York: Oxford University Press.

Sebeok, T. A. (1973). Between animal and animal. *Times Literary Supplement, 3734*(5 October 1973), (p. 1187–1189).

Sebeok, T. A. (1976). *Contributions to the doctrine of signs*. Bloomington: Indiana University Press.

Sebeok, T. A. (1998). The Estonian connection. *Sign Systems Studies, 26*, 20–41.

Simpson, J., & Weiner, E. (Eds.). (1989). *Oxford English dictionary*. Oxford/New York: Oxford University Press.

Darwin's Ethology and the Expression of the Emotions: Biosemiotics as a Historical Science

Thomas Robert

Abstract Because of the reduction of his theory to *The Origin of Species* (1859) and its slogan "descent with modification by means of natural selection", Darwin's contribution to the study of language is largely overlooked. However, in later works, such as *The Descent of Man and Selection in Relation to Sex* (1871) and *The Expression of the Emotions in Man and Animals* (1872), Darwin develops his theory of language and of signs in general. These considerations are contained within Darwin's ethology, which is different from the theory of instinct of *The Origin* based on natural selection. Respecting the idea of continuity between non-human animals and human beings, the Darwinian animal appears as a hermeneutical subject that constructs its own world and behaves accordingly by taking into account both its structure and the surrounding conditions. Moreover, the Darwinian animal is able to emit both voluntary and involuntary signs that can be recognised as such by the animal or an observer (human or non-human). *The Expression of the Emotions* is dedicated to the study of sign emission, which has to be understood in the context of Darwinian ethology. In this article, I argue that both Darwinian ethology and biosemiotics (represented by the theory of the expression of the emotions) correspond to Saussure's definition of historical sciences. Darwin's ethology and biosemiotics are composed of contingent facts that have to be studied historically.

Keywords Ch. Darwin • F. de Saussure • Ethology • Expression • Emotions • Signs • Biosemiotics

Introduction

Darwin's contribution to the study of language may not appear central to his theory of evolution or to the history of linguistics. Indeed, the English naturalist is sometimes reduced to the author of a single book, *On the Origin of Species* (1859), which is summarised in a slogan: descent with modification by means of natural selection.

T. Robert (✉)
Independent Scholar, Geneva, Switzerland
e-mail: thomas-robert@live.fr

E. Velmezova et al. (eds.), *Biosemiotic Perspectives on Language and Linguistics*, Biosemiotics 13, DOI 10.1007/978-3-319-20663-9_13

Although the classification of languages is analogous to the classification of species,[1] the origin and development of language is not treated in *The Origin*. The fact that Darwin is not interested in such aspects of language in *The Origin* is not really surprising. Human beings are strategically absent of the book, except for a sentence in its conclusion, probably the most famous understatement of the history of science: "Light will be thrown on the origin of man and his history".[2] Moreover, even animal behaviour is practically left untouched in *The Origin*, excepting some considerations in the context of the law of use and disuse and the seventh chapter of the first edition dedicated to instinct.[3]

Darwin's silence on human faculties in *The Origin* does not mean that they cannot be explained by means of natural selection. The faculty of language makes no exception, as it is argued by Steven Pinker's popularisation of such a view in *The Language Instinct* (1994). Darwin's contribution to the study of language can be easily identified in two closely related books: *The Descent of Man and Selection in Relation to Sex* (1871) and *The Expression of the Emotions in Man and Animals* (1872). With respect to language, both these books directly contradict an adaptationist, neo-Darwinian approach of faculty. More precisely, Darwin is not interested in the faculty of language itself but in its expression through diverse kinds of languages. In other words, Darwin develops a semiotic theory that Sarah Winter justly qualifies as biosemiotics.[4] In order to explain Darwin's biosemiotics, which is drastically different from a neo-Darwinian approach to language such as the one defended by Pinker (1994), it is necessary to understand its basis that can be associated with the ethology developed in *The Descent of Man* and *The Expression*.

In order to illustrate the historical and theoretical value of Darwin's biosemiotics,[5] I will firstly explain the principles of the ethology developed in *The Descent of Man* and *The Expression*. Secondly, I will illustrate Darwin's biosemiotics through the analysis of the three principles of expressive movements given in *The Expression*. This study will lead to a Saussurean definition of both the naturalist's ethology and biosemiotics as historical sciences.

---

[1] Darwin 1859, pp. 422–423; Alter 1999.

[2] Darwin 1859, p. 488.

[3] *Ibid.*, pp. 134–139, 207–244.

[4] Winter's definition of biosemiotics will be applied in this article: "By biosemiotics I mean not only a theory that reads biological systems in semiotic terms but also one that shows how such systems function at all levels through signaling and thus through producing nonlinguistic biological signs" (Winter 2009, p. 130).

[5] While I will adopt Winter's definition of biosemiotics, I will focus on Darwinian ethology and on its compatibility with Saussure's epistemology of the science of language, which constitutes an extension of Winter's work.

Principles of Darwinian Ethology

Although several authors have emphasised the importance of animal behaviour in Darwin's thought,[6] this subject remains largely unexplored. Two complementary reasons explain why Darwinian ethology has been generally overlooked. First of all, the seventh chapter of *The Origin* seems to reduce most animal behaviour to the complex phenomenon of instinct.[7] Far from being a hasty solution, Darwin's theory of instinct is subtly articulated and is the result of a long maturation on the subject, the naturalist's path towards the theory of *The Origin* being identifiable in his manuscripts.[8] Indeed, confronted to the problem of explaining the different structures and behaviour of neuter insects, i.e. insects in a given community that are unable to reproduce, Darwin has to abandon the Lamarckian theory of the heredity of habits, which necessitates direct reproduction.[9] Using for the first time community selection, Darwin reduces most complex behaviour to the selection of accidental/spontaneous variations that present an advantage for the individual or its community in the context of a general struggle for existence.[10] Given the solution to the question of instinct in the seventh chapter of *The Origin*, it would be superfluous to look for a more developed ethology, especially if this book is considered as a summary of Darwin's thought.

Secondly, the theory of instinct contained in *The Origin* is based on principles that are in line with the natural sciences. By contrast, Darwin's ethology developed in his manuscripts, in *The Descent of Man* and in *The Expression* can be accused of relying on dubious principles and methodology.[11] For instance, the laws of heredity used by Darwin in the context of sexual selection are modelled on the hypothesis of pangenesis,[12] while *The Expression* is entirely based on the heredity of habits.[13]

[6] Cf., e.g., Durant 1985; Burkhardt 1985; Richards 1987; Townshend 2009.

[7] Darwin does not give a definition of instinct, which he seems to reduce to an innate tendency to accomplish more or less complex actions in accordance with the external circumstances.

[8] The entries in the *M* and *N* notebooks are numerous, the evolution of Darwin's thought on instinct is also particularly well illustrated in his 1842 *Sketch*, his 1844 *Essay* and his 1856–1858 *Natural Selection*. Cf. Darwin, quoted in Barrett et al. 1987 [2008, pp. 517–596]; Darwin, quoted in F. Darwin 1909, pp. 17–21, 112–132; Darwin, quoted in Stauffer 1975, pp. 466–527. I have treated this issue elsewhere, cf. Thomas 2013.

[9] Richards 1987, pp. 142–152.

[10] Darwin 1859, p. 242.

[11] Cf., e.g., Ghiselin 1969 [2003, pp. 187–213].

[12] The "provisional hypothesis of pangenesis" is Darwin's theory of heredity developed in *The Variation of Animals and Plants under Domestication* (1868). This complex and fascinating theory states that each part of an organism emits gemmules that are transmitted through reproduction. Changes in the organism lead to similar changes in the gemmules and to the possibility of the inheritance of new traits. However, complex rules preclude a systematic heredity of new characters. It has to be noted that the hypothesis of pangenesis can be considered as a theory of reproduction compatible with the heredity of habits. Cf. Darwin 1868, vol. II, pp. 357–432; Ghiselin 1969 [2003, pp. 181–186]; Ruse 1979 [1999, pp. 212–213]; Hodge 1985, pp. 227–237; Endersby 2009, pp. 82–86.

[13] The extensive use of the heredity of habits renders *The Expression* strangely un-Darwinian for readers that consider *The Origin* as a summary of Darwin's thought (cf. Radick 2010).

Moreover, animal behaviour is described through anecdotes and explained in anthropomorphic terms. Both pangenesis and the heredity of habits have been contested by Friedrich Leopold August Weismann's theory of the impermeability between *soma* and *germen* and the development of genetics. Anthropomorphism and anecdotes have been criticised in ethology since Conwy L. Morgan's canon: "In no case is an animal activity to be interpreted in terms of higher psychological processes if it can be fairly interpreted in terms of processes which stand lower in the scale of psychological evolution and development".[14] In sum, Darwinian ethology can be viewed as outdated.[15]

In order to attest the historical importance of Darwin's ethology and to recognise its potential usefulness in the study of signs, it is necessary to understand its context of development. Between 1859 and 1871, Darwin's silence on man in *The Origin* has allowed the most diverse extrapolations, such as Herbert Spencer's social Darwinism, William Rathbone Greg and Francis Galton's eugenics or Alfred Russel Wallace's surprising limitation of natural selection to exclude human higher faculties. However, such interpretations do not represent Darwin's theory of man.[16] Prompted by such wrong applications or limitations of natural selection, Darwin chose to end his silence on man, realising his project of writing a treatise on this subject that had been abandoned during his work on the different editions of *The Origin* and *Variations*. More precisely, Darwin wrote *The Descent of Man* in reaction to "The limits of natural selection as applied to man", in which Wallace, the co-discoverer of natural selection who had recently converted to spiritualism, states that natural selection cannot be responsible of the higher faculties of man and that an intelligent force must be reintroduced in evolution.[17] In short, while *The Origin* was written to challenge appeal to independent creation and natural theology, *The Descent of Man* is mostly directed against creative design with a focus on the case of man.

Although *The Descent of Man* rectifies the applications and limitations of natural selection with respect to man, Darwin does not argue for a vision of evolution based uniquely on natural selection. Indeed, the naturalist recognises a certain instance of limitation of natural selection:

> Thus a large yet undefined extension may safely be given to the direct and indirect results of natural selection; but I now admit, after reading the essay by Nägeli on plants, and the remarks by various authors with respect to animals, more especially those recently made by Professor Broca, that in the earlier editions of my "Origin of Species" I perhaps attributed too much to the action of natural selection or the survival of the fittest. I have altered the fifth edition of the "Origin" so as to confine my remarks to adaptive changes of structure; but I am convinced, from the light gained during even the last few years, that very many structures which now appear to us useless, will hereafter be proved to be useful, and will therefore come within the range of natural selection. Nevertheless, I did not formerly con-

[14] Morgan 1895, p. 53.

[15] Ghiselin 1969 [2003, pp. 187–213]; Durant 1985, pp. 291–292, 302–303; Burkhardt 1985, pp. 328, 348–349, 351.

[16] Cf. Tort 2010, pp. 63–152.

[17] Cf. Wallace 1871, pp. 332–371. Cf. also Kottler 1974 and 1985, pp. 420–421; Richards 1987, pp. 186–187.

> sider sufficiently the existence of structures, which, as far as we can at present judge, are neither beneficial nor injurious; and this I believe to be one of the greatest oversights as yet detected in my work. I may be permitted to say [...], that I had two distinct objects in view; firstly, to shew that species had not been separately created, and secondly, that natural selection had been the chief agent of change, though largely aided by the inherited effects of habit, and slightly by the direct action of the surrounding conditions. I was not, however, able to annul the influence of my former belief, then almost universal, that each species had been purposely created; and this led to my tacit assumption that every detail of structure, excepting rudiments, was of some special, though unrecognised, service. Any one with this assumption in his mind would naturally extend too far the action of natural selection, either during past or present times.[18]

As a manifesto in favour of transmutationism, *The Origin* is prone to exaggeration. As Darwin emphasises, natural selection is not the only operative principle in evolution. Indeed, as already mentioned in the first edition of 1859, he gave weight to other principles. These principles, such as the heredity of habits or the influence of conditions, complement natural selection in cases in which behaviour lacks adaptive value. Darwin is relatively careful on such non-adaptive characteristics, underlining that an adaptive value could be inaccessible due to the advancement of science. However, the second part of *The Descent of Man*, dedicated to sexual selection and written before the first part,[19] is more radical on this issue. Indeed, sexual selection is not a particular case of natural selection[20] but, rather, leads to the development of useless and injurious structures and behaviour. In other words, the observation of animal behaviour, particularly in the context of courtship, makes Darwin open what could be considered as a domain of the useless,[21] in which non-adaptive and anti-adaptive structures and behaviour can be explained. The domain of the useless, though never identified as such by Darwin, constitutes the naturalist's answer to the arguments against natural selection or its misuse.

Darwin's ethology is directly opposed to Morgan's canon since the naturalist states that "the more the habits of any particular animal are studied by a naturalist, the more he attributes to reason and the less to unlearnt instincts".[22] This generosity towards animals is opposed to the reduction of animal behaviour to complex instincts in *The Origin*. Persuaded of the continuity between man and other animals, Darwin develops a double argument representing his global ethology: zoomorphic anthropology and anthropomorphic zoology.[23] Every human faculty is identifiable in animals, despite quantitative differences, while animal behaviour can be described and explained by anecdotes related in anthropomorphic terms. Sexual selection is certainly the most pregnant illustration of Darwinian ethology. During courtship, males try to seduce females by singing or displaying their ornaments. Such behaviour is at

[18] Darwin 1874, pp. 61–62.

[19] Burkhardt 1985, pp. 349–350.

[20] In other words, Darwin is not a precursor of sociobiology (cf. Mayr 1972, p. 88). For a conciliation between Darwin and sociobiology, cf. Cronin 1991, pp. 113–249.

[21] The adjective *useless*, although prone to be criticised, is here chosen on purpose as opposed to *useful*, i.e. to advantageous behaviour and structures in the context of the struggle for existence.

[22] Darwin 1874, p. 75.

[23] These terms are used as they have been established by Durant (1985).

first conscious but can become an instinct thanks to the heredity of habits.[24] Females choose consciously their favourite males by using their sense of beauty. Anthropomorphism reaches its climax in the comparison of such seductive behaviour and human cultural rituals:

> With respect to female birds feeling a preference for particular males, we must bear in mind that we can judge of choice being exerted, only by analogy. If an inhabitant of another planet were to behold a number of young rustics at a fair courting a pretty girl, and quarrelling about her like birds at one of their places of assemblage, he would, by the eagerness of the wooers to please her and to display their finery, infer that she had the power of choice. Now with birds, the evidence stands thus; they have acute powers of observation, and they seem to have some taste for the beautiful both in colour and sound. It is certain that the females occasionally exhibit, from unknown causes, the strongest antipathies and preferences for particular males. When the sexes differ in colour or in other ornaments the males with rare exceptions are the more decorated, either permanently or temporarily during the breeding-season. They sedulously display their various ornaments, exert their voices, and perform strange antics in the presence of the females. Even well-armed males, who, it might be thought, would altogether depend for success on the law of battle, are in most cases highly ornamented; and their ornaments have been acquired at the expense of some loss of power. In other cases ornaments have been acquired, at the cost of increased risk from birds and beasts of prey. With various species many individuals of both sexes congregate at the same spot, and their courtship is a prolonged affair. There is even reason to suspect that the males and females within the same district do not always succeed in pleasing each other and pairing.[25]

At the opposite of Darwin's anthropomorphic account of courtship among animals, Wallace, respecting Morgan's canon, denies both the conscious seductive action of males and the choice exerted by females. According to Wallace, the fittest males develop ornaments thanks to a surplus of energy and are able to captivate females.[26] While animal behaviour can be explained by natural selection operating on mechanical variations, human structures and behaviour need the intervention of an intelligent force since they are far too developed in primeval men and savages or are obviously injurious to them to be accounted for by natural selection, which has immediate utility as criterion.[27] An ultra-adaptive theory leads Wallace to maintain a discontinuity between animals and man. Darwin softens natural selection, letting the domain of the useless emerge, in order to allow a strict continuity in the animal reign.

The opposition between Wallace's discontinuous theory and Darwin's insistence on continuity is mostly obvious in the account given to the useless and the injurious. Indeed, prefiguring the Baldwinian theory of evolution based on organic selection,[28]

---

[24] Darwin 1874, p. 402. In more modern terms, the epigenetic level, represented by the heredity of habits, is the source of the evolution of behaviour. This explanation of the evolution of behaviour is opposed to the theory of instinct defended in *The Origin*, which would correspond to a genetic account of behaviour.

[25] *Ibid*., pp. 420–421.

[26] Wallace 1889, pp. 268–300.

[27] Wallace 1871, pp. 332–371 and 1889, pp. 445–478; Kottler 1980 and 1985, pp. 417–425; Cronin 1991, pp. 123–164.

[28] Baldwin 1896.

Darwin transfers the selective power to the animal in the case of non-adaptive or anti-adaptive structures and behaviour. Not only are females able to choose their mate thanks to high intellectual faculties and aesthetical sense, but also the animals in general can be considered as reacting intelligently to their structure, taking into account their surrounding conditions. For example, female birds that have inherited conspicuous colours, acquired by the males in the context of sexual selection, can intelligently modify their habits of nidification.[29] The useless and the injurious can be conserved thanks to the transfer of selective power to the animal. The domain of the useless is ever-expanding and leads to a modification of the definition of natural selection. From the positive and creative principle of the *Origin* that acts on spontaneous variations, natural selection becomes a negative and eliminative principle in *The Descent of Man* due to the transfer of the selective power to the animal. Darwin's insistence on continuity does not simply consist of zoomorphic anthropology, recognising the fact that man is only an animal, but is also based on anthropomorphic zoology, emphasising the mental powers of the entire reign.

The Darwinian animal does not correspond to the animal described by mainstream ethology, which has followed both René Descartes and Conwy L. Morgan's path.[30] Indeed, Darwin's animal is not at all a machine deprived of surprises that can be described by an ethogram. On the contrary, the Darwinian animal is always interpreting and constructing its own world influenced by a triple history: phylogenetic, cultural, historical.[31] The fact that the animal is at the crossroads of three histories is particularly well illustrated by the emission of signs, which is explained in *The Expression* where Darwin considers the semiotic character of the animal. The study of expression, completing Darwin's ethology, has to follow the same principles developed in *The Descent of Man*. In short, understanding the animal consists of considering how and why it constructs its own world as it is.[32]

---

[29] Wallace argues that coloration is always under the realm of natural selection. According to him, conspicuous females having the habits of hatching unprotected are eliminated while less conspicuous females are selected. Darwin reverses this process by stating that when females become conspicuous, they alter their habits of nidification. While it could be possible that such new habits are the results of the selection of a spontaneous variation, Darwin's anthropomorphic zoology suggests that such behaviour is the product of intelligence and can become instinctive by the heredity of habits. It has to be noted that this alteration of the hatching habits echoes the acquisition of instinctive fear in *Natural Selection*, which is certainly the best example of the use of the heredity of intelligent habits in Darwin's manuscripts (cf. Darwin, quoted in Stauffer 1975, pp. 495–496; Darwin 1874, pp. 452–453; Wallace 1871, pp. 249–263 and 1889, pp. 277–281).

[30] Dominique Lestel considers that ethology is mainly realist-Cartesian: "Contemporary ethology emphasizes an approach to the animal which could be characterized as realistic and Cartesian. It combines fundamental description of the world with stipulation of the legitimate ways of studying it. It supposes that there is a world which is separated from the subject, and that we can provide a genuine description of the animal by investigating the causal and mechanical procedures determining animal behaviour. The possibility of observations without observers, and the description of an animal as a machine, therefore fundamentally define this approach" (Lestel 2011, pp. 83–84).

[31] Cf. *ibid.*, pp. 84, 89.

[32] This approach corresponds to bi-constructivism, which is Lestel's alternative to the realist-Cartesian paradigm (*ibid.*, pp. 83–102).

Darwin's Biosemiotics: *The Expression of the Emotions in Man and Animals*

Darwin wanted to include his considerations on expression in *The Descent of Man*. However, due to the large amount of data accumulated by the naturalist, it was published separately in 1872.[33] Therefore, *The Expression* has to be understood in the context of the argument for continuity.[34] Darwin's main antagonist is no more Wallace but Charles Bell:

> All the authors who have written on Expression, with the exception of Mr. Spencer – the great expounder of the principle of Evolution – appear to have been firmly convinced that species, man of course included, came into existence in their present condition. Sir C. Bell, being thus convinced, maintains that many of our facial muscles are "purely instrumental in expression"; or are "a special provision" for this sole object. But the simple fact that the anthropoid apes possess the same facial muscles as we do, renders it very improbable that these muscles in our case serve exclusively for expression; for no one, I presume, would be inclined to admit that monkeys have been endowed with special muscles solely for exhibiting their grimaces. Distinct uses, independently of expression, can indeed be assigned with much probability for almost all the facial muscles.[35]

As a reaction to Bell's design theory, Darwin states that the expression of emotions is not adaptive, which places *The Expression* within the domain of the useless.[36] Both zoomorphic anthropology and anthropomorphic zoology are at play in the explanation of expressive movements.

Having gathered data on animal and human expressions from a large diversity of sources, Darwin is able to give three main principles that, combined, should explain most expressive movements:

I. *The principle of serviceable associated Habits*. Certain complex actions are of direct or indirect service under certain states of the mind, in order to relieve or gratify certain sensations, desires, etc.; and whenever the same state of mind is induced, however feebly, there is a tendency through the force of habit and association for the same movements to be performed, though they may not then be of the least use. Some actions ordinarily associated through habit with certain states of the mind may be partially repressed through the will, and in such cases the muscles which are least under the separate control of the will are the most liable still to act, causing movements which we recognise as expressive.

---

[33] Browne 1985, pp. 308–309; Richards 1987, p. 230.

[34] It is also important to note that the study of expression constitutes an important argument for monogenism, cf. Winter 2009; Desmond and Moore 2009 [2010].

[35] Darwin 1890, pp. 10–11.

[36] Although numerous scholars have noted this particularity of *The Expression* (cf., e.g., Browne 1985; Burkhardt 1985; Richards 1987 and more recently Radick 2010), others, such as Michel T. Ghiselin (cf. Ghiselin 1969 [2003]), have completely distorted Darwin's text by introducing natural selection within *The Expression*. Only four occurrences of natural selection appear in *The Expression* and none of them considers natural selection as an essential explanation of expressive movements (Darwin 1890, pp. 44, 110, 113, 381).

In certain other cases the checking of one habitual movement requires other slight movements; and these are likewise expressive.

II. *The principle of Antithesis*. Certain states of the mind lead to certain habitual actions, which are of service, as under our first principle. Now when a directly opposite state of mind is induced, there is a strong and involuntary tendency to the performance of movements of a directly opposite nature, though these are of no use; and such movements are in some cases highly expressive.

III. *The principle of actions due to the constitution of the Nervous System, independently from the first of the Will, and independently to a certain extent of Habit*. When the sensorium is strongly excited, nerve-force is generated in excess, and is transmitted in certain definite directions, depending on the connection of the nerve-cells, and partly on habit: or the supply of nerve-force may, as it appears, be interrupted. Effects are thus produced which we recognise as expressive. This third principle may, for the sake of brevity, be called that of the direct action of the nervous system.[37]

The principle of serviceable associated habits is certainly the most complex and the most important. Indeed, it presupposes the Darwinian concepts of habit and instinct. In order to explain this first principle, therefore, it is necessary to address its physiological side, its psychological side and the relations between habits and reflex actions. The physiology of the first principle allows Darwin to explain the potential transmission of expressive actions, which is necessary to understand their innate character. Using Max Müller and Herbert Spencer's neo-Lamarckian physiological theory, Darwin states that habits can alter the nervous system by repetition.[38] Under the same emotion, nerve-force follows such modified nervous channel, which leads to an instinctive expressive movement. The link between the first and the third principle is evident when the physiology of associated habits is considered.

The psychological side of the first principle is based on the principle of association. An emotion leads to the unconscious accomplishment of a certain action or a chain of actions that can be highly complex. Despite their very different origins, such actions, which are at first conscious and intelligent, are no more distinguishable from instinct, i.e. from the selection of spontaneous variations.

The difference between habit and instinct is central in the consideration of reflex actions. Such actions are "due to excitement of a peripheral nerve, which transmits its influence to certain nerve-cells, and these in their turn excite certain muscles or glands into action".[39] Unlike instincts that are centred in the brain,[40] reflex actions are localised in an affected area of the nervous system. Moreover, reflex actions are mostly not the product of natural selection and are generally issued from habits,

[37] *Ibid.*, pp. 29–30.

[38] *Ibid.*, pp. 9–13, 30, 41–43, 49–50, 71–75, 80, 209, 358–364.

[39] *Ibid.*, p. 36.

[40] Or in the centre of the nervous system.

which are at first conscious and voluntary.[41] Darwin's double argumentation against Bell and an adaptive explanation of the expression appears. Indeed, the animal accomplishes complex actions, that are at first voluntary and that allow him to obtain satisfaction. Repeated by association, these actions alter the nervous system of the animal, which can then be transmitted to its offspring through the heredity of habits. It is particularly important to underline that the useful actions leading to habits and reflex actions are not selected but correspond to an intelligent reaction[42] of the animal, taking into account both its structure and the surrounding conditions. Therefore, Darwin disavows the adaptive perspective with respect to the origin and development of the expression of the emotions. Moreover, once the emotion and its corresponding action are associated, which constitutes the expressive character of such movements, the utility criterion becomes superfluous. In sum, the first principle contains Darwin's argument of continuity leading to the consideration of the expression of the emotions as a part of the domain of the useless.

At first sight, the mechanism explaining the principle of antithesis seems trivial. According to the first principle, useful actions are associated to a mental state and become automatic. Such actions correspond to expressive movements. An opposed emotion logically leads to equally opposed movements. However, important implications of the principle of antithesis have to be addressed. Indeed, the limits of convention in the context of expression appear with this second principle of expression. The expressive actions triggered by the principle of antithesis are useless. Movements contrary to other movements initially issued from the first principle cannot be useful, except with respect to communication.[43] Although Darwin doubts of the possibility of the voluntary development, by animals, of such complicated movements in order to communicate,[44] the naturalist recognises that the will can replace simple muscular expressive movements.[45] With the second principle explaining the expression of the emotions, a possibility of negotiation through communication emerges for the animal.

The name given to the third principle explaining the expression of the emotions seems to isolate it from the two previous principles. Indeed, only the nervous system, independently of the will and habits is considered. In other words, the third principle seems to account for pure useless characters within the domain of the useless. Indeed, the actions explainable by the third principles are not, at first, useful actions as it is the case in the first principle. Moreover, no voluntary action seems to be interfering in the context of the third principle since it concerns parts of the nervous system inaccessible to the will. However, the link with the two other principles

[41] Darwin 1890, pp. 41–44.

[42] Contrarily to instinctive actions, intelligent actions are issued from conscious choices (from the animal).

[43] For example, the position that the cat adopts in order to attack its prey is useful. On the contrary, the movement of the loving cat, opposed to the position of the attacking cat, is of no use. Still, the cat can, by this position, show that it is not going to attack.

[44] Darwin 1890, pp. 67–68.

[45] *Ibid.*

of the expression of the emotions is not completely eliminated. All the principles can interact with each other during expressive movements and have Müller and Spencer's neo-Lamarckian physiological theory as common ground.[46]

Certain essential characters of Darwin's theory of expressive signs appear from this (too) brief exposition of the three principles of the expression of the emotions. Reflecting the double argumentation of Darwinian ethology (zoomorphic anthropology and anthropomorphic zoology), expressive movements are issued from an encounter between purely physiological mechanisms and intelligent reactions to structure and surrounding conditions. Moreover, expressive signs are not adaptive and must be studied in the context of the domain of the useless, as it is proved by the extensive use of the heredity of habits, which links all three principles of the expression of the emotions. Finally, the possibility to act on one's expressive movements opens a space in which the animal can negotiate. Thanks to such actions, the animal can wilfully communicate and try to convince its interlocutor. For example, the seductive antics of the males during courtship are entirely based on such actions on expressive movements. Since Darwin thinks that sexual selection is the context of the emergence of language,[47] allowing one to seduce, convince, praise and blame, and that moral societies are in part built on such exercise of communicative power,[48] the minimal convention observable in the control of expressive movements appears as a key phenomenon in evolution.

A Saussurean Definition of Ethology and Biosemiotics

The studies trying to conciliate Charles Darwin and Ferdinand de Saussure are few, despite several fruitful possibilities. For instance, it is possible to make an analogy between the Saussurean dichotomy diachrony/synchrony and Darwin's successive considerations of variations and natural selection.[49] Closer to the perspective of this article, Darwin and Saussure's theories can be compared both historically and theoretically. For example, Winter states that Darwin's theory of expressive signs respects the Saussurean definition of the arbitrariness of the linguistic sign since there is "no semantic relation"[50] between the expressive movements and the emotion. In sum, an exhaustive comparative work linking Darwin's theory with the Saussurean tradition remains to be done.

An agreement between Darwin and Saussure's epistemology can be found. In the first conference given by Saussure in 1891 for his return to the University of Geneva after years spent in Paris, the linguist defines the science of language as an historical science:

---

[46] *Ibid.*, pp. 86–87.

[47] Darwin 1874, pp. 84, 92 and 1890, pp. 88–100.

[48] Darwin 1874, pp. 130–134.

[49] Variations can be studied independently through time while natural selection represents a differential system (cf. Röllin 1980).

[50] Winter 2009, p. 145.

> À mesure <qu'on> a mieux compris la véritable nature <des faits de> langage, <qui sont si près de nous,> mais d'autant plus difficile à saisir dans leur essence, il est devenu plus évident que la science du langage est une science historique et rien d'autre qu'une science historique. C'est de cette qualité de science historique que se réclamera toute espèce d'études linguistiques pour figurer dans une Faculté des Lettres. Comme c'est particulièrement aussi sur cette idée d'histoire qu'il est insisté dans le titre de ce cours – alors que d'autres dénominations comme *Grammaire comparée* sont plus usitées – je crois devoir essayer de faire le commentaire, nécessairement très abrégé et incomplet, du sens qu'a ce mot *histoire* pour le linguiste. C'est sur ce sujet que j'aurais voulu solliciter votre attention presque sans autre préambule, car il contient tout: plus on étudie la langue, plus on arrive à se pénétrer de ce fait que *tout* dans la langue *est histoire*, c'est-à-dire qu'elle est un objet <d'analyse> historique, et non <d'analyse> abstraite, qu'elle se compose de *faits*, et non de *lois*, que tout ce qui semble *organique* dans le langage est en réalité *contingent* et complètement accidentel.[51]

Linguistics is not a natural science, contrarily to what could have been inferred from August Schleicher's *Die Darwinische Theorie und die Sprachwissenschaft* (1863).[52] The study of language is based on the analysis of different languages that are historical realities influenced by the circumstances of human history.

Both Darwin's ethology and linguistics can be considered as historical, and not natural, sciences. Indeed, animal behaviour is studied through anthropomorphic anecdotes. Therefore, animal behaviour is a collection of historical facts and cannot be reduced to abstract laws, except if the model of the seventh chapter of *The Origin* is used. Each animal possesses its own individual history, influenced by a phylogenetic history (itself composed of the sum of individual histories), and a cultural history.[53] With the transmission of selective power to the animal, historical contingency is at its peak. Although the expressive movements seem to be triggered by organic laws, due to their physiological determination, Darwin's biosemiotics has to be considered as a historical science and corresponds to the epistemology that Saussure tried to establish in 1891.[54] The three principles of the expression of the emotions have something to do with animal voluntary actions.[55] The expressive movements due to the principle of associated serviceable habits are initially the

---

[51] Cf. Saussure 1891 [1967–1974, IV, p. 5] and 2002, pp. 148–149.

[52] Cf. Tort 1980.

[53] Behaviour can be considered as cultural when the animal actions are not determined by their biology and their environment. It is necessary to add the importance of meaning for the animal considered as a subject (cf. Lestel 2001, p. 368).

[54] A common mistake has to be avoided here. Saussure uses *historical* as opposed to *natural*. Linguistics is a historical (and not natural) science. With respect to the later distinction of diachronic and synchronic linguistics, both aspects have to be studied by a historical, i.e. not natural, science. In other words, the definition of linguistics as a historical science must not be reduced to diachronic considerations. It is via this very broad sense of *historical* as opposed to *natural* that Darwin's ethology and linguistics can be linked with Saussure's epistemology of the science of language.

[55] Historical sciences, according to Saussure, study voluntary actions. However, the voluntary character of actions can be more or less pregnant. With respect to language, the voluntary character of linguistic acts is reduced to its minimum by Saussure (cf. Saussure 1891 [1967–1974, pp. 5–6] and 2002, p. 150).

result of actions done on purpose by the animal in order to obtain satisfaction. The principle of antithesis is indirectly the result of voluntary actions. The movements due to the principle of the direct action of the nervous system can be voluntarily used by the animals for their communicative value, as the ones issued from the second principle. As any animal action, the emission of signs is influenced by the three histories of the animal. The individual history determines what kind of emotion the animal is led to feel and what kind of expressive movements it is used to display. The phylogenetic history determines the range of possible movements that the animal is capable of. Finally, the cultural history determines the minimal convention intervening in expressive signs. Therefore Darwin's biosemiotics has to be considered as a historical science based on a historical analysis of expressive contingent facts issued, in their origins, from the voluntary actions of animals.

Conclusion

Darwin's theory of behaviour is far more developed than its limited expression in the seventh chapter of *The Origin*. In order to study the full range of Darwin's ethology, it is necessary to concentrate on his manuscripts, on *The Descent of Man* and on *The Expression of the Emotions*. The Darwinian animal appears as a hermeneutical subject that constructs its own world and that behaves accordingly, taking into account both its structure and the surrounding conditions. Moreover, the Darwinian animal is capable of emitting both voluntary and involuntary signs that can be recognised as such by the animal or the observer (human or non-human). Despite an important physiological determination, most of the expressive movements of the animal are due to the automatisation of voluntary actions through the heredity of habits. Even movements apparently completely determined by the nervous system can be used on purpose by the animal.

Darwin's non-adaptive ethology and theory of expression is compatible with the Saussurean definition of historical science. Ethology is the historical analysis of behavioural, contingent actions of the animals. Such contingent actions constitute the facts on which ethology (defined as a historical science) is founded. Biosemiotics is the historical analysis of expressive contingent facts issued, in their origins, from the voluntary actions[56] of animals that are studied by ethology. In sum, biosemiotics is a particular aspect of a global non-adaptive ethology maintaining the continuity between non-human animals and human beings by a historicisation of nature.

[56] The voluntary character of actions is one of the "conditions" of historical sciences as defined by Saussure.

References

Alter, S. G. (1999). *Darwinism and the linguistic image, language, race and natural theology in the nineteenth century*. Baltimore: John Hopkins University Press.

Baldwin, J. M. (1896). A new factor in evolution. *American Naturalist, 30*, 441–451, 536–553.

Barrett, P. H., Gautrey, P. J., Herbert S., Kohn D., & Smith, S. (Eds.). (1987 [2008]). *Charles Darwin's notebooks 1836–1844*. Cambridge: Cambridge University Press.

Browne, J. (1985). Darwin and the expression of the emotions. In Kohn (Ed.), 1985, pp. 307–326.

Burkhardt, R. W. Jr. (1985). Darwin on animal behaviour and evolution. In Kohn (Ed.), 1985, pp. 327–365.

Cronin, H. (1991). *The ant and the peacock*. Cambridge: Cambridge University Press.

Darwin, C. R. (1859). *On the origin of species by means of natural selection, or the preservation of favoured races in the struggle for life*. London: John Murray.

Darwin, C. R. (1868). *The variations of animals and plants under domestication* (Vol. II). London: John Murray.

Darwin, C. R. (1874). *The descent of man and selection in relation to sex* (2nd ed.). London: John Murray.

Darwin, C. R. (1890). *The expression of the emotions in man and animals* (2nd ed.). London: John Murray.

Darwin, F. (Ed.). (1909). *The foundation of the origin of species, two essays written in 1842 and 1844 by Charles Darwin*. Cambridge: Cambridge University Press.

Desmond, A., & Moore, J. (2009 [2010]). *Darwin's sacred cause: Race, slavery and the quest for human origins*. London: Penguin.

Durant, J. R. (1985). The ascent of nature in Darwin's *Descent of Man*. In Kohn (Ed.), 1985, pp. 283–306.

Endersby, J. (2009). Darwin on pangenesis, generation and sexual selection. In M. J. S. Hodge & G. Radick (Eds.), *The Cambridge companion to Darwin* (pp. 73–95). Cambridge: Cambridge University Press.

Ghiselin, M. T. (1969 [2003]). *The triumph of the Darwinian method*. New York: Dover Publications.

Hodge, M. J. S. (1985). Darwin as a lifelong generation theorist. In Kohn (Ed.), 1985, pp. 207–243.

Kohn, D. (Ed.). (1985). *The Darwinian heritage*. Princeton: Princeton University Press.

Kottler, M. (1974). Alfred Russel Wallace, the origin of man and spiritualism. *Isis, 65*(2), 145–192.

Kottler, M. (1980). Darwin, Wallace and the origin of sexual dimorphism. *Proceedings of the American Philosophy Society, 124*(3), 203–226.

Kottler, M. (1985). Charles Darwin and Alfred Russel Wallace: Two decades of debate over natural selection. In Kohn (Ed.), 1985, pp. 367–432.

Lestel, D. (2001). *Les origines animales de la culture*. Paris: Flammarion.

Lestel, D. (2011). What capabilities for the animal? *Biosemiotics, 4*(1), 83–102.

Mayr, E. (1972). Sexual selection and natural selection. In B. Campbell (Ed.), *Sexual selection and the descent of man* (pp. 87–104). Chicago: Adline.

Morgan, C. L. (1895). *An introduction to comparative psychology*. London: Walter Scott.

Pinker, S. (1994 [2007]). *The language instinct, how the mind creates language*. New York: Harper Perennial.

Radick, G. (2010). Darwin's puzzling expression. *Comptes rendus de l'Académie des Sciences: Biologies, 333*(2), 181–187.

Richards, R. J. (1987). *Darwin and the emergence of evolutionary theories of mind and behaviour*. Chicago: Chicago University Press.

Robert, T. (2013). The Darwinian theories of instinct: From Lamarckism to selection. In P. Hanna (Ed.), *An anthology of philosophical studies* (Vol. 7, pp. 121–132). Athens: ATINER.

Röllin, K. (1980). Parola e riproduzione : un confronto epistemologico tra il *Corso di linguistica generale* di Ferdinand de Saussure e *L'origine delle specie* di Charles Darwin. *Cahiers Ferdinand de Saussure, 34*, 99–115.
Ruse, M. (1979 [1999]). *The Darwinian revolution, science red in tooth and claw*. Chicago: The University of Chicago Press.
Saussure, F. de (1891 [1967–1974]). *Conférences inaugurales à l'Université de Genève*. In *Cours de linguistique générale* (Bzoch. IV). Wiesbaden: Harrasowitz.
Saussure, F. de (2002). *Écrits de linguistique générale*. Paris: Gallimard.
Schleicher, A. (1863). *Die Darwinische Theorie und die Sprachwissenschaft*. Weimar: Bölau.
Stauffer, R. C. (Ed.). (1975). *Charles Darwin's natural selection, being the second part of his big species book written from 1856 to 1858*. Cambridge: Cambridge University Press.
Tort, P. (1980). *Évolutionnisme et linguistique*. Paris: Vrin.
Tort, P. (2010). *Darwin n'est pas celui qu'on croit*. Paris: Le cavalier bleu.
Townshend, E. (2009). *Darwin's dogs, how Darwin's pets helped form a World-changing theory of evolution*. London: Frances Lincoln.
Wallace, A. R. (1871). *Contributions to the theory of natural selection*. London: Macmillan.
Wallace, A. R. (1889). *Darwinism*. London: Macmillan.
Winter, S. (2009). Darwin's Saussure: Biosemiotics and race in the *Expression*. *Representations, 107*(1), 128–161.

Darwin's Biosemiotics: The Linguistic Rubicon in the *Descent of Man*

Deana Neubauer

Abstract In this essay I shall attempt to show how Darwin's work on, and contribution to, linguistics has often charted a proto-biosemiotic trajectory of thought. While modern linguistic studies have tended to explore Darwin's contribution to linguistics adopting a Saussurean view I argue that such views undermine Darwin's notion of continuity between animals and humans and I propose to look at Darwin's theory of language, which stems from the Romantic thought, by adopting a biosemiotic perspective which clarifies Darwin's own ideas on the origin of language and animal evolution as well as their inter-relations.

Keywords Biosemiotics • Origin of language • F. de Saussure • Ch.S. Peirce • Sign • Evolutionary theory • Ch. Darwin • German Romanticism • Philology

Introduction

Literary criticism[1] has seen an increased interest in Charles Darwin's work in relation to his views on language and its origins, and his contribution to the development of linguistics. However, his theory of language has more often than not been interpreted through a twentieth century Saussurean tradition based on the arbitrariness of signs. Although such interpretations shed light on what has been defined as Darwin's theory of signs,[2] they seem to fall into what Hans Aarslef[3] identifies as an error, namely that of confusing the formal criteria of a discipline in its maturity, with the motives and influences that brought it into being. In a similar way, it could be argued that Darwin's theory of language has often been interpreted from the standpoint of how his work was conceptualised in its maturity, rather than understood in the light of the influences that lead to its emergence. By building on the account of the rise and proliferation of comparative philology in nineteenth century England

[1] Alter 1999; Herbert 2001; Winter 2009.

[2] Winter 2009.

[3] Aarsleff 1983, p. 7.

D. Neubauer (✉)
London Metropolitan University, London, UK
e-mail: d.neubauer@londonmet.ac.uk

E. Velmezova et al. (eds.), *Biosemiotic Perspectives on Language and Linguistics*, Biosemiotics 13, DOI 10.1007/978-3-319-20663-9_14

and by outlining the legacy of the German Romantic thought as a model for Darwin's own understanding of language theory, the aim of this essay is to show how Darwin's semiotic project, which emerges in the *Descent of Man, and Selection in Relation to Sex* (1871), could be argued to be closer to a proto-biosemiotic perspective of language than a Saussurean. In order to do so I shall endeavour to show how Darwin's postulation of the link between animal cognition and human cognition could be understood from a biosemiotic perspective as the result of sign interpretation or semiosis.

Doctrines Concerning the Theory and Origin of Language: A Background to Darwin's Evolutionary and Language Theory[4]

The study of language has traditionally been seen as the central question about the nature of man and as such linked to questions concerning intelligence, reason, thought, and progress of knowledge. In its eighteenth century formulation, as Aarsleff notes,[5] the origin of language and speech was the key to the history of thought and mankind. John Locke, for instance, had noted that any inquiry into the human race would necessarily involve an inquiry into the origin of language. This point was taken up and quoted, as Gillian Beer[6] observes, by Lord Monboddo in *Of the Origins and Progress of Language*[7] where he devoted two chapters to epistemological questions and the nature of man before reverting to the origin of language as the central issue to the understanding of human mind and its progress.

A central preoccupation in the eighteenth century was to show the separateness of man's natural endowments from his artificial accomplishments and language was not only a prime example of man's art, but also the foundation of the progress of knowledge and thought. It is within this context that Horne Tooke's (1736–1812) publication of his first volume of *Diversions of Purley* appeared in 1786, almost simultaneously with Sir William Jones's famous discourse "On Hindus" in England. These publications marked the beginning of the decline of one tradition in the study of language, namely the philosophical, based on a general reasoning **a priori** and etymology,[8] and the emergence of another, the comparative and historical, based on the study of grammar and kinship of languages. The fact that both works were published in the same year reflects of the controversies which were to capture the imagination of two generations in England and which were to form the basis for

[4] Part of this argument has been developed in Neubauer 2013.

[5] Aarsleff 1976.

[6] Beer 1996.

[7] Monboddo 1779.

[8] Beer notes that etymology was the oldest form in which linguistic change was recorded, however it didn't allow to establish an authoritative sequence of change, since it was possible to produce more than one convincing etymology for one word (Beer 1996, p. 109).

discussions among linguists on the necessity to institute a study of language upon a more scientific footing.

As Aarsleff points out, the reputation of Tooke's *Diversion of Purley* is "one of the most remarkable phenomena in the intellectual and scholarly life of England during the first third of the 19th century"[9] as it kept England immune to the new philology which had to be imported from the Continent by German linguists and which was adopted and popularised in England by Max Müller. At the root of such a success lay Tooke's proposed integration of philosophy and philology. He set out to demonstrate the dependence of thought on language and to create a system of language which, applied to metaphysical domains, could transcend them. He based his views largely on theories proposed by the empiricist John Locke (1632–1704) and the French philosopher, epistemologist and psychologist Étienne Bonnot de Condillac (1715–1780).

Locke's *Essay Concerning Human Understanding* is a philosophical landmark devoted not only to the understanding of the nature and limits of human knowledge in terms of concepts and ideas, but also to the discussion of the role language plays in human cognition. The *Essay* is divided in four books, where after a critique of innate ideas and an extensive discussion of the origin and classification of ideas in *Books I* and *II* respectively, Locke turns to his discussion on language. *Book III* starts with a chapter entitled "Of words or language in general" followed by "On the signification of words". In the latter he emphasises the arbitrary nature of words as well as their importance in communication as he writes:

> Words are sensible signs, necessary for communication of ideas. [...] in their immediate signification, words are the sensible signs of his ideas who uses them, how imperfectly soever or carelessly those ideas are collected from the things which they are supposed to represent. [...] *Words* come to be made use of by Men, as *the Signs of* their *Ideas* [...] not by any natural connexion, that there is between particular articulate sounds and certain *Ideas*, [...] but by a voluntary Imposition, whereby such a Word is made arbitrarily the Mark of such an *Idea.*[10]

As the above passage clearly explains, Locke sees the relation between words and ideas as a human artefact and a product of voluntary activity. He further states that it is only "the arbitrary imposition of Men" that connects words and ideas because he believes there is no "natural connexion between sound and Idea".[11] By stating that there is no natural connection between sound and idea, Locke dismissed the then predominant view of natural language largely associated with the Biblical description of Adam's language.

It is important to point out here that with his discussion on language and his classification of science into three domains whereby the third was occupied by what he termed "σημειωτική [*sēmeintikē*], or *the Doctrine of Signs*; the most usual whereof being words, it is aptly enough termed λογική [*logikē*], Logick" Locke introduced the formal study of signs into philosophy. The task of this doctrine was "to consider

[9] Aarsleff 1983, p. 73.

[10] Locke 1690 [1975], Book III, p. 405.

[11] *Ibid.*, p. 477.

the nature of signs, the mind makes use of for the understanding of things, or conveying its knowledge to others".[12] The type of signs which should be studied by his doctrine are not only words, but also ideas which are signs of external objects, or as he writes: "[S]ince the things the mind contemplates are none of them, besides itself, present to the understanding, it is necessary that something else, as a sign or representation of the thing it considers, should be present to it: and these are ideas".[13] By postulating the link between words that signify ideas and ideas that represent objects, Locke showed the intrinsic relation between representation and knowledge.[14]

Similarly to Locke, Condillac based his explanation of the operation of mind and the origin of human knowledge on a theory of signs. This theory was founded on two principles namely on the Lockean doctrine of the origin of ideas in sensation and on the rational principles of the universal grammar whereby language was supposed to have a single origin. Condillac presented his theory in *Essai sur l'origine des conaissances humaines*,[15] which was translated into English with the title *An Essay on the Origin of Human Knowledge, Being a Supplement to Mr. Locke's Essay on the Human Understanding*. Due to the parallelisms between Condillac and Locke's work, critics initially believed that Condillac's *Essai* was just an extension of Locke's *Essay*. Although Condillac admired Locke greatly, he didn't agree with Locke's reduction of ideas to a dual origin in sensation and reflection. Instead Condillac wished to demonstrate that reflection could be derived from sensation and to do so he postulated a new principle, namely that of the connection of ideas which depended on the use of signs.

Condillac's *Essai* is divided into two parts; the first discusses the operations of mind and postulates the importance of an active and deliberate use of signs which he divided into three categories – the accidental, natural and instituted or as he put it: "I distinguish three sorts of signs: 1. accidental signs, or the objects which particular circumstances have connected with some of our ideas, so as to render the one proper to revive the other. 2. Natural signs or the cries which nature has established to express the passions of joy, of fear or of grief. 3. Instituted signs or those we have chosen ourselves, and bear only an arbitrary relation to our ideas".[16] As Aarsleff explains,[17] all knowledge, according to Condillac, is stocked on these three internalised sign categories and its progress depends on the sign capability to open the way to reflection which is an expression of reason. Human beings are capable of higher degrees of reflection in proportion to their reason. Progress in knowledge and language is possible only from this. However, to be able to use the third type of signs, the instituted or conventional ones, human beings need to have control over the first two sign-types. In order to explain this, Condillac reverted to the study of the origin of language, which represented the second part of his essay. He argued that

[12] *Ibid.*, Book IV, p. 720.

[13] *Ibid.*

[14] Cf. Losonsky 2007 for an in-depth discussion of Locke's essay.

[15] Condillac 1746 [2001].

[16] *Ibid.*, p. 51.

[17] Aarsleff 1982.

language developed from animal cries or what he called natural signs which human beings used to communicate in situations of danger and fear. It was the repetition of the same gestures and cries over a long period of time that enabled man to recall specific signs at will rather than use them instinctively. This way, Condillac believed that mind and the use of signs would interact to the mutual advantage of both.

Although Condillac's view of linguistic signs is limited to the notion that they are a special category outside the mind and that they are arbitrary (a term he uses consistently in his *Essay*) what is important is that, he reverted to nature and natural signs in order to understand the origins and nature of language by paving the way for a possible correspondence between the natural world and the cultural one. Yet, Condillac's assertion that animals do not have reason (although he conceded that they have rudimentary forms of thought) meant that man and animal were separated by the higher capacity of human beings to use arbitrary signs in language and speech. In the nineteenth century this view was challenged by Darwin's evolutionary theory and his suggestion in the *Descent of Man* about a similar genealogy for human beings and language and the shared intellectual capacities of human and animals. However, it is fair to say that Darwin's interest in the workings of language stemmed from his preliminary readings of Monboddo and of Tooke whom Darwin held up as "one of the founders of the noble science of philolology".[18] The importance of Monboddo's and Tooke's influence on Darwin should not be underestimated as Darwin's initial concern to show that language had a natural origin and that it developed over time in a genealogical progression found expression in the following note he wrote in his notebook *M*[19] after returning from the Beagle Voyage: "Origins of Man now proved. Metaphysics must flourish. He who understands baboon would do more towards metaphysics than Locke",[20] and a few pages later, referring to Locke's *tabula rasa* of human knowledge he writes: "[T]he monkey understand the affinities of man better than the boasted philosopher himself".[21] The reference to baboons which we find in his notebook relate to Monboddo's view that there is a clear relation between human beings and orangutans not less so because they exhibit "exactly [...] the same human form; walking erect [...] they use sticks

[18] Darwin 1871 [1981, p. 87].

[19] Darwin's notebooks represent an important testimony to the development of his thought and theories. According to Jonathan Hodge (2009) they have helped in transforming the understanding of Darwin's entire life and work since 1960s. They reveal the vast range of Darwin's readings which contributed to his elaboration of the origin of species, his theory of the origin of moral sense in man from ancestral animal instinct, as well as language theory. The notebooks were written between 1831 and 1839. He started recording his observations during his voyage on the *Beagle* in the Field Notebooks which were followed by the Red Notebook. Darwin labelled each notebook with a letter. For instance, notebook *A* was written in July 1837 and was devoted to geology whereas notebook *B* was headed *Zoonomia* and was devoted to the laws of life. By July 1838 he had filled notebook *C* devoted to transmutation which was followed by notebook D on the same subject. In 1838 he started filling notebooks *M* and *N* on metaphysics and moral expression. For a detailed commentary and Darwin's notebooks cf. Barrett et al. 1987.

[20] Darwin, Notebook *M*, p. 84e, quoted in Wyhe 2002.

[21] Darwin, quoted in Degler 1991, p. 7.

for weapons; […] and have the organs of pronunciation as perfect as we have".[22] Although Monboddo affirms that *Homo sapiens* and orangutans share the same organs of pronunciation, and that *Homo sapiens* began articulating in the imitation to natural cries of animals, he is adamant that only human beings "had received the disposition"[23] for developing speech.

Monboddo's theory of the origin of language is similar to Condillac's view; the difference between the two lies in the fact that for Condillac the development of language started at a later stage than that proposed by Monboddo. Monboddo's work received mixed reviews in England, however his reputation was better in Germany where his work was translated by Johann Gottfried von Herder (1744–1803). In the preface of the translation Herder explained that the German audience would not object to Monboddo's attack on Locke and Isaac Newton because of the developing view of nature and language which stemmed from the German philosophy of Naturphilosophie. I shall address this shift of perspective in the next section before turning to discussing Darwin's indebtedness to such views in the formulation of his own theory.

The German Legacy: Romanticism and the Rise of Philology in England

As Condillac, in France, and Monboddo in England were elaborating their view on language and its origins, the German philosopher Johann Georg Hamann (1730–1788) also addressed the origins of language and criticised responses offered by both the Enlightenment and its critics. The issue rested, as Andrew Bowie[24] points out, on the fact that language was either conceived as the result of consciousness coming to make animal cries into meaningful signs, as proposed by Condillac, or, alternatively, it was in the nature of humankind to establish social conventions that gave agreed meanings to signs. Instead Hamann saw language as a creative force. To sustain his view of the creative force of language, Hamann presented a series of texts from antiquity to the present in order to demonstrate a kind of continuity of thought and to establish new contexts and meanings for his carefully selected material. By so doing, he introduced a historical dimension to language. In contrast to the rationalist tradition of the Enlightenment, he believed that primary contact with the world is in terms of feeling or sensation and not ideas. For Hamann, as Bowie explains, "human beings have a fundamental conviction of the reality of things which are prior to any abstract philosophical attempt to establish the nature of that reality".[25] Such belief is supported not by reason but by the immediate or

[22] Monboddo 1774, pp. 187–189.

[23] *Ibid.*, p. 481.

[24] Bowie 2003.

[25] *Ibid.*, p. 46.

non-inferential thinking. In other words, the world is revealed as something that is always already intelligible since the intelligibility of language and things are inseparable, because they are created by God's word. While it is God's word that brings utterances concretely into existence, it is human language which reveals how his word can be translated into new forms.

Hamann's views on language are important for two main reasons: firstly because the connection he posited between the creativity of language and art prefigured developments in early Romantic thought and, specifically, its re-animation of nature that replaced a mechanistic model of the natural world with an organic one. Secondly, the introduction of a historical dimension of language helped the institutionalisation of philology and anthropology by allowing language to change and grow or evolve over time. Philology held that the meaning of words is not fixed and immutable, but rather alters in history as a result of adaptation. Understanding the meaning and evolution of words became a way of interpreting past societies and cultures. Thus, both historians and anthropologists reverted to the study of words. As we shall see shortly, it was the interest in language and classical, later comparative, philology that initiated an interest and furnished Darwin with a key metaphor and an example or illustration of an evolutionary process.

Similarly to Hamann, Herder attributed a historical dimension to language. By asserting that language works in the manner of nature he equated the development of language with stages of the development of man. For instance, in the childhood stage language is determined by the affective reaction to the environment, and it is based on feeling and instinct. In the next stage, when human beings move to a more developed stage of thought, language becomes more able to deal with abstract concepts, until it reaches its youth or the poetic stage when a direct link is visible between man and nature. In the final, mature phase, language reaches what he termed the era of prose and philosophy where language "loses the pure poetry of nature".[26] Herder's equation of the development of language to the various phases of human evolution prefigures concepts in Darwin's evolutionary theory of species.[27]

John Wyon Burrow[28] notes that the development of comparative philology in England was closely linked to the central doctrines of German Romanticism, specifically in its understanding of language as something, not made, but natural and thus growing and evolving. The historical-comparative study of language in England sprung from the work of the aforementioned Sir William Jones who in 1789, delivered his famous lecture "On the Hindus" to the Asiatic Society of Calcutta which

---

[26] Herder 1767 [1985, p. 441].

[27] It is important to highlight, that Herder was familiar with Condillac's work, specifically with the part that deals with the origin of language and that his ideas may have influenced the writing of Herder's essay *Über den Ursprung der Sparche* (1772) or *On the Origin of Language* (Aarsleff 1982).

[28] Burrow 1967, p. 189.

was a part of his wider contribution "The Sanskrit Language",[29] in which he suggested that that classical languages, such as Greek and Latin, had a common root with Sanskrit, and that they may be further related to Gothic and Celtic as well as Persian. Jones put it as follows:

> The Sanskrit language, whatever be its antiquity, is of a wonderful structure; more perfect than the Greek, more copious than the Latin, and more exquisitely refined than either, yet bearing to both of them a stronger affinity, both in the roots of verbs and in the forms of grammar, than could possibly have been produced by accident; so strong indeed, that no philologer could examine them all three, without believing them to have sprung from some common source, which, perhaps no longer exists; there is a similar reason, though not quite so forcible, for supposing that both the Gothic and the Celtic, though blended with a very different idiom, had the same origin with the Sanskrit; and the old Persian might be added to the same family, if this were the place for discussing any question concerning the antiquities of Persia.[30]

Jones's statement about the kinship and affiliations between Indo-European languages was based on a comparative study of grammar which, he argues, was far more accurate than etymology. In their account of the importance of language studies in England, Burrow and Aarsleff agree that Jones' insight caused a revolution in the study of language because "the classification and derivation of languages could be systematised and that linguistic change could be studied on a comparative basis".[31] The consequences were various, but by far the most important one was that comparative philology became a model for different kinds of inquiry into the remote past and an ethnological tool or means of classifying racial families and even finding a single origin of the human race. It was this endeavour, i.e. the attempt to trace phenomena in an unbroken line to a remote past, which appealed to nineteenth century scholars working along these lines in biology and geology. Charles Lyell and Charles Darwin, for instance, found in comparative philology a consonant analogue.

While Jones' foremost Oriental scholarships – in particular his declaration of affiliation between languages was a fundamental contribution to the development of philology and Darwin's language theory lay in the work of Wilhelm von Humboldt (1767–1835), Jacob Grimm (1785–1863) and August Schleicher (1821–1868) who were all in various ways and to various degrees indebted to German Romanticism and who based the historical-comparative study of language on grammar alone.[32]

---

[29] Cf. Aarsleff 1983.

[30] Jones 1799, p. 26.

[31] Hoenigswald 1963, p. 7. Hoenigswald notes, however, that term *comparative* as used in the eighteenth century does not refer to comparison at large, but to a process whereby original features can be separated from recent ones and where the aim of classification is subordinated to the aim of reconstruction (cf., e.g., his argument in Hoenigswald 1963).

[32] Linda Dowling argues that Englishmen have contributed relatively little to the development of this new science, so much so, that when M. Müller arrived in England in 1846, he believed to have set foot in a country rich of philological resources, yet relatively poor in philological achievements (cf. for instance Dowling 1982).

In his *On the Comparative Study of Language and its Relation to the Different Periods of Language Development* (1820), for instance, Humboldt argued that language is creative. Its function was not limited simply to representing or communicating existing ideas and concepts but it was a "formative organ of thought",[33] and as such was capable of creating and not just reflecting the existing world. It is precisely because of its creative aspect that the structure and organization of a language could not be gathered from actual verbal forms of its construction or its grammar. Rather, they had to be obtained from an analysis of the procedures language employs in speech, hence through sounds.[34] In his earlier empirical essay, *Thinking and Speaking: Sixteen Theses on Language* of 1795, he drew a clear distinction between the "physical sound of nature"[35] (similar to Condillac's animal cries) on the one hand and the "articulated sounds"[36] that constitute language on the other. In his studies he argued that the latter alone could form discernible units and were thus capable of embodying features to allow these sounds to enter into specific relationships with each other and any other sound. In other words, for Humboldt, the individual sound of a given language can be formed only "in relation to the others"[37] that make up the entire "sound system" of that language.[38]

He endeavoured to compile a phonetic system of different languages cataloguing them either according to their phonetic affinities or their oppositions. Humboldt's emphasis on language's creativity and its productive ability represents not only his critique of the rationalist (Descartes) and empiricist (Locke and Condillac) views on language whereby it was assumed that signs constituted a special class of objects outside an independently existing mind, to which convenient labels agreed upon by society had been attached.[39] As Robert J. Richards[40] observes, Alexander von Humboldt not only conveyed a conception of living nature, which Darwin later incorporated into his evolutionary theory, but Humboldt also suggested that language helped to create human intellect, an idea that became predominant in Darwin's language theory.

On the other hand, others such as J. Grimm in his *Deutsche Grammatik* (or *German grammar*) (1819) set out to illustrate the resemblance and kinship between languages based on the *Lautverschiebung* 'sound shift' or Grimm's law where, over a period of time, sets of consonants displace each other in a predictable and regular fashion.[41] Grimm presented the development of a single mother tongue, in this case Sanskrit, through a series of natural transformations through sound inheritance. Grimm's work was so revolutionary, that the translator of the English version of his

[33] Humboldt 1820 [1986, p. 100].

[34] Mueller-Vollmer 2011.

[35] *Ibid.*

[36] *Ibid.*

[37] *Ibid.*

[38] Mueller-Vollmer 1989.

[39] Mueller-Vollmer 2011.

[40] Richards 2002, p. 26.

[41] Gamkrelidze and Ivanov 1990, p. 111.

work remarked that it had "created a new epoch in the science of comparative philology"[42] and he compared it to Newton's *Principia in mathematics* and Bacon's *Novum Organum.* It is important to note here that Hensleigh Wedgwood, Darwin's cousins, published a technical exposition of Grimm's work and expressed enthusiasm for his achievement: so such work was familiar to Darwin who was in contact with his cousin when he was writing the *Descent of Man*. As Beer notes, Grimm's achievement lied in the fact that he did not draw on an even spread of evidence, but on the discovery of a law which could trace with precision the various shifts in language, or as Wedgewood himself wrote: "The illustrious scholar Grimm, has here given us, under the modest title of German Grammar, a thorough history not only of his own language, but of that of every descendant of the Gothic stock throughout Europe, tracing at the same time every inflection in every dialect through every intermediate stage up to the earliest period of which any literary monuments remain".[43] In the *Descent of Man* Darwin referred to Wedgwood and Schleicher as sources for ideas about the evolutionary descent of language. Richards argues that it was Schleicher's "thorough-going naturalism that Darwin depended on for his theory of the constructive effect of language on mind".[44] Schleicher held that contemporary languages had gone through a process in which simpler languages or *Ursprachen* had given rise to descendant languages. Schleicher maintained that this fact was perfectly in line with Darwin's theory and that the linguistic model was a repeated analogue for the biological one. In his *Darwinsche Theorie und die Sprachwissenschaft* (1863) he identified four areas which he thought would advance Darwin's theory based on a linguistic model. Among the most interesting of these seems to be Schleicher's point that languages are natural organisms, yet they have an advantage over natural organisms as far more transitional forms of language have survived as compared to animal's fossilised remains. In his discussion on Schleicher's argument on the correspondence between pattern of language descent and human descent, Richards explains that Schleicher found a justification for such a claim in his belief in monism. In fact Schleicher states that:

> Thought in the contemporary period runs unmistakably in the direction of monism. The dualism, which one conceives as the opposition of mind and nature, content and form, being and appearance, or however one wishes to indicate it this dualism is for the natural scientific perspective of our day a completely unacceptable position. For the natural scientific perspective there is no matter without mind [*Geist*] (that is, without that necessary power determining matter), nor any mind without matter. Rather there is neither mind nor matter in the usual sense. There is only one thing that is both simultaneously.[45]

As Richards goes on to explain, the doctrine of monism provided Schleicher with a metaphysical ground for his theory that the organism of language simply represented the material side of mind. What this implied, was that the evolution of language carried the evolution of mind and vice versa. This idea itself had its roots

[42] Aarsleff 1983, p. 160.

[43] Beer 1996, p. 103.

[44] Richards 2002, p. 31.

[45] Schleicher, quoted in Richards 2009b, p. 126.

in German Romanticism which, as noted above, dispensed with the mechanistic understanding of nature and propounded the concept of organism as the fundamental principle "in terms of which human mentality and all natural phenomena were ultimately to be understood".[46] Schleicher's view became an important aspect of Darwin's theory which I consider next.

Darwin's *Descent of Man* and the Linguistic Rubicon

In the period leading to his elaboration of the theory of species transmutation, Darwin became increasingly interested in the workings of language. As early as 1839, Darwin had been fascinated by Lord Henry Brougham's *Dissertations of Subjects of Science Connected with Natural Theology* which insisted that both animals and humans shared the capacity for abstraction because they could understand signs. Beer[47] notes, that what Darwin did not understand about this work is that Brougham thought, and, I suggest, much as Condillac before him, that signs are to be understood as arbitrary, in a view later developed by the Swiss linguist Ferdinand de Saussure (1857–1913) in the *Course in General Linguistics* (1916). Beer explains that Brougham argued that the relation between signifier and signified (to use Saussure's terminology) is as arbitrary in animal communication as it is in human language. Brougham states:

> Have not animals some kind of language? At all events they understood ours. A horse knows the encouraging or chiding voice or whip, and moves and stops accordingly. [...] But they seem to have some knowledge of conversational signs. If I am to teach a dog or a pig to do certain things on a given signal, the process I take to be this. I connect his obedience with reward, his disobedience with punishment. But this only gives him the motive to obey, the fear of disobeying. It in no way can give him the means of connecting the act with the sign. Now connecting the two together (action and sign), whatever be the manner in which the sign is made, is Abstraction; but it is more, it is the very kind of abstraction in which all language has its origin – the connecting the sign with the thing signified; for the sign is purely arbitrary in this case as much as in human language.[48]

Although Darwin could have used Brougham's suggestion of the common origins between man and animal and animal intelligence, it was the move from the idea of abstraction to that of language which Darwin found difficult to grasp as he found no evidence for it in Brougham's work. In a passage in the *Descent of Man*, Darwin questions the claim that *animals* do not have the power of abstraction or that of forming general concepts and he states that: "[W]hen a dog sees another dog at a distance, it is often clear that he perceives that it is a dog in the abstract; for when he gets nearer his whole manner suddenly changes if the other dog be a friend".[49]

---

[46] *Ibid.*, p. 30.

[47] Beer 1996.

[48] Brougham 1839, vol. 2, pp. 195–196.

[49] Darwin 1871 [1981, p. 64].

Yet Beer suggests that Darwin's concern was the result of his preoccupation at the time with ideas of continuity and connections and that the idea of semiotic arbitrariness as the prototype of abstraction would have undermined Darwin's primary concerns.[50] Beer makes a valid point here, since as it is known from Darwin's very early theorising in his notebooks *M* and *N* and from the *Descent of Man*, he believed in the non-arbitrary understanding of the relation between words and things at the origin of language. Darwin came to believe that there was a necessary connection between "things and voices" or rather he believed in the musical basis of language which implied either a mimetic or an abstract relation between a thing and a voice. In the *Descent of Man*, in fact, Darwin states that:

> With respect to the origin of articulate language, after having read on the one side the highly interesting works of Mr. Hensleigh Wedgwood, the Rev. F. Farrar, and Prof. Schleicher, and the celebrated lectures of Prof. Max Müller on the other side, I cannot doubt that language owes its origin to the imitation and modification of various natural sounds, the voices of other animals, and man's own instinctive cries, aided by signs and gestures. It is, therefore, probable that the imitation of musical cries by articulate sounds may have given rise towards expressive of various complex emotions. The strong tendency in our nearest allies, the monkeys, and in the barbarous races of mankind, to imitate whatever they heard deserves notice, as bearing on the subject of imitation. Since monkeys certainly understand much that is said to them by man, and when wild, utter signal-cries of danger to their fellows; * (3) and since fowls give distinct warnings for danger on the ground, or in the sky from hawks (both, as well as a third cry, intelligible to dogs), * (4) may not some unusually wise apelike animal have imitated the growl of a beast of prey, and thus told his fellow-monkeys the nature of the expected danger? This would have been a first step in the formation of a language. As the voice was used more and more, the vocal organs would have been strengthened and perfected through the principle of the inherited effects of use; and this would have reacted on the power of speech.[51]

In other words, Darwin thought that it was through natural selection that the primitive vocal efforts of animals and human beings had evolved into a vast array of songs, sounds and cries and ultimately into speech. In his notebook *N*, some 30 years before the publication of the *Descent of Man*, as Richards points out,[52] Darwin already supposed that our aboriginal ancestors began imitating the sounds of nature and that language developed from these simple beginnings. What is particularly interesting, however, is Darwin's focus on imitation since it shows parallelisms with a biosemiotic perspective in animal communication or zoosemiotics which is based on the Peircean sign model. According to zoosemiotics, imitation could be based either on an iconic (i.e. based on similarity or resemblance) or an indexical (i.e. based on spatio-temporal contiguity) interpretation of signs. For instance, iconic imitation could have developed for purposes of predator deception or self-protection. An example of iconic imitation is alarm calls whose loudness is proportional to the degree of threat felt by the animal, as Darwin also suggested.[53] In line with Thomas

[50] Beer 1996.

[51] Darwin 1871 [1981, p. 68].

[52] Richards 2009a, p. 109.

[53] Nöth 1995, p. 163.

Sebeok's view of biosemiotics,[54] Darwin also sees that communication among animals is based on nonverbal signs and that human beings share this capacity with animals. Iconic and indexical signs are most often seen as being nonverbal.[55]

However, the difference between Sebeok and Darwin here is that Sebeok makes a distinction between language and speech whereas Darwin doesn't. In Sebeok's view,[56] language evolved as an adaptation much *earlier* than speech in humans and it did so not for communicative purposes, but for what Sebeok will call modelling. In other words, for Sebeok, language is a communicative device, so the specific function of language is neither to give information nor to transmit it. Sebeok,[57] instead, describes language as a modelling device, and although every species is endowed with a model that produces its own world, language is the specific model belonging to the human species. Speech, like language, made its appearance as an adaptation later than language and for the sake of communication. In its form "speech" (and later script language), it enabled humans to attain an enhanced nonverbal capacity, which they already possessed in less developed form. Darwin and Sebeok are in agreement on this point, as Darwin in the *Descent* similarly acknowledges that "articulate speech" (by which he means vocalization augmented by controlled movement of the lips and tongue[58]) is "peculiar to man",[59] but differently from Sebeok he denies that this mere power of articulation suffices to distinguish human language from animal vocalisations, "for as everyone knows, parrots can talk".[60] Translated into a biosemiotic perspective we could suggest that they both agree on continuity between animals and human beings and that this continuity has later been elaborated by Sebeok as being based on the iconicity and indexicality of signs.

The correspondence between animal language and human language and its origins postulated by Darwin in the *Descent of Man* encountered severe criticism by M. Müller who in *The Science of Language* (1861) presented the implications of Darwin's theory of natural selection as an overt attack on humankind. He argued that the use of language implied the ability to form concepts and, since animals cannot do that, there must be an impassable barrier between the two. Müller's point of view emerged from his conviction that language and thought coincide and, as Dowling argues, since Müller believed there is an exact coincidence between the two, "all language becomes meaningful, with reason transpiercing its apparent opacities and formal elements from within".[61] Given the inherent meaningfulness of words, Müller also believed that language could never arise conventionally as a system of external signs and as Saussure would later assert of arbitrary signs,

---

[54] Sebeok 2001.

[55] Cf., e.g., Martinelli 2010; Sebeok 1990 and 1972.

[56] Sebeok 1994.

[57] *Ibid.*

[58] Darwin 1871 [1981, p. 59].

[59] *Ibid.*, p. 55.

[60] *Ibid.*

[61] Winter 2009, p. 128.

because he held that humans would have needed words to hold the convention. Instead he portrayed it as internal and expressive in origin. Also, given the fact that Müller argued for a perfect identity between thought and language, he retorted that language stood in opposition to the evolutionary view proposed by Darwin. In fact he declared that: "One of the great barriers between the brute and man is *Language*. Man speaks and no brute had ever uttered a word. Language is our Rubicon and no brute will dare to cross it. […] It admits of no cavilling, and no process of natural selection will ever distil significant words out of the notes of birds and the cries of beasts".[62] Although there are differences between Darwin and Müller's views, the fact that they both believe in the non-arbitrariness of language is an important element which they concur. As John Deely points out, Saussure's definition of a sign rests on the notion that a sign is linguistic in essence and dyadic in character, and is arbitrary in the sense that it rests upon a stipulation.[63] In other words, Saussure postulates the relationship between form and meaning, arbitrarily restricting signs to the human sphere thus "severing their connection with the motivating history of the sign users as embodied in their language".[64] The severing of this connection also serves to separate human beings from animals, contrary to Darwin's view.

In the Peircean model of signs, the investigation of signs is not based upon an arbitrary dyadic model. Instead, it is based on a relational, triadic model. Where Saussure began positing a stipulated definition of a linguistic, thus cultural, sign, Peirce began with a descriptive definition of any sign, not only of linguistic signs.[65] Peirce emphasised the importance of investigating and interpreting signs rather than positing arbitrary meanings to them, and upon this he based his interdisciplinary science of the study of signs, which biosemiotics is based upon as a fundamental principle. The way he defines signs is according to the type of relations they have, where iconic and indexical signs (non-arbitrary signs) are shared between the human species and animals.

Many twentieth century linguists have followed Saussaure and thus neglected the importance of iconic and indexical signs. Often, they have concentrated their attention on cultural and conventional signs. It is therefore significant, for example, to observe that Sarah Winter,[66] in her article on Darwin's semiotic project on the expression of the emotions in man and animals, states that his project can be satisfactorily understood or investigated through a Saussurean lens. Winter states that, "Darwin's theory of expression falls into place within Saussure's disciplinary chronology"[67] yet immediately goes on to assert, "[w]hat I will characterise as Darwin's biosemiotic thinking in Expression also has important implications for clarifying our understanding of the status of race in Darwinian theory".[68] Winter

[62] Müller 1861 [1862, p. 354].

[63] Deely 2010.

[64] *Ibid.*, p. 20.

[65] *Ibid.*, p. 21.

[66] Winter 2009.

[67] *Ibid*, p. 131.

[68] *Ibid*.

thus argues that Darwin engaged in proto-biosemiotic thinking, yet within a Saussurean perspective on signs. In an endnote to the article, Winter quotes from Jesper Hoffmeyer and Thomas Sebeok to account for her perspective on biosemiotics, stating that they "favour a Peircean perspective"[69] of sign. It is not clear how they could have favoured or adopted a non-Peircean perspective, as the Saussurean view contradicts the non-arbitrary nature of the phenomena Darwin was investigating, namely (to use today's terms) the iconic and indexical signs of animal expression. In addition since iconic and indexical signs are not part of *langue*, Saussure was simply not concerned with the non-arbitrary signs of Darwin, Peirce and Hoffmeyer.

To conclude, in this paper, I have attempted to show how Darwin's work on, and contribution to, linguistics has often charted a proto-biosemiotic trajectory of thought which cannot be adequately underpinned, I have argued, by a Saussurean tradition. However, modern linguistic theorists often favour a Saussurean perspective, apparently overlooking and undermining Darwin's notion of continuity between animal and human. It is striking that this is a legacy of Romantic thought which influenced him and the evolution of his own ideas on the origin of language and animal evolution, and their inter-relations.

References

Aarsleff, H. (1976). An outline of language origin theory since the Renaissance. In S. R. Harnad et al. (Eds.), *Origins and evolution of language and speech* (pp. 4–13). New York: Annals of the New York Academy of Sciences.

Aarsleff, H. (1982). Wilhelm von Humboldt and the linguistic thought of the French ideologues. In H. Aarsleff (Ed.), *From Locke to Saussure: Essays on the study of language and intellectual history* (pp. 335–355). London: Athlone.

Aarsleff, H. (1983). *The study of language in England 1780–1860*. Princeton: Princeton University Press.

Alter, S. G. (1999). *Darwinism and the linguistic image: Language, race and natural theology in the nineteenth century*. Baltimore/London: Johns Hopkins University Press.

Barrett, P. H., Gautrey, P. J., Herbert, S., Kohn, D., & Smith, S. (Eds.). (1987). *Charles Darwin's notebooks 1836–1844: Geology, transmutation of species metaphysical enquiries*. London: British Museum (Natural History)/Cambridge University Press.

Beer, G. (1996). *Open fields: Science in cultural encounter*. Oxford: OUP.

Bowie, A. (2003). *Introduction to German Philosophy: Form Kant to Habermas*. Cambridge: Polity Press.

Brougham, H. (1839). *Dissertations on the subject of science connected with natural theology*, 2 vols. London: C. Knight.

Burrow, J. W. (1967). The uses of philology in Victorian England. In R. Robson (Ed.), *Ideas and institutions of Victorian England. Essays in honour of George Kitson Clark* (pp. 108–204). London: G. Bells and Sons.

Condillac, É. B. de (1746 [2001]). *Essai sur l'origine des connaissances humaines*. Cambridge: Cambridge University Press.

Darwin, C. R. (1871 [1981]). *The descent of man, and selection in relation to sex*. Princeton: Princeton University Press.

[69] *Ibid.*, p. 156.

Deely, J. (2010). *The semiotic animal: A postmodern definition of "human being" transcending patriarchy and feminism*. South Bend: St. Augustine Press.
Degler, C. N. (1991). *In search of human nature. The decline and revival of Darwinism in American social thought*. Oxford: Oxford University Press.
Dowling, L. (1982). Victorian Oxford and the science of language. *PMLA, 97*(2), 160–178.
Gamkrelidze, T. V., & Ivanov, V. V. (1990). The early history of Indo-European languages. *Scientific American, 262*(3), 110–116.
Herbert, C. (2001). *Victorian relativity: Radical thought and scientific discovery*. Chicago: University of Chicago Press.
Herder, J. G. von (1767 [1985]). *Über die neuere deutsche Literatur. Fragmente*. Berlin: Aufbau.
Hodge, J. (2009). The notebook programmes and projects of Darwin's London years. In J. Hodge & G. Raddick (Eds.), *The Cambridge companion to Darwin* (2nd ed., pp. 44–72). Cambridge: Cambridge University Press.
Hoenigswald, H. M. (1963). On the history of comparative method. *Anthropological Linguistics, 5*(1), 1–11.
Humboldt, W. von (1820 [1986]). The nature and conformation of language. In K. Mueller-Vollmer (Ed.), *The hermeneutics reader: Texts from the German tradition from the Enlightenment to the present* (pp. 99–105). Oxford: Basil Blackwell.
Jones, W. (1799). The Sanskrit language. In L. Teignmouth (Ed.), *The works of Sir William Jones* (Vol. III, pp. 19–34). London: John Syockdale and John Walker.
Locke, J. (1690 [1975]). *An essay concerning human understanding*. Oxford: Clarendon Press.
Losonsky, M. (2007). Language, meaning and mind in Locke's essay. In L. Newman (Ed.), *The Cambridge companion to Locke's essay concerning human understanding* (pp. 286–313). Cambridge: Cambridge University Press.
Martinelli, D. (2010). *A critical companion to zoosemiotics, people, paths, ideas*. New York: Springer.
Lord Monboddo, J. B. (1774). *The origins of language*. London/Edinburgh: J. Balfour and T. Cadell Publishers.
Lord Monboddo, J. B. (1779). *Of the origins and progress of language*. Edinburgh: J. Balfour and T. Cadell.
Mueller-Vollmer, K. (1989). Thinking and speaking: Herder, Humboldt and Saussurean semiotics. A translation and commentary on Wilhelm von Humboldt's "Thinking and speaking: Sixteen theses on language". *Comparative Criticism, 11*, 159–214.
Mueller-Vollmer, K. (2011). Wilhelm von Humboldt. In E.N. Zalta (Ed.), *The Stanford encyclopedia of philosophy (fall 2011 edition)* (http://plato.stanford.edu/archives/fall2011/entries/wilhelm-humboldt/; website consulted on 15 June 2013).
Müller F. M. (1861 [1862]). *Lectures on the science of language, delivered at royal institution of Great Britain in April, May and June* (2nd rev. ed.). New York: Scribners.
Neubauer, D. (2013). Lady Welby and the interpretive mind. *Semiotica, 196*, 243–260.
Nöth, W. (1995). *Handbook of semiotics*. Bloomington: Indiana University Press.
Richards, R. J. (2002). The linguistic creation of man: Charles Darwin, August Schleicher, Ernst Haeckel, and the missing link in the nineteenth-century evolutionary theory. In M. Doerres (Ed.), *Experimenting in tongues: Studies in science and language* (pp. 20–48). Stanford: Stanford University Press.
Richards, R. J. (2009a). Darwin on mind, morals and emotions. In J. Hodge & G. Raddick (Eds.), *The Cambridge companion to Darwin* (2nd ed., pp. 96–119). Cambridge: Cambridge University Press.
Richards, R. J. (2009b). *The tragic sense of life: Ernst Heckel and the struggle over evolution*. Chicago: Chicago University Press.
Sebeok, T. A. (1972). *Perspectives in zoosemiotics*. The Hague: Mouton de Gruyter.
Sebeok, T. A. (1990). *Essays in zoosemiotics*. Toronto: Toronto Semiotic Circle.

Sebeok, T. A. (1994). *Signs: An introduction to semiotics*. Toronto: University of Toronto Press.
Sebeok, T. A. (2001). *Global semiotics*. Bloomington: Indiana University Press.
Winter, S. (2009). Darwin's Saussure: Biosemiotics and race in expression. *Representations, 107*(1), 128–161.
Wyhe, J. van (2002). *The complete work of Charles Darwin online* (http://darwin-online.org.uk; website consulted on 15 May 2013).

The Bakhtinian *Dialogue* Revisited: A (Non-biosemiotic) View from Historiography and Epistemology of Humanities

Ekaterina Velmezova

Abstract One of the key concepts and categories of Bakhtin's philosophy, that of the *dialogue*, was perceived by Bakhtin in different ways. Even if this category acquires its typically "Bakhtinian" sense in his works beginning the 1950s, already Bakhtin's early writings contained some germs of his future "dialogical" thoughts, the category of dialogue being connected with other important notions of Bakhtin's theories.

Keywords Dialogue • M. Bakhtin • History of ideas • Linguistics and philosophy

> *Everything [...] can be reduced to a dialogue.*
>
> (Baxtin 1929b [1997–2012, vol. II, p. 157])[1]

There already exist researches where the notions of *dialogue* and *dialogism* worked out by Russian historian of literature and philosopher Mikhail Bakhtin (1895–1975) are referred to as concepts having a certain importance (or at least relevance) for biosemiotic studies: let us refer, first of all, to the works by Augusto Ponzio and Susan Petrilli.[2] Though inspired by these scholars, we cannot claim to be biosemioticians or semioticians, thus, in the following article we set ourselves a much more modest task: on the basis of primary sources, to try to reconstruct in a historiographical and epistemological perspective what Bakhtin himself meant by

---

[1] Extracts from Bakhtin's work are translated by ourselves. They are far from being as distinguished as already existing texts of Bakhtin's translations into English; here we give preference to the fidelity of translation, sometimes at the expense of language or stylistic elegance (it also concerns the titles of Bakhtin's translated works in the References).

[2] Among the latest studies, let us refer to the article Petrilli and Ponzio 2013 (cf. also Ponzio 2004; Kull 2007; Ponzio 2012 and the corresponding bibliographical references in these articles, especially in the first and in the third ones).

E. Velmezova (✉)
Centre for Linguistics and Language Sciences / Department of Slavic and South Asian Studies, University of Lausanne, Lausanne, Switzerland
e-mail: ekaterina.velmezova@unil.ch

E. Velmezova et al. (eds.), *Biosemiotic Perspectives on Language and Linguistics*, Biosemiotics 13, DOI 10.1007/978-3-319-20663-9_15

dialogue.[3] We hope that despite this simple goal, our text could be useful to researchers (including, maybe, biosemioticians), interested in Bakhtin (as he certainly deserves it) and wishing to be guided by his thoughts in their own reflections.[4]

Today the name of Mikhail Bakhtin is immediately associated with the notions of *dialogue* and *dialogism*. In this article, without claiming any completeness, we are going to try to briefly answer the following questions:

- What did Bakhtin mean by *dialogue* and in which way was this category connected with other key-notions of Bakhtin's work?
- In which way have Bakhtin's ideas on dialogue evolved with time?
- Who were the forerunners of his reflections on dialogue? (The answer to this question seems particularly important given the actual tendency to present Bakhtin as an unparalleled genius whose work had nothing in common with the research conducted at his time.)

The idea of a "dialogical Bakhtin" has attracted scholars' attention for a long time. However the limited volume of this article requires to minimize references not only to secondary sources, but also to the works written by researchers who had presumably been Bakhtin's forerunners in his "dialogic" reflections (though the names of some of them will appear in this study). For this reason, we shall have to limit our analysis to some presumed sources of Bakhtin's ideas on dialogue and to some of Bakhin's works which are currently available. First of all, the works that are sometimes attributed to Bakhtin will be excluded from the analysis, because their authorship remains questionable – such as, for instance, the article "Contemporary vitalism" (1926) (written by Bakhtin or Ivan Kanaev?), the books *Freudianism* (1927) (written by Bakhtin or by Valentin Voloshinov?), *The Formal Method in Literary Scholarship: A Critical Introduction to Sociological Poetics* (1928) (Bakhtin or Pavel Medvedev?), *Marxism and the Philosophy of Language* (1929) (Bakhtin or Voloshinov?), likewise some other articles with debatable authorship.[5]

[3] Similar attempts have also been undertaken in the past. However, the range of reliable sources that one can use, increases with time, that is why returning to this issue does not seem superfluous.

[4] The article which follows was published in French for the first time (*Cahiers de praxematique*, 2011, 57, pp. 31–50), as the text of our plenary paper presented at the Conference "Dialogisme: langue, discours" (Université Montpellier-III/CNRS), organized in Montpellier in September 2010. The English translation is a slightly revised version of the original text. – *E.V.*

[5] Though some of these works contain certain ideas connected with the subject of our article.

Trying to Define Bakhtinian *Dialogue*

Although the category of dialogue was one of the most important for Bakhtin, we do not find any strict definition of *dialogue* in his work.[6] Its reconstruction on the basis of the quasi totality of his work[7] shows that the notion of dialogue was understood by Bakhtin in at least two different ways. In its narrow sense[8] (we shall designate it as "linguistic"), the dialogue was understood by Bakhtin as a particular organisation of speech, opposed to monologue[9]; Bakhtin also perceived dialogism as a discussion or a polemic.[10] Another interpretation of dialogue in Bakhtin's work is, on the contrary, extremely broad; here already it seems to be possible to insist on a typically "Bakhtinian" sense of this category.[11] In this sense, Bakhtin analyzed dialogue at different levels:

- social and psychological (dialogue was connected with the problems of developing consciousness and its origin, etc.),
- religious (each utterance presupposed at least two receivers, and not the only one [real and particular][12]). Although Bakhtin refuses to reduce the "third" participant in question to a "mystical or metaphysical entity", the following series of synonyms is present in his work: "[...] dialogue, asking [*voprošanie*], prayer"[13],
- culturological (dialogue being considered as a universal means, even as a *sine qua non* condition for the existence of culture and, at the same time, as one of the key facilitators of the permanent renewal of culture),
- existential and ethical (dialogue as an instrument of the "accomplishment" of a human being as a person: it is through dialogue that a human being "opens himself" not only to others, but also to himself, learning to know himself as a unique being),
- philosophical (dialogue as a premise of the existence of ideas, each idea originating in a dialogue of several types of consciousness), etc.[14]

The dialogic for Bakhtin is connected with the sense (understood in a large way but obviously with reference to human beings) and its transmission (from the trans-

---

[6] In general, Bakhtin did not like definitions and was the first to recognize it, emphasizing his "love for variations and for a multitude of terms referring to one and the same phenomenon" (Baxtin 1971–1974 [1997–2012, vol. VI, p. 431]).

[7] Some of Bakhtin's studies (or their parts) have been lost forever.

[8] In this study, we shall not distinguish *sense* and *meaning*.

[9] Baxtin 1959–1960 [1997–2012, vol. V, p. 325], etc.

[10] Baxtin 1961 [1997–2012, vol. V, p. 332, etc.].

[11] Unless otherwise stated, speaking of dialogue in this article we shall refer to this broad sense.

[12] Baxtin 1961 [1997–2012, vol. V, p. 338 *sq*.].

[13] Baxtin, early 1920s–1974 [1986, p. 515].

[14] The content of Bakhtinian *dialogue* does not allow to define any exact number of these levels. On the other hand, Bakhtin's "dialogic" concerns were not limited to his theoretical researches, cf. for instance Bakhtin's criticism of the "monological" direction in the teaching of Russian at school (Baxtin 1945 [1997–2012]).

fer of most intimate verbal interventions to that of collective knowledge from one generation to another)[15] – including the dialogue in the narrowest, linguistic sense. That is why, even though the Russian word *dialogue* was already three centuries old by Bakhtin's time, used as it was in his works, this lexeme gained if not a terminological,[16] at least a categorical novelty.

Speaking of dialogue in the broad sense, Bakhtin apparently counted on the more or less common and ordinary interpretation and understanding of this word by his readers. But today this mixture of two semantic levels of the word's use (in their meaning of everyday life and at the same time in another sense, less widespread and more particular one) creates difficulties for the reception of Bakhtin's ideas, because it is not always easy to understand which of the two *dialogues* is discussed in one or another of his works.

From the Relation "*I (Self) vs* the *Other*" to the *Dialogue*

For most of his life, Bakhtin remained an unofficial figure of Soviet culture. Likewise, one of the forms of his work was also unofficial *par excellence*: his preparatory notes. The scholar wrote them throughout his life (from the 1920s to 1974) to outline his future major projects. The words *dialogical*, *dialogism* and *dialogue* (in the both above-mentioned senses, but especially in the second, broader one) are much more frequent in his notes and, generally in his work, dating from the 1950s–1970s.[17] Therefore, the notion of dialogue retained Bakhtin's attention far more in his later works than at the beginning of his intellectual career. However, already in Bakhtin's early works the seeds of his future "dialogic" ideas could be found: at the heart of Bakhtin's "dialogical" reflections going back to the last period of his intellectual activity were his earlier ideas on the relationship between *I (Self)* and the *Other*, together with several other categories of Bakhtin's philosophy related to these reflections.

In the first third of the twentieth century, the problem of the relationship between *I (Self)* and the *Other* was discussed very intensely both in Russia and in Western Europe; some of Bakhtin's forerunners in this field were mentioned in his work.[18]

[15] Cf. Baxtin 1963 [1997–2012, vol. VI, p. 51].

[16] The absence of terminological strictness in the case of Bakhtin and his "conceptual plasticity" (cf. Brès and Rosier 2007, p. 437 *sq.*) do not allow such a formulation.

[17] In Bakhtin's intellectual activity several phases could be distinguished. During the 1920s, Bakhtin was interested in the problems of general (especially literary) aesthetics, seen through the prism of philosophy. In the 1930s, he studied, first of all, historical poetics of literary genres. Finally, in his research of the 1950s–1970s, the scholar came back to a number of subjects of his philosophy of aesthetics in the 1920s, such as the problem of text in general, the study of utterances, of speech genres, etc.

[18] Let us note, however, that Bakhtin, who did not like definitions, did not like references to particular sources either: there are not many references in his work for at least three important reasons. First, some of his works (especially his early texts), anyway, remained unfinished (including at the

Among others, there were Max Scheler (in particular, it is with a reference to Scheler that Bakhtin mentions in his book on Fyodor Dostoevsky in 1929, the "criticism of monologism as a specifically Kantian form of idealism"[19] which began in the West in the 1920s[20]), Ludwig Feuerbach, whose philosophy (where the category of the *Other* was a central one) was well-known in Russia at that time (Bakhtine refers to Feuerbach several times[21]), Hermann Cohen. In the early twentieth century, Cohen was sometimes seen in Russia as one of very few philosophers who seemed to have understood the importance of the connection between *I (Self)* and the *Other* as a fundamental category of ethics and of the "first philosophy": unlike Martin Buber, Cohen acknowledged *der Andere* as (**a priori**) particular and different compared to *Ich*, therefore the relation *Ich – Du* was neither symmetrical nor reversible in his view. Bakhtin's interest for Cohen, the head of the neo-Kantian school of Marburg, also reflects his belief in this current: Bakhtin considered it the only one capable of solving the problem "*I (Self) vs* the *Other*" (or, in any case, oriented towards solving this problem). On the contrary, Bakhtin criticized the philosophy of his time for not having worked enough on the problem of "the *other I (Self)*", of "*I (Self) vs* the *Other*", and in this Bakhtin was not alone. Apart from Bakhtin, the category of the *Other* was discussed at that time in Russia by many other scholars. Among them were not only philosophers and historians of philosophy (Boris Vysheslavcev, Ivan Lapshin, Alexander Vvedensky, Nikolai Lossky, etc.), but also psychologists (Vladimir Bekhterev, Lev Vygotsky, etc.). These discussions also constituted an important source of Bakhtinian theories.

It is in the light of the category "*I (Self) vs* the *Other*" that, already in the 1920s, Bakhtin assumed the dialogical character (even if he did not always use the same word) of knowledge and cognition in general. According to Bakhtin, in the field of knowledge and cognition, there are neither actions nor works "isolated" from one another,[22] which presupposes that the "objective unity" of knowledge and cognition has neither beginning nor end.[23] Speaking about knowledge and cognition, Bakhtin discusses ***dialogic exchanges*** that take place, among others, ***between individuals***. In Bakhtin's philosophy, one of the important concepts which appears already in his early works is the *sobytie bytija*, literally the 'event of being'. It presupposed the perception of being [*bytie*] by (individual) consciousness[24] and was connected to the

level of references). Secondly, Bakhtin always counted on a sufficiently high level of his potential readers (*sapienti sat*) and sometimes consciously refused to give precise references. Finally, in the late nineteenth and early twentieth century, there existed a particular genre of philosophical treatise that did not involve references in general.

19 Baxtin 1929b [1997–2012, vol. II, p. 60].

20 This note disappeared from the second edition of the book (Baxtin 1963 [1997–2012]).

21 Cf. for instance Baxtin 1923–1924 [1997–2012, vol. I, p. 125], etc.

22 Baxtin 1924 [1997–2012, vol. I, p. 285].

23 *Ibid.*, p. 318.

24 Baxtin 1923–1924 [1997–2012, vol. I, p. 246 *sq.*].

phenomenology of Edmund Husserl[25] and to the philosophy of being of Vyacheslav Ivanov (discussed, for instance, in Bakhtin's book[s] on Dostoevsky[26]).

In particular, in order for an aesthetic[27] event to be fulfilled, at least two participants are necessary – as well as two types of consciousness, different and therefore capable of entering into dialogic relations with one another. For Bakhtin, the event of being does not occur if one of the consciences dissolves completely into the other – this case would be that of empathy [*včuvstvovanie*] rather than of dialogue.[28]

The *Other* as a category constituted a point of intersection between the notions of dialogue and of the event of being. This way, it was considered as the organizing force of all aesthetic forms, therefore, some events (above all, the "creatively productive",[29] unique and irreversible ones) could not, by definition, occur at the level of a single consciousness. The being, *bytie*, was not an abstract category for Bakhtin, but rather a "live" event, presupposing a (dialogic) interaction of a particular human life with the existential universe of others: the search of one's own voice would be, in reality, that of a *Word* [*slovo*][30] which is greater than *Me* (or any *Self*) and which is connected with the *Other*. Because as *(my)Self, I* can never be a "primary author" either of my life or of my works: "One needs to stop being only oneself in order to enter history", says Bakhtin.[31]

One could distinguish the premises of this idea already in Bakhtin's earliest article to have "survived" to the present day.[32] Although Bakhtin does not use the word *dialogue* here, the idea of dialogue is still there in his text. For already in this work, Bakhtin explicitly manifests his system of values, opposing what is "mechanic" [*mexaničeskoe*] and superficial or external [*vnešnee*] (seen negatively) to what is, instead, "impregnated with the interior unity of sense"[33] (judged positively). It is the

---

[25] *Ibid.*, p. 246, etc. It is also the deep Bakhtinian antipsychologism (cf. for instance Baxtin 1918–1924 [1997–2012, vol. I, p. 15 *sq.*]) that brings him closer to Husserl; as for the dialogue, it has never been studied by Bakhtin on a purely psychological level.

[26] Baxtin 1929b [1997–2012] and 1963 [1997–2012].

[27] For Bakhtin, *aesthetic* implied phenomena related to the humanities, unlike natural sciences. In this opposition a reference to Wilhelm Dilthey (cf. Baxtin 1966–1967 – ? (a) [1997–2012, vol. VI, p. 403, 407]) and to Heinrich Rickert (*ibid.*, p. 407) could be distinguished, even though, for Bakhtine, the boundaries between these two types of knowledge were not always impenetrable (*ibid.*).

[28] In the early twentieth century, the reception in Russia of the theory of *Einfühlung* 'empathy' (immediately associated with the name of Theodor Lipps, to whom Bakhtin referred several times [Baxtin 1923–1924 (1997–2012, vol. I, p. 94, 138, 140), etc.]), took place to a large extent through Lapshin; in the 1910s, the concept of *Einfühlung* was already widespread in the Russian humanities.

[29] Baxtin 1923–1924 [1997–2012, vol. I, p. 159].

[30] The Russian word *slovo* refers not only to the 'word', but also to the 'discourse', to the 'speech', etc. (cf. in Bakhtin's work [Baxtin 1953–1954 (1997–2012, vol. V, p. 171)]) and sometime has religious connotations (*In the beginning was the Word…*).

[31] Baxtin 1924 [1997–2012, vol. I, p. 280].

[32] Baxtin 1919 [1986].

[33] *Ibid.*, p. 3.

interaction of the different parts of a whole (their dialogue) that allows to overcome the mechanical nature of such links – let us emphasize here an implicit reference of Bakhtin to Auguste Comte and his principle of *solidarity*. Likewise, speaking in this article about the human personality, Bakhtin echoes discussions on the isolation of art from life: this problem was one of the most crucial in the early twentieth century both in Russia (cf. the works by Bakhtin, Gustav Shpet, etc.) and in Western Europe (cf. Rickert, Husserl and others).

In Bakhtin's idea of interaction and of interpenetration of different parts of the whole (for example, of the human personality[34]), a reference to the category of dialogue could be presumed for the following reason. One of the important notions that appears already in Bakhtin's early works is *vnenaxodimost'*, 'outsideness' presupposing an inability of *Others* to be at the same time and in the same place as *I*. This notion implies the category of the *Other*, who would be the only one capable of seeing *Me* as *I* am.[35] Therefore a human being alone could never be the author of his own "value", since he needs to be "realized" – "impregnated with the interior unity of sense" – through the prism of the "evaluating soul" [*ocenivajuščaja duša*] of the Other,[36] which also presupposes the category of responsibility (answerability) [*otvetstvennost'*] with regard to the Other.[37]

Dialogue as a Cultural Unity: Between Literature and Philosophy

According to Bakhtin, dialogic exchanges exist not only between individuals, but also

- ***between particular ideas***, the dialogue being at the very origin of human sciences and every idea being considered as an echo to other thoughts[38];
- ***between texts[39] and their parts.***[40] In particular, in respect to literary texts, the origins of literary works for Bakhtin sometimes go back to very ancient times, to the folk culture (he shows it while analyzing the works of François Rabelais, Nikolai Gogol, etc.[41]). In turn, every work is reflected in the later texts;

[34] *Ibid.*

[35] Cf. similar ideas connected with the metaphors of reverberation and reflection in the works of other Russian philosophers in the early twentieth century (in particular, Lapshin and Shpet who analyzed, among others, Paul Natorp's related ideas).

[36] Baxtin 1923–1924 [1997–2012, vol. I, p. 111].

[37] Baxtin 1918–1924 [1997–2012].

[38] Baxtin 1959–1960 [1997–2012, vol. V, p. 306 *sq.*].

[39] In the broad sense of the word, Bakhtin understood the *text* as a "coherent whole complex of signs" (*ibid.*, p. 308), that is, as a semiotic unit *par excellence* (on this subject cf. Ponzio 2007).

[40] Baxtin 1959–1960 [1997–2012, vol. V, p. 308].

[41] Baxtin 1940 [1997–2012], 1940/1970 [1997–2012] and 1965 [1997–2012].

- dialogues also exist ***between literary genres and between languages***. This phenomenon, according to Bakhtin, is typical especially for the modern era, but it began already during the Renaissance, when languages came into active interaction with one another. This process promoted the development of linguistic ideas (any language could be better studied and understood through the prism of another one)[42];
- one could also speak about dialogue ***between different fields of art and culture:*** for instance, this is how Bakhtin speaks about analogies between the compositional forms of different arts, such analogies being determined by the common character of their architectonic goals.[43] In the same way, for example, Russian poetics as a discipline, says Bakhtin, would gain a lot if it were to connect with other arts, with the unity of art in general (otherwise, it would lead to an extreme simplification of its tasks and to a superficial and incomplete study of its object),[44] etc.;
- dialogic exchanges also exist ***between cultures*** as such, says Bakhtin, entering into controversy with Oswald Spengler[45] whose philosophy was well known and often discussed by other Russian thinkers (Semyon Frank, Fyodor Stepun, Nikolai Berdyaev, Yakov Bukshpan) at the time. Refusing to consider culture as a "closed circle", Bakhtin offers an opposite conception of culture, perceived as an open unit interacting with other cultures. In addition, a real understanding of other cultures is only possible in a dialogue with them.[46] This way, according to Bakhtin, the novel was born at a particular era thanks to a (dialogic) interaction of languages and cultures and began to develop intensively as a specific genre. Among various types of the novel's origins (rhetorical, erotic, satirical, autobiographical, utopian ones, etc.), Bakhtin mentions its "dialogic" roots pointing out that they had not been properly studied yet.[47]

Regarding Bakhtin's own work on the novel, it is his book on Dostoevsky which is the best known today in relation to his "dialogical" ideas. But already before launching an analysis of Dostoevsky's novels, Bakhtin had thought about the "dialogical" problems[48] in the light of literature and aesthetics in general, discussing the following issues:

[42] Baxtin 1940 [1997–2012, vol. IV(1), p. 489 *sq*.].

[43] Baxtin 1924 [1997–2012, vol. I, pp. 278–279 *sq*.].

[44] *Ibid.*

[45] Cf. Baxtin 1918–1924 [1997–2012, vol. I, p. 51] and especially Baxtin 1970 [1997–2012, vol. VI, p. 455].

[46] Baxtin 1970 [1997–2012].

[47] Baxtin, early 1920s–1974 [1986, p. 514].

[48] In his early studies, Bakhtin also touches upon the problem of dialogues (in the narrow sense of the word) in literary works – for example, speaking about dialogue in drama (Baxtin 1923–1924 [1997–2012, vol. I, p. 75 *sq*.]), etc.

1. *Dialogues between the author and his character(s)*: author and character meet in the literary work, entering into relations of various types (depending on particular writers and genres, etc.)[49];
2. The creation of a literary character was sometimes tantamount, in Bakhtin's view, to a response that the author gave to his own. It is through the complex dialogical relationship between author and character in literary works that one could explain, among other things, the fact that the language of literature is not always the same, but changes from one passage to another (cf. also *hereroglossia*). Therefore even *the relationship between form and content in a literary work could be perceived dialogically*;
3. Speaking of the literary work, one could also distinguish *a dialogue between a work and its readers*: reading a text, we do not perceive it "outside" ourselves, but we appropriate it, making it "ours", in order, subsequently, to respond and/or to react to it in a certain way.[50] Here once again, Bakhtin emphasizes the role of dialogue in the process of cognition;
4. Finally, Bakhtin insists on the importance of *dialogue(s) between the literary work and* the context, above all, historical, of *its time*. The writer always determines his position in relation to the foregoing culture and events; therefore, in order to understand a literary text, we must place it in the particular context in which it appeared (what Bakhtin did himself when analyzing, for example, Lev Tolstoy's fiction,[51] etc.).

But it is especially about Dostoevsky's novels[52] that Bakhtin was thinking in the light of his interest in dialogue and in dialogism in literature and in the "whole ideological culture" of that time.[53] In the early 1960s, Bakhtin reworked the first edition of his book on Dostoevsky, being directed by his ever growing interest in the problems of not only historical poetics, but also of dialogue and dialogism. However, already in the first edition of his book he mentions criticism against the "monological" paradigm of so-called classical philosophy (Bakhtin traces this paradigm to the Kantian idealism) and the gradual replacement of this paradigm by the "dialogic" principle of thought.[54] Regarding the dialogue in the narrow sense, according to Bakhtin, in literature before Dostoevsky, replicas of dialogues in novels had been of a monological character: each character-participant had his own universe, their worlds being closed to one another. Dostoevsky, on the contrary, created a particular kind of novel – the polyphonic[55] or dialogic one (for Bakhtin, *dialogue* couples with *polyphony* and these words are often used in his work as synonyms). In addition, in the dialogues in Dostoevsky's novels, one of the charac-

[49] Baxtin 1923–1924 [1997–2012].

[50] Cf. also Baxtin 1953–1954 [1997–2012, vol. V, p. 170].

[51] Baxtin 1929a [1997–2012] and 1929c [1997–2012].

[52] Baxtin 1929b [1997–2012] and 1963 [1997–2012].

[53] Baxtin 1929b [1997–2012, vol. II, p. 59].

[54] *Ibid.*, p. 60.

[55] Baxtin 1963 [1997–2012, vol. VI, p. 7].

ters sometimes embodies the inner voice of another hero. This thesis implicitly reflects the discussions of Russian psychologists about inner speech. For instance, Lev Vygotsky's works explore transformation, in a child's life, of external speech into internal one. Bakhtin could hardly ignore these discussions, even though, speaking of Dostoevsky's novels, he reverses the process described by Vygotsky: the inner speech of his characters "externalizes" itself in others.[56] What is more, dialogues between characters in Dostoevsky's novels often seem unfinished, so that nobody "wins". It allows, in principle, to continue these dialogues indefinitely, like any real dialogue in the broad sense of the word. Here, according to Bakhtin, a border passes between dialogue and rhetoric[57]: if the purpose of rhetoric is to defeat an opponent, in dialogue, on the contrary, the metaphorical death of one participant would be equivalent to the end of the dialogue as such. In contrast to the rhetoric, the purpose of dialogue is to find the truth and to get closer to the truth. This interactionist side of the dialogue is also related to its social dimension, which could be illustrated in Bakhtin's work through a comparison of two editions of his book on Dostoevsky. There are some contexts where the word *social* in the 1929 edition is replaced by *dialogic* in the edition of 1963.[58]

Therefore the two aforementioned senses of the word *dialogue* in Bakhtin's work obviously "meet" in his book on Dostoevsky[59] and we can consider his book published in 1929 as a kind of intermediary between Bakhtin's early and later writings (devoted, among others, to the historical poetics).

Bakhtin could not bypass the notion of dialogue speaking of his other hero, François Rabelais: although in his work on Rabelais the notion of dialogue understood as the basis of any culture is far from being at the center of attention, the idea of dialogue is here nevertheless present. First, Bakhtin discusses dialogues in Rabelais' work in the usual (linguistic) sense of this word: deprived of his own inner world, the human being here manifests himself through his "exterior" behavior – including the verbal behaviour, or more particularly the dialogues in which he participates.[60] On the other hand, in Rabelais' work the most unexpected things and phenomena can enter into dialogues: Rabelais breaks off the ordinary semantic links by establishing, in their place, much less predictable connections.[61] The world-

[56] *Ibid.*, Chapter 5, Part 4. Cf. at the same time the notion of "microdialogue" which implies, on the contrary, the internalization of dialogic replicas (*ibid.*, p. 51). This way, Bakhtin insisted on the dialogic nature of even interior "monologues", emphasizing their importance in his book on Dostoevsky.

[57] Cf. in particular Baxtin 1966–1967 – ? (b) [1997–2012, vol. VI, p. 413] and early 1920s–1974 [1986, p. 528]. In his book on Dostoevsky, dialogue is also opposed to the (Hegelian) dialectics, the latter implying the process of formation and growth (Baxtin 1963 [1997–2012, vol. VI, p. 33 *sq.*]). This opposition is also present in Bakhtin's other works (including in Baxtin 1959–1960 [1997–2012]).

[58] Cf. Baxtin 1929b [1997–2012, vol. II, p. 99] *vs* 1963 [1997–2012, vol. VI, p. 225].

[59] In addition to those mentioned above, this book also contains reflections on some other aspects of dialogues in literature.

[60] Baxtin 1937–1938 [1986, p. 272].

[61] *Ibid.*, pp. 203–204.

view that had been typical for the Middle Ages was in the process of disintegration at the Renaissance, and the task of Rabelais consisted in constructing another vision of the world, the one that would reflect better a "new material basis"[62] of the society in question. Finally, with the example of Rabelais, Bakhtin tries to connect two of his key concepts: *dialogue* and *carnival.* He emphasizes the carnivalesque character of the Rabelaisian universe: carnival eliminates distance between people who, in this way, become able to start a true contact (a dialogue) with one another.[63] In general, Bakhtin's emphasis on the idea of dialogue and interaction has allowed him to put in a new way *the problem of borders*, which was particularly important for the whole of Soviet culture in the first half of the twentieth century and also discussed outside the USSR. For Bakhtin, the most interesting phenomena in the field of culture occur precisely *on the borders.*[64] In addition, culture itself, for Bakhtin, has no boundaries, it "lies entirely *on* the borders, the borders go anywhere, penetrating all its elements".[65] Therefore, no cultural action, no phenomenon of culture could ever be neutral, they are always defined in relation to something else. This way, the "unity of culture"[66] is ensured.

For example, it is by the disappearance of particular boundaries that Bakhtin explains the emergence of new literary genres such as parody, this "intentioned 'dialogized' hybrid" (or, in other words, a result of mixing styles, languages, dialects, etc.).[67] For the same reason, among characters of folk culture[68] who were obviously very dear to Bakhtin, there are jesters, cheaters and fools, that is, those "on the borders" between several worlds, who are able to enter "dialogues" with different universes. That is why, in literature, these characters often become those expressing not only "the absolute truth", but also the author's position.[69]

As it happens, Bakhtin himself could be seen as someone who worked *on* the boundaries between different areas of culture, these areas entering into dialogue with each other. One of his favorite subjects of reflection, already in his early works, was the link between the history of literature and the history of philosophy, between literary and philosophical phenomena in general. If, speaking of Friedrich Nietzsche

---

[62] *Ibid.*, p. 239.

[63] Baxtin 1965 [1997–2012, vol. IV(2), p. 25].

[64] The same thing would be true in case of individuals (Baxtin 1961 [1997–2012, vol. V, p. 344]).

[65] Baxtin 1924 [1997–2012, vol. I, p. 282]. Here Bakhtin echoes philosophy of Rickert who was well known in Russia at that time. For Rickert, the main methodological problem of philosophy concerned the distinction between different fields of culture, and the boundaries between them. In general, Bakhtin uses the word *boundary* in different contexts – in particular, reflecting in his early works on philosophical problems, with references to Friedrich Schleiermacher, Georg Wilhelm Friedrich Hegel, Paul Natorp, Arthur Schopenhauer.

[66] Baxtin 1924 [1997–2012, vol. I, p. 267 *sq.*].

[67] Baxtin 1940 [1986, p. 385].

[68] It is speaking of dialogue (among others) that Bakhtin opposes the official culture (which reflects only the "small experience" of a particular society) to folk culture (reflecting a "great experience" of humanity), unlimited and infinite, in which everything is alive, everything speaks, everything is dialogical (Baxtin, early 1920s–1974 [1986, pp. 518–520]).

[69] Baxtin 1937–1938 [1986, p. 194 *sq.*].

and Arthur Schopenhauer, Bakhtin defines their theoretical conceptions as "mid-philosophical", "mid-literary",[70] one could say the same thing about Bakhtin's own work.[71]

This is how Bakhtin defines his own work: "We shall be obliged to name our analysis a philosophical one, failing to find a better definition: for it is neither a linguistic analysis, nor a philological, nor a literary, nor any other one. [...] our research is on the border of all these disciplines [...]"[72] – this statement seems to be a quintessence of the very logic of interdisciplinary research, which is such a current issue at present.[73]

Dialogue, First and Primary Category

There is nothing surprising that, with such a credo, Bakhtin was sometimes opposed to the very existence of particular branches of knowledge – such as, among others, linguistics: he expresses this point of view in his work "The problem of speech genres ",[74] which is particularly important for a better understanding of the evolution of categorical values in the scholar's work.

[70] Baxtin 1923–1924 [1997–2012, vol. I, p. 86].

[71] Already in Bakhtin's early works, namely in "Author and hero in aesthetic activity", his thoughts on the literary works serve as a pretext for him to talk about philosophy: there are very few examples from literature in this work; even the word *author* here refers not only to writers, but also to the creative acts always being in need of the *Other*. This way, Bakhtinian ideas about the relationship between author and character are transformed into thoughts about the relationship *I (Self) vs* the *Other* in general. In the work about the "philosophy of the act", some examples from literature seem to be lost in Bakhtin's philosophical reflections. Afterwards, in his book on Dostoevsky, it is philosophy that Bakhtin blames for its monological nature, discussing polyphony and dialogism in literature (Baxtin 1929b [1997–2012, vol. II, p. 59 *sq.*]). Similarly, in some of his later texts, it is not always easy to understand whether Bakhtin discusses literature or philosophy, dialogues in literary works or dialogical relations in a much broader sense. Apart from Bakhtin, many other Russian scholars of the early twentieth century also pondered the problems of relationship between author and character not only in literary, but also in philosophical contexts: among them, were Ivan Lapshin, Alexander Lappo-Danilevsky, Timofey Rainov, Lev Pumpyansky, etc.

[72] Baxtin 1959–1960 [1997–2012, vol. V, p. 306]. In his dialogues with Victor Duvakin Bakhtin defines himself in the following way: "[I am] a philosopher, rather than a philologist. [...] I am a thinker" (*Besedy* 1973 [1996, p. 42]). Recorded two years before his death, these dialogues can now serve as Bakhtin's memoirs not only of his own life, but also of his contemporaries and of a whole era.

[73] One of the inspirations of Bakhtin's reflections was certainly Buber, with his discovery of the domain of *das Zwischenmenschliche*.

[74] Baxtin, early 1950s [1997–2012]. Several issues discussed in this article by Bakhtin are also present in his preparatory texts, eloquently titled "Dialogue(s)", even if in one of these texts appears a seemingly unexpected, for Bakhtin's work, topic, that of the *relative* character of the opposition *dialogue vs monologue* (*ibid.*, p. 209).

According to Bakhtin, speech genres consist of particular and relatively stable types of utterances[75] that each sphere of use of one or another language develops. Unlike sentences or propositions, utterances had not been properly studied by linguistics yet, according to Bakhtin, and this was for several reasons. First, the communicative (or dialogical) function – that is, according to Bakhtin, which is essential to language, had not attracted linguists' attention very much. Second, utterances are very heterogeneous: from replicas of a dialogue (dialogue in the narrow sense of this word was, according to Bakhtin, the simplest and the most typical form of verbal communication) to great novels.[76]

This criticism of linguistics (though very few linguists are named in Bakhtin's studies) explains the fact that Bakhtin even proposes a new term for the future science which would study utterances: metalinguistic, or "translinguistic".[77] Once again, this science would exist "on the borders" of several branches, because the relationships between utterances, the *dialogical* relations would be neither of a linguistic nor psychological, nor philological, (etc.) character. Rather those are relations implying a "transmission of sense".

For Bakhtin, the utterance represents a real and genuine unit of *communication*, unlike the proposition, the main unit of *language*. In addition, every utterance (unlike propositions) has an immediate contact with reality, thus being unique.[78] Moreover, unlike boundaries between propositions, those between utterances are determined by the alternation of speakers (which is easiest to see in a dialogue in the narrow sense of this word). Finally, in the Bakhtinian sense, the utterance must necessarily be addressed to someone, and the scholar especially insists on the fact that the receiver is not passive, but active, in the same way as the person producing utterances. For, besides the fact that he understands, the receiver of any utterance is supposed to react to what he hears afterwards: "The word wants to be listened, to be understood, to be answered to, and it wants, in its turn, to answer to another answer, and so *ad infinitum*. It enters into a dialogue where sense has no end".[79]

Therefore, the speaker, in turn, answers to previously produced utterances, so that the exchange of utterances, the dialogue becomes infinite, like science, art and culture, these particular forms of human activity presupposing the transmission of sense. It means that every utterance could/should be considered as a link in an unfinite chain of statements, and all our utterances (all our ideas, works, etc.) are

[75] The notion of utterance [*vyskazyvanie*] in the linguistic sense of the word appears in Bakhtin's work as from 1924 (Baxtin 1924 [1997–2012]), visibly under the influence of Lev Yakubinsky. In some other Bakhtin's texts (Baxtin 1918–1924 [1997–2012], 1923–1924 [1997–2012], etc.), the word *utterance* is, in addition, used as a synonym of *judgment*.

[76] As early as the 1920s, Bakhtin spoke about utterances as very heterogeneous units, which made them particularly difficult to be studied (Baxtin 1924 [1997–2012, vol. I, pp. 300–301]).

[77] The idea of metalinguistics as a particular discipline is already outlined in 1929, in Bakhtin's book on Dostoevsky, even if this word is still not used there (cf., on the contrary, Baxtin 1963 [1997–2012, vol. VI, p. 203]).

[78] That is why, in particular, no translated text would be completely adequate to its original (Baxtin 1959–1960 [1997–2012, vol. V, p. 310]).

[79] Baxtin 1961 [1997–2012, vol. V, p. 338].

penetrated by the utterances of others.[80] Therefore, the speaker, in the same way as the receiver, *I* (Self) and the *Other*, are formed in a dialogue, rather than preceding its realisation. This change seems crucial in Bakhtin's "dialogic" conception, compared to his early works.

If *dialogic* was, for Bakhtin, everything which implied sense and its transmission, already in the early 1920s he stated that "sense cannot be born, sense cannot die – in the same way as continuity of sense in life can be neither initiated nor completed".[81] That is why, "each sense will one day celebrate its resurrection, nothing will be forgotten".[82] This way, the celebration of *dialogue* in Bakhtin's work seems to become a hymn to life itself: "Life is inherently dialogical. To live means to participate in a dialogue".[83]

Summarizing what has been stated above, we come to the following conclusions:

1. Without ever defining dialogue, Bakhtin uses this word in (at least) two ways: in its narrow sense (linguistic) and its broad sense (referring to the idea of the transmission of sense).
2. It is mainly from the 1950s onwards that dialogue in the broad ("Bakhtinian") sense of the word draws the attention of the scholar. However these thoughts go back to the ideas of his youth about the relationship between *I (Self)* and the *Other*; between the whole and its parts, between art and science on the one hand, and life on the other, etc. Reflecting on these constant subjects of his philosophy, Bakhtin had forerunners not only in Russia but also in Western Europe (primarily among German philosophers). Read in this way, through the prism of the intellectual context of his time, Bakhtin's work appears less original than it could, seen at first glance.
3. Compared to his own early works, in his later research, Bakhtin explicitly changes his priorities. If at the beginning, at least two participants – *I (Self)* and the *Other* – were thought to be necessary for a dialogue, with time it is the dialogue that appears as the first and primary category, the *sine qua non* condition of the formation of categories such as (in particular) *I (Self)* and the *Other*.

Finally, although by now, Bakhtinian *dialogue* has turned into an epistemological obstacle for the study of the scholar's work (the notion of dialogue in Bakhtin's work being too general, it no longer is operational, even if its semantico-semiotic nature remains indisputable), what at first sight seems a defect of Bakhtinian work (the absence of clear definitions or rigor) could also be seen positively. It cannot be excluded that it is precisely the non-rigorous style of Bakhtin's work and the eclec-

[80] Baxtin 1953–1954 [1997–2012, vol. V, p. 193]. This thesis allows to raise the problem of the *author* of texts in a new way and can explain, in part, the complicated situation around the authorship of certain works composed by the members of the so-called "Bakhtin's circle" (cf. Ivanov 1973 *vs* Ivanov in Velmezova and Kull 2011).

[81] Baxtin 1923–1924 [1997–2012, vol. I, p. 182].

[82] Baxtin, early 1920s–1974 [1986, p. 531].

[83] Baxtin 1961 [1997–2012, vol. V, p. 351].

tic nature of his philosophical language (a kind of terminological polyphony) that makes his work change depending on the demands and interests of its readers, inviting the latter, inspired by the variety of subjects treated by Bakhtin, to a "Bakhtinian" dialogue, with Bakhtin himself.

References

Primary Sources

Baxtin, M. M. (1918–1924 [1997–2012]). K filosofii postupka. In Baxtin 1997–2012, Vol. I, pp. 7–68. [Towards a philosophy of the act].

Baxtin, M. M. (1919 [1986]). Iskusstvo i otvetstvennost'. In Baxtin 1986, pp. 3–4. [Art and answerability].

Baxtin, M. M. (early 1920s–1974 [1986]). Zametki. In Baxtin 1986, pp. 509–531. [Notes].

Baxtin, M. M. (1923–1924 [1997–2012]). Avtor i geroj v èstetičeskoj dejatel'nosti. In Baxtin 1997–2012, Vol. I, pp. 69–263. [Author and hero in aesthetic activity].

Baxtin, M. M. (1924 [1997–2012]). K voprosam metodologii èstetiki slovesnogo tvorčestva. I. Problema formy, soderžanija i materiala v slovesnom xudožestvennom tvorčestve. In Baxtin 1997–2012, Vol. I, pp. 265–325. [Towards the methodology of aesthetics of verbal creation. I. The problem of form, content and material in verbal artistic creation].

Baxtin, M. M. (1929a [1997–2012]). Ideologičeskij roman L.N. Tolstogo. Predislovie. In Baxtin 1997–2012, Vol. II, pp. 185–204. [Tolstoy's ideological novel].

Baxtin, M. M. (1929b [1997–2012]). Problemy tvorčestva Dostoevskogo. In Baxtin 1997–2012, Vol. II, pp. 5–175. [The problems of Dostoevsky's works].

Baxtin, M. M. (1929c [1997–2012]). Tolstoj-dramaturg. Predislovie. In Baxtin 1997–2012, Vol. II, pp. 176–184. [Tolstoy the playwright. Preface].

Baxtin, M. M. (1937–1938 [1986]). Formy vremeni i xronotopa v romane. Očerki po istoričeskoj poètike. In Baxtin 1986, pp. 121–290. [Forms of time and of the chronotope in the novel].

Baxtin, M. M. (1940 [1986]). Iz predystorii romannogo slova. In Baxtin 1986, pp. 353–391. [Of the prehistory of the novelistic Word]

Baxtin, M. M. (1940 [1997–2012]). Fransua Rable v istorii realizma. In Baxtin 1997–2012, Vol. IV(1), pp. 11–505. [François Rabelais in the history of realism].

Baxtin, M. M. (1940/1970 [1997–2012]). Rable i Gogol' (Iskusstvo slova i narodnaja smexovaja kul'tura). In Baxtin 1997–2012, Vol. IV(2), pp. 509–521. [Rabelais and Gogol (The art of the Word and the folk culture of laughter)].

Baxtin, M. M. (1945 [1997–2012]). Voprosy stilistiki na urokax russkogo jazyka v srednej škole. In Baxtin 1997–2012, Vol. V, pp. 141–156. [Questions of stylistics in teaching the Russian language at secondary school].

Baxtin, M. M. (early 1950s [1997–2012]). Iz arxivnyx zapisej k rabote "Problema rečevyx žanrov". In Baxtin 1997–2012, Vol. V, pp. 207–286. [From the archive notes to the work "Problem of speech genres"].

Baxtin, M. M. (1953–1954 [1997–2012]). Problema rečevyx žanrov. In Baxtin 1997–2012, Vol. V, pp. 159–206. [The problem of speech genres].

Baxtin, M. M. (1959–1960 [1997–2012]). Problema teksta. In Baxtin 1997–2012, Vol. V, pp. 306–326. [The problem of text].

Baxtin, M. M. (1961 [1997–2012]). 1961 god. Zametki. In Baxtin 1997–2012, Vol. V, pp. 329–360. [1961. Notes].

Baxtin, M. M. (1963 [1997–2012]). Problemy poètiki Dostoevskogo. In Baxtin 1997–2012, Vol. VI, pp. 5–300. [The problems of Dostoevsky's poetics].

Baxtin, M. M. (1965 [1997–2012]). Tvorčestvo Fransua Rable i narodnaja kul'tura srednevekov'ja i Renessansa. In Baxtin 1997–2012, Vol. IV(2), pp. 7–508. [The work of François Rabelais and folk culture of the Middle Ages and Renaissance].

Baxtin, M. M. (1966–1967 – ? (a) [1997–2012]).“Rabočie zapisi 60-x – načala 70-x godov. Tetrad' 2. In Baxtin 1997–2012, Vol. VI, pp. 385–410. [Working notes made in the 1960s – early 1970s. Notebook 2].

Baxtin, M. M. (1966–1967 – ? (b) [1997–2012]). Rabočie zapisi 60-x – načala 70-x godov. Tetrad' 3. In Baxtin 1997–2012, Vol. VI, pp. 411–416. [Working notes made in the 1960s – early 1970s. Notebook 3].

Baxtin, M. M. (1970 [1997–2012]). Otvet na vopros redakcii “Novogo mira”. In Baxtin 1997–2012, Vol. VI, pp. 451–457. [Answer to the question of “Novyj mir” editors].

Baxtin, M. M. (1971–1974 [1997–2012]). Rabočie zapisi 60-x – načala 70-x godov. Tetrad' 4. In Baxtin 1997–2012, Vol. VI, pp. 416–431. [Working notes made in the 1960s – early 1970s. Notebook 4].

Baxtin, M. M. (1986). *Literaturno-kritičeskie stat'i*. Moskva: Xudožestvennaja literatura. [Literary criticism. Articles].

Baxtin, M. M. (1997–2012). *Sobranie sočinenij v semi tomax*: Vol. I (2003), II (2000), III (2012), IV(1) (2008), IV(2) (2010), V (1997), VI (2002). Moskva: Russkie slovari (vol. I, II, V, VI)/ Jazyki slavjanskix kul'tur (vol. I, III, IV [I], IV [II], VI). [Collected works in seven volumes].

Besedy. (1973 [1996]). *Besedy V.D. Duvakina s M.M. Baxtinym*. Moskva: Izdatel'skaja gruppa “Progress”. [V.D. Duvakin's conversations with M.M. Bakhtin].

Secondary Sources

Brès, J., & Rosier, L. (2007). Réfractions: polyphonie et dialogisme, deux exemples de reconfigurations théoriques dans les sciences du langage francophone. In B. Vauthier (Ed.), Bakhtine, Volochinov et Medvedev dans les contextes européen et russe [*Slavica Occitania*, 25], pp. 437–461.

Ivanov, V. V. (1973). Značenie idej M.M. Baxtina o znake, vyskazyvanii i dialoge dlja sovremennoj semiotiki. *Trudy po znakovym sismemam*, 6, 5–44. [The significance of M.M. Bakhtin's ideas on sign, utterance and dialogue for modern semiotics].

Kull, K. (2007). Biosemiotic conversations: Ponzio, Bakhtin, Kanaev, Driesch, Uexküll, Lotman. In S. Petrilli (Ed.), *Philosophy of language as the art of listening: On Augusto Ponzio's scientific research* (pp. 79–89). Bari: Edizioni dal Sud.

Petrilli, S., & Ponzio, A. (2013). Modelling, dialogism and the functional cycle: Biosemiotic and philosophical insights. *Sign Systems Studies, 41*(1), 93–115.

Ponzio, A. (2004). Dialogism and biosemiotics. *Semiotica, 150*, 39–60.

Ponzio, A. (2007). Dialogue, intertextualité et intercorporéité dans l'œuvre de Bakhtine et du Cercle. In B. Vauthier (Ed.), Bakhtine, Volochinov et Medvedev dans les contextes européen et russe [*Slavica Occitania*, 25], pp. 181–202.

Ponzio, A. (2012). Alterity. In D. Favareau, P. Cobley, & K. Kull (Eds.), *A more developed sign: Interpreting the work of Jesper Hoffmeyer* (pp. 21–24). Tartu: Tartu University Press.

Velmezova, E., & Kull, K. (2011). Interview with Vjačeslav Vsevolodovič Ivanov about semiotics, the languages of the brain and history of ideas. *Sign Systems Studies, 39*(2–4), 290–313.

Index

E. Velmezova et al. (eds.), *Biosemiotic Perspectives on Language and Linguistics*, Biosemiotics 13, DOI 10.1007/978-3-319-20663-9

MIX
Papier aus verantwortungsvollen Quellen
Paper from responsible sources
FSC® C105338
FSC
www.fsc.org

If you have any concerns about our products,
you can contact us on
ProductSafety@springernature.com

In case Publisher is established outside the EU,
the EU authorized representative is:
Springer Nature Customer Service Center GmbH
Europaplatz 3, 69115 Heidelberg, Germany

Printed by Libri Plureos GmbH
in Hamburg, Germany